West Asia
in Turmoil

Published by Academic Foundation
in association with the
Institute for Defence Studies
and Analyses (IDSA), New Delhi

West Asia in Turmoil

Implications for Global Security

Editors

N.S. Sisodia

Ashok K. Behuria

ACADEMIC FOUNDATION

NEW DELHI

First published in 2007
by

ACADEMIC FOUNDATION
4772-73 / 23 Bharat Ram Road, (23 Ansari Road),
Darya Ganj, New Delhi - 110 002 (India).
Phones : 23245001 / 02 / 03 / 04.
Fax : +91-11-23245005.
E-mail : academic@vsnl.com
www. academicfoundation.com

in association with :

Institute for Defence Studies and Analyses, New Delhi
www.idsa.in

West Asia in Turmoil: Implications for Global Security
editors: N.S. Sisodia and Ashok K. Behuria

ISBN 13: 978-81-7188-626-5
ISBN 10: 81-7188-626-4

Typeset by Italics India, New Delhi.
Printed and bound in India.

Contents

Section II

Regional Security: Different Perceptions

Section III

Energy and Security: The Political Economy of Energy

Section IV

WMD and the Nuclear Dynamic in West Asia

Section V

Political Reforms and Popular Participation: Regional and Extra-regional Approaches

Section VI

Non-Traditional Challenges to Security: The West Asian Experience

Section VII

India and West Asia

Editors/Contributors

Editors:

N.S. Sisodia is Director, Institute for Defence Studies and Analyses, and a member of the National Security Advisory Board. He holds an Honours degree in History from University of Delhi and a Master's Degree in Public Policy and Management from Harvard University, USA, where he was a Mason Fellow. He joined the Indian Administrative Service in 1968 and served as Additional Secretary, National Security Council Secretariat, and Secretary to Government of India in the Ministries of Finance and Defence. He was a member of the Task Force constituted to recommend measures for reforming the Management of Defence. He has also been Vice-Chancellor, University of Udaipur.

He has co-edited *Emerging India: Security and Foreign Policy Perspectives* (2005), *India and the World* (2005), *Changing Security Dynamic in Eastern Asia: Focus on Japan* (2005) and *India-Japan Relations: Partnership for Peace and Security in Asia* (2006).

Ashok K. Behuria is a Research Fellow at the Institute for Defence Studies and Analyses, New Delhi. He has done his Ph.D. on International Relations from Jawaharlal Nehru University and has published chapters in books and written many articles on issues related to national and international politics and security in various research journals and magazines of repute. His recent publications on West Asia include occasional papers on West Asia, titled *Reconstructing Iraq: The Challenges Ahead, President Khatami's India Visit: Reconstructing India-Iran Relations.*

Contributors:

Marian Abisheva is a Deputy Director for Science at the Kazakh Institute for Strategic Studies under the President of the Republic of Kazakhstan. She was earlier Academic Secretary of the Kazakhstan Institute for Strategic Studies under the President of the Republic of Kazakhstan. She is a Ph.D. in Political Science. She wrote her Ph.D. dissertation on "Factors of Social Security in the Context of Provision of Internal Political Stability of Kazakhstan Society" in 2004. She has published more than 20 articles on issues relating to national security, globalisation and intercultural cooperation, and the internal policy of the Republic of Kazakhstan.

Talmiz Ahmad is the Director General of Indian Council of World Affairs, New Delhi, India. He is a career diplomat and has been the Official Spokesman of the Ministry of External Affairs, Joint Secretary (Gulf/Hajj) and Ambassador of India to Saudi Arabia and the Sultanate of Oman. He was also Additional Secretary in the Ministry of Petroleum and Natural Gas. Mr. Talmiz Ahmad has a special interest in political Islam and Islamic affairs in general, as also the politics of West Asia. He has published a monograph titled: *An Introduction to Contemporary Islamic Groups and Movements in India*. His book titled: *Reform in the Arab World External Influences and Regional Debates* was released in July 2005.

Jaffar Mirghani Ahmed is the Director General, Sudan Civilization Institute. He has been the Head of Department, Linguistics, in the International University of Africa. He has also been a Visiting Professor at the University of Ben Saud in Riyadh, University of Chad, Islamic University of Omdurman, Institute of Ethiopian Studies in Addis Abbaba, Sudan Higher Institute and Burgeiba Institute in Tunisia. He has published many papers in different fields of studies, i.e., Linguistics, Phonetics, Language Studies, Social History, Islamic Studies and Cultural Aspects.

Nobumasa Akiyama is a Senior Research Fellow at CPDNP, Japan Institute of International Affairs, Tokyo. He has been the Advisor to the Japanese delegation to the May 2005 NPT Review Conference; Research Advisor to Ambassador Endo, the Japanese Representative to IAEA's expert group on Multilateral Nuclear Approaches (October 2004–March 2005); and Member of Non-Proliferation Study Group of Japan Atomic Industry Forum (November 2004). He has book chapters and various articles in journals of repute to his credit.

Ahmed Ali Salem Al-Atrash is a Faculty and Researcher at the Political Science Department, Al-Fateh University, Libya. He is also the Deputy Director, Centre for International and Strategic Studies, Academy for Graduate Studies, Libya. He obtained his Ph.D. in International Relations, University of Kent at Canterbury, UK (2000). Dr. Al-Atrash has published articles in English as well as in Arabic. His major publications are on *Peacemaking in Africa as a Challenge, Resolving Conflicts: A Non-Western Perspective, Libya's Peacemaking Efforts in Africa and Future of Peaceful Management of Regional Conflicts in the Context of Afro-Arab Space, Arab Summits: An Arena for Inter-Arab Conflicts or Conflict Management?* (2003) and *Diplomacy and Arab-China Relations: From Theory to Practice* (2005).

Sahar A.M. Al-Majali is the Vice President of the Gandhi Center for Strategic Peace Studies and Lecturer at the Faculty of Arts, Al-Zaytoonah University Jordan, and Lecturer/General Administration in the Royal Jordanian Army. Dr. Al-Majali has also served as a Colonel (Cornelle) in the Jordanian Armed Forces from 1986 till 2004. Dr. Al-Majali's publications include *The Arab Legion: 1921-1951: Its Role in the Arab-Israeli Conflict,* Volume I, II. (ed.), Amman, Jordan (Arabic Version), and *The Development of the Arabian Army in Andalusia,* (138-422AH/756-1031 AD), (Ist. ed.), Amman-Jordan (Arabic Version).

Faisal Odeh Al-Rfouh is working as Professor at the Department of Political Science, University of Jordan. He has also served as Minister of Culture (1999-2000), Minister of Social Development (1999) and was also Acting Minister of Labour as well as Health in the same year. He was awarded "Ambassador for Peace" by the Interreligious and International Federation for World Peace, "Man of the Year 1999" by the American Biographical Institute and its Board of International Research, and "The Arab Historian Medal" by the Union of Arab Historians. He has several publications to his credit in various journals like, *Defense Journal, Middle East Peace Initiative* and *Journal of Political Science*.

A.F. Alhajji is an Associate Professor of Economics in the College of Business Administration, Ohio Northern University. He previously worked at two universities that are strongly affiliated with the oil industry, the University of Oklahoma and Colorado School of Mines. Dr. Alhajji has more than 300 articles and columns to his credit. He focuses on issues relating to international economics, natural resource economics, environment, and oil industry. His research interest also includes trade policy and globalisation, education policy, and human capital. He is a contributing editor for *World Oil*. He is an Associate Editor for *Oil, Gas, and Energy Law*. In addition, he was the energy columnist for *Alwatan* daily newspaper in Saudi Arabia. Dr. Alhajji developed PetroTrade, a computer programme to assist oil producing countries in determining their trade policies in today's volatile market.

Mohammad Hamid Ansari is the Chairman of the National Minority Commission, India. Mr. Ansari had joined the Indian Foreign Service in 1961. He was Ambassador to the United Arab Emirates, Afghanistan, Iran, Saudi Arabia, High Commission to Australia, and Permanent Representative to the United Nations in New York. He was awarded the Padma Shri in 1984. He has been the Vice-Chancellor of the Aligarh Muslim University and a Visiting Professor at the Centre for West Asian and African Studies, Jawaharlal Nehru University and the Academy for Third World Studies, Jamia Millia Islamia, New Delhi. Mr. Ansari has edited *Iran Today: Twenty-five Years After the Islamic Revolution* (New Delhi, 2005), and has written a number of academic papers and newspaper articles on West Asian politics.

Smail Benamara is currently Algerian Ambassador to Canada. He has also been Director of the Institute of Diplomacy and International Relations, Algiers and Ambassador of Algeria to Senegal, Guinea-Bissau, Gambia and Cap-Vert (1996-2003). Ambassador Benamara is a member of several international associations and has published various articles on law and international relations.

Cameron S. Brown is the Deputy Director of the Global Research in International Affairs (GLORIA) Center, a part of the Interdisciplinary Center in Herzliya, Israel. Mr. Brown focuses on Israel, US Middle East policy, the Arab World, Turkey, and Azerbaijan. His commentary has been published in papers such as the *Chicago Sun-Times*, *Newsday* and Canada's *National Post*. Mr. Brown lectures occasionally at the NATO School in Germany and has spoken at research centres and universities in the United States, Sweden, Italy, Turkey, and Azerbaijan.

Gulshan Dietl is a Professor at Centre for West Asian and African Studies, School of International Studies, Jawaharlal Nehru University, New Delhi, India. She also serves as the Director, the Gulf Studies Programme at the University. She has been a Fulbright Scholar-in-Residence at the Mount Saint Mary College, Newburgh, New York (1993-94), a Guest Research Fellow at the Copenhagen Peace Research Institute (1998-99). She is Joint Secretary of the Indian Association of Central and West Asian

Studies, Member and Member of the Editorial Advisory Board of *Human Rights Global Focus*. Her publications include *The Dulles Era: America Enters West Asia* (Lancer International, New Delhi, 1985), *Through Two Wars and Beyond: A Study of the Gulf Cooperation Council* (Lancer Books, New Delhi, 1991).

Mohamed Noman Galal is an advisor for Strategic International Studies and Dialogue of Civilizations, Bahrain Center for Studies and Research. He has been in the Egyptian Foreign Service for 38 years and occupied many posts including Deputy Permanent Representative to the UN, Permanent Representative to the League of Arab States, Ambassador to Pakistan, Ambassador to China and Assistant Foreign Minister for Policy Planning. Dr. Galal has authored around 26 books on Arab, International and Asian issues as well as Human rights with a special study on human rights in Islam. Among his recent books are: *Neo-realism in Arab Thought; Strategy, Diplomacy and Protocol and Diplomacy of International Dialogue*. He writes regularly in *Alahram* and *Algoumhoria* newspapers in Egypt as well as occasionally in other Arab and Chinese papers and journals. He is a part-time professor of international relations and international organisations.

Siret Hürsoy is an Assistant Professor in the Department of International Relations at the Ege University, Turkey. Dr. Hürsoy served as a research and teaching assistant between 1998-2000 at Ege University–Faculty of Economics and Administrative Sciences. Later, he completed his Ph.D. at the end of 2002 by receiving his degree from *Philipps-Universität Marburg–Gesellschaftswissenschaften und Philosophie, Fachbereich Politikwissenschaft* (Germany). His main fields of specialisation and research interests are: international relations theories, European Union (integration theories), foreign and security policy analysis, and war and defence studies. He has many publications to his credit including a book titled *NATO's Transformation and the Position of Turkey*, (Ege University Press, Izmir, 2004).

Akmal Hussain is a Professor of International Relations in the Dhaka University and he did his Ph.D. from the Institute of Law, Academy of Sciences, Prague. He has been a Visiting Research Fellow at the Institute of Developing Economies (IDE), Japan and a Visiting Professor at the University of Rajasthan. Prof. Hussain has about 25 publications in national and international journals and chapters in edited volumes as well. In addition to these articles, he has one research monograph and two books in vernacular on current issues. He is also a regular contributor to leading newspapers on current national and global issues. In addition, he has edited the *Journal of International Relations* (until 2004) and the *Journal of the Asiatic Society of Bangladesh* (1996 and 1997). He is also a regular speaker at the National Defence College of Bangladesh.

N. Janardhan is the Editor of the Gulf in the Media at the Gulf Research Center in Dubai, UAE. He was formerly with *The Asian Age* daily in New Delhi and *The Gulf Today* in Sharjah, UAE, and was the Gulf correspondent for the Inter Press Service (Asia-Pacific), Bangkok. His doctoral thesis was "The Al-Sabahs and the Kuwaiti National Assembly–The Legitimacy factor" (Jawaharlal Nehru University, New Delhi, 2001). Dr. Janardhan has had over 300 articles published in Indian and Middle East newspapers.

Dietrich Jung is a Senior Research Fellow at the Danish Institute for International Studies in Copenhagen and an External Associate Professor at the Institute for Sociology, University of Copenhagen. He holds an M.A. in Political Science and Islamic Studies, as well as a Ph.D. in Political Science from University of Hamburg, Germany. Dr. Jung has been teaching Political Science, Sociology and Middle Eastern Studies at Aarhus University (Denmark), Bilkent University (Ankara), and the Universities of Copenhagen, Hamburg and Southern Denmark. He is editor of the Palgrave Series on Governance, Security and Development and he has published on causes of war, theories of world society and on conflicts in the Middle East. His most recent books are *Shadow Globalization, Ethnic Conflicts and New Wars: A Political Economy of Intra State Wars* (Routledge, 2003), *Contemporary Security Analysis and Copenhagen Peace Research* (Routledge, 2004: co-edited with Stefano Guzzini), and *The Middle East and Palestine, Global Politics and Regional Conflict* (Palgrave, 2004).

Abdusamat Khaydarov is at present Faculty, Tashkent State Institute of Oriental Studies. He has had a distinguished diplomatic career ever since he completed his studies from Moscow State Institute of International Relations, Ministry of Foreign Affairs of the former USSR. He also has a degree from Diplomatic Academy, Ministry of Foreign Affairs of the former USSR. He has been Visiting Fulbright Scholar at the University of Georgia, USA (2004-05), Deputy Minister of Foreign Affairs of the Republic of Uzbekistan (1999-2000), Ambassador Extraordinary and Plenipotentiary of the Republic of Uzbekistan in the Islamic Republic of Iran and Permanent Representative of the Republic of Uzbekistan in Economic Cooperation Organization (ECO)

(1996-1999). He has a number of scientific works and publications on international relations, Afghanistan, Iran, Central and South Asia.

Gurpreet S. Khurana is a Research Fellow at the Institute for Defence Studies and Analyses, New Delhi, India. Commander Khurana is a specialist in Missile Warfare and Gunnery and a qualified Ship's Diving Officer. His research interests include maritime security issues, particularly in the Indo-Pacific region and Laws of the Sea. His published papers have encompassed issues like sea-line security, sea-piracy, maritime terrorism, other non-traditional maritime threats, imperatives for naval sealift and airlift capabilities, Proliferation Security Initiative (PSI), Container Security Initiative (CSI) and security in the Malacca Straits.

Joseph McMillan is a Senior Research Fellow at the National Defense University's Institute for National Strategic Studies, focusing on issues related to terrorism, the greater Middle East, and South Asia. He has been associated with NDU since August 2001, first as a Visiting Research Fellow at INSS and later as academic chairman of the Near East-South Asia Center for Strategic Studies. He has published a number of articles and book chapters on foreign and security affairs.

Bakhtiyar Mirkasymov is the Head of the Asia-Pacific Department in Russia's Institute for Strategic Studies (RISS) which does research and analytical work for different government organisations. He has held this position since 2002. Prior to his current appointment, Bakhtiyar Mirkasymov worked as Senior Research Fellow in RISS. He began his career in Tashkent in the Uzbek branch of the Union of Soviet Friendship Societies having graduated from the Moscow State Institute of International Relations (international economic relations faculty) in 1974. He has authored several articles on the current political problems pertaining to religious extremism and security in the Persian Gulf which were published in various Russian and foreign publications.

Pranab Mukherjee, currently the External Affairs Minister of India, is the President of the Executive Council of the Institute for Defence Studies and Analyses (IDSA). He was educated at Vidyasagar College, University of Calcutta, where he received his Bachelor of Law (LLB), Masters (History and Political Science) and D.Litt (*Honoris Causa*) degrees. He has held various important portfolios in the Union Cabinet. He was Minister of Defence, Deputy Minister for Industrial Development (1973-74); Deputy Minister for Shipping and Transport (January 1974-October 1974); Minister of State for Finance (1974-75); Commerce and Steel and Mines (1980-1982); Minister of Finance (1982-1984); Deputy Chairman, Planning Commission (1991-1996); and Minister for External Affairs (1995-96). He was also on the Board of Governors of the International Monetary Fund (1982-85), the World Bank (1982-1985), and the Asian Development Bank (1982-1985). Mr. Mukherjee is also the Leader of the House (Lok Sabha), from June 2004 till date.

Sari Nusseibeh is a Professor of Philosophy and President of the Arab Al-Quds University in Jerusalem. He has a Ph.D. in Philosophy from Harvard University (1978). Formerly Professor at Birzeit University (1978-1990) and until December 2002, the representative of the Palestinian Authority in Jerusalem. Along with Ami Ayalon, he is a co-founder of *The People's Voice*, an Israeli-Palestinian civil initiative, which aims to advance the process of achieving peace between Israel and the Palestinians. He is also the Head and Founder of the Palestinian Consultancy Group (PCG), Jerusalem, and is a co-founder and member of several Palestinian institutions. He is the author of numerous articles and a book co-authored with Mark Heller, *No Trumpets, No Drums: A Two-State Solution of the Israeli-Palestinian Conflict* (Hill & Wand, New York, 1991).

Girijesh Pant is the Chairperson and Professor at the Centre of West Asian and African Studies at Jawaharlal Nehru University, New Delhi, India Dr. Pant is a renowned academician and specialist in issues related to economics and energy. He has authored and edited over 8 books, and has published more than 50 articles in various journals. His specialisation includes Political Economy of Development, West Asian Economies and Hydrocarbon Economics.

A.K. Pasha is a Professor at the Centre for West Asian and African Studies, School of International Studies, Jawaharlal Nehru University, New Delhi. He has served as Director, Gulf Studies Programme, and Director, Maulana Azad Centre for Indian Culture, Cairo. A prolific writer, Dr. Pasha has published 10 books, edited 8 and co-edited 2 books. He has also contributed chapters to over 17 edited books, and articles in various academic journals. His latest books include, *Iraq: Sanctions and Wars* (New Delhi, 2003), and *Egypt in a Changing World* (New Delhi, 2003).

Basant Repswal, a serving Colonel in the Indian army, was a Research Fellow, Institute for Defence Studies and Analyses, New Delhi, India. Colonel Basant Repswal was commissioned in the Indian Army on December 15, 1984 and during his service career served as logistic officer with United Nations (UNIFIL) in Lebanon (1998-99) and his areas of expertise include Terrorism, Military Logistics and Strategy, and Implications of International Terrorism on India.

Chen Shuangqing is at present Research Fellow in the West Asian and African Division, China Institute of Contemporary International Relations (CICIR). He was in charge of research on the Middle East Peace Process, the Foreign Policy of Israel and Palestinian Authority. He has published about 200 articles on Middle Eastern issues in some famous journals and newspapers in China, such as *World Affairs, Contemporary International Relations, Beijing Review, People's Daily,* etc. He has also co-authored 15 books.

S.K. Singh, formerly Foreign Secretary of India, has a firm understanding of the West Asian politics. He has been India's longest-serving Ambassador to Pakistan from 1985 until 1989. He also served as Ambassador to Austria, Afghanistan, Jordan, Lebanon and Cyprus. At the moment, he is the Governor of the Indian province of Arunanchal Pradesh.

Emmanuel Sivan is a Professor of Islamic History at the Hebrew University of Jerusalem and has authored many books on contemporary Islam. His publications include, *Radical Islam, The 1948 Generation,* and *Mythes Politiques Arabes*. In the late 1980s, he was among the first scholars to begin to look at radical forms of Islamic movements that were generated from the Iranian revolution. He has also written more than 150 articles and chapters, and one of his areas of expertise is Islamic conceptions of the sanctity of Jerusalem. Sivan is a regular commentator on Islamic movements for Israeli television and writes prolifically for the Israeli press.

Premarani Somasundram is a Senior Fellow with the Contemporary Islam Programme at the Institute of Defence & Strategic Studies, Nanyang Technological University, Singapore. She has worked at Singapore's Ministry of Defence and also served a term as Counsellor in the Singapore Embassy in Washington, DC. Her research interests include Islamism in the Middle East and Southeast Asia.

Ryoji Tateyama is working at Graduate School of Security Studies and Department of International Relations, National Defense

Academy, Japan. He is also Visiting Professor, Aoyama-Gakuin University. He was previously Chief Researcher at the Japanese Institute of Middle Eastern Economies, Head of Administration and Special Project Officer United Nations Relief and Works Agency for Palestine Refugees in the Near East (UNRWA) HQs, Vienna and Amman and Special Assistant at the Embassy of Japan, Tel Aviv, Israel. He has published many books which include, *A Jewish State and Post-Zionism* (Bungei-shinju-sha, Tokyo, 2000), *Prospects for Peace in the Middle East* (Chuo-Koron-sha, Tokyo, 1995), *Jerusalem* (Shincho-sha, 1993, Tokyo), *Israel and Palestine: Searching for a Settlement of Israeli-Palestinian Conflict* (Chuo-Koron-Sha, Tokyo, 1989) and has edited a book titled *Middle East* (Jiyu-Kokumin-sha, Tokyo, 1998). He has written numerous articles in various international journals on matters concerning the Middle East and West Asia.

Mostafa Zahrani is a member of Iranian foreign ministry since 1981 and has been teaching in the School of International Relations since 1993. He has also been the Director General of the Institute for Political and International Studies (IPIS) in Tehran. His major recent publications include, *Theories of Economic Sanction* and *New International Trends*; others such as *State Sovereignty in the Age of Globalization, Second Order Change and Decision Making in War Termination: Iran-Iraq War* are under publication.

Preface

West Asia is in a state of ferment. Since the end of the Cold War, the entire region is experiencing pressures from within and without. There are many assertive political constituencies within the states seeking reform or overhaul of the existing systems of governance. At another level, the world outside, dependent on energy resources of the region, is seeking to hedge the prospect of chronic instability by advocating incremental change in favour of participatory and responsible governance in almost all the states in the region. There has also been a conscious effort by external powers with capacity to influence the region's politics to prevent the rise of any regional hegemon, powerful enough to shift the balance to the disadvantage of their energy interests.

As the case of Saddam Hussein demonstrates, the Western powers as well as the regional ones apprehensive of Saddam's intentions have not hesitated to use force to deal with a defiant regime. However, Saddam's fall and final execution has not led to any improvement in the security situation of the region. Iraq is burning and Iraqis are yet to find their initial experiments with democracy too assuring. Nevertheless, the presence of external forces has changed the security dynamics in the region unalterably.

Following the attack on the World Trade Centre on September 11, 2001, the principal problems agitating the minds of security analysts the world over, are terrorism and religious extremism. Some analysts believe that the nerve centre of Islamic extremism lies in the region, and that some of the states, unable to contain such phenomenon at home, are exporting it to other countries and regions. There are also the issues of proliferation of weapons of mass destruction in the region and that of Palestine which need to be addressed to make the region more stable and secure.

The evolving situation in West Asia warrants careful analysis. In fact, there has been a surge of interest in the politics of the region in the West ever since the First Gulf War. Given the dependence of their economies on hydrocarbon supplies from the Gulf, the Western states have watched developments in the region with great anxiety and concern. There has also been a great deal of intellectual churning taking place in the West, which is focused on devising remedies to the chronic turbulence in the region. The prescriptions range from sustained democratisation and partnership initiatives to even 're-colonisation'. It was interesting indeed, to find Robert Cooper, one-time strategic foreign policy advisor to Tony Blair, advocating aggressive recolonisation of the troubled world for the betterment of the mankind.

In this wider debate, on how to stabilise West Asia and bring peace and prosperity to the people there, the voice from the region is either grossly underrepresented or ignored. At another level, there is an argument that intellectual capital is in short supply in the region and whatever inputs one lays access to are irremediably biased, highly emotive and policy wise, unacceptable. Even then, the importance of a dialogue with academics, thinkers, journalists, policy makers and leading opinion makers is emphasised by all concerned.

Against this backdrop, the Institute for Defence Studies and Analyses (IDSA), New Delhi, had organised a three-day international conference on January 30 till February 1, 2006, titled "Changing Security Dynamics in West Asia: Implications for the Post-9/11 International Systemic." This compilation is the outcome of the conference. The Institute had given special attention to participation from the region and its immediate neighbourhood, i.e., from North Africa and Central Asia, apart from participants from other regions of the world and many eminent researchers, diplomats and policy makers presented their views in the conference. Some of them were in the shape of papers and others were transcribed and given the shape of papers, subsequently.

The papers were later updated by the authors keeping in view the rapidly changing security situation in the region. They were edited and thoroughly revised by the authors to sharpen their arguments in light of the comments and queries made by the participants in the conference.

The contributors in this compilation have made an effort to situate the region in the global security context, dwell on critical issues of our time like religious extremism, democratisation, external intervention and their impact on regional and global security, and suggest alternatives for securing peace and prosperity for the region and the world.

The book is divided into seven sections. Each focuses on a specific dimension of the evolving security scenario and discusses it in detail. There is a happy blend of perspectives from within the region and outside in the first two sections of the book. The regional understanding, as would appear from the papers presentations by Prof. Sari Nusseibeh, Dr. Mohamed Noman Galal, Prof. Faisal Al-Rfouh and others, has a critical sympathy for the causes that give rise to religious extremism, even if there is a consensus that 'terror' in all its forms needs universal condemnation. Prof. Nusseibeh would hint at the limits of military power and argue that intermeshing of critical elements like religion, identity and allegation of injustice can be a lethal combination but to seek an alternative through the use of force is certain to aggravate the situation and legitimise the use of violence. A solution to the unresolved issue of Palestine, he argues, has to be sought outside the paradigm of military coercion. Dr. Cheng Shuangqing's paper also corroborates the argument that the Palestine issue remains one of the principal issues fomenting wider disaffection against the West, encouraging extremism and even contributing to the proliferation of the WMDs in the region.

The responses of the scholars from Central Asia and North Africa show their concern regarding the unstable situation in West Asia after the Iraq War of 2003. The apprehension of instability in the region seeping into their societies emerges clearly from their discourses. The other factor that most of them have drawn attention to is the understanding that 'societies in the region are steeped in traditional values and would perceive changes brought about under external pressure or persuasion as corrosive and destabilising'. Prof. Jaffar Mirghani Ahmed would even go one step further and argue that the modern security discourse on Middle East or West Asia is driven by Euro-centric or western idioms and vocabularies, which continually reinforce the hold of the 'West' on the 'mind' of the world. He would quote from F. W. Fernau's book published in 1954 to drive home his point that the Western world deeply believes that "lack of stability characterises the world of Islam" and stability in this "intermediate world is dependent more on external power-plays than on internal transformation, reform or revolution."

The issues of democratisation, non-traditional threats to security are covered in two sections (V & VI) of the book. Focusing on *jihadist* insurgency in his paper, Joseph McMillan makes a forceful argument that such insurgency is the outer manifestation of a deep running crisis of legitimacy in the Islamic world, which needs to be addressed properly through a mixed strategy of force and persuasion. Most of the scholars from the region and beyond tend to agree that there is a crisis of legitimacy. However, they point to the reflexive opposition among the people of the region to political ideas or institutions being transplanted by Western states. By so doing the desired aim of segregating the terrorists from the people will not be possible, they would argue. Prof. Gulshan Dietl analyses the Saudi adaptation of representative governance at a rudimentary level and says that it marks a good beginning but lot of changes remain to be initiated before democracy can take roots there. Dr. Siret Hürsoy from Turkey presents the case of democracy in Turkey but says that it is born out of a specific sociopolitical situation and may not be replicable in other states. Dietrich Jung dwells on the issue of democratisation in great detail in his paper and says that "one-size-fits-all solutions for democratic and economic reform in the Middle East do not exist" and would only invite more trouble. Marian Abisheva concurs that the result of such strategy would lead to situations like Iraq, which may have already become a "laboratory for global jihad."

In his paper on the role of religion in politics, Prof. Emmanuel Sivan writes that the debate on whether any religion is compatible with democracy ought to be put to rest. He would rather argue—quoting instances from the early interactions between the Church and the democratic State—that no institutionalised religion would favour democracy. Thus the road to democracy in Islamic societies in the region will have to be sought through broadening of social bases of the existing regimes and incremental promotion of values like pluralism, diversity, transparency and public reasoning.

The discussions on Energy in Section III reinforce the critical dependence of all growing economies on energy resources of the region and reiterate the importance of the region for global security. In the following section, the question of WMD proliferation and especially the efforts of Iran to acquire nuclear weapons have been dealt with in detail. While Premarani Somasundram analyses the rise of Ahmadinijad to power and the challenge it could pose for the regional security, Basant Repswal deals with Iran-US relations. Both of them argue that Iran has to be handled with lot of care and caution. Premarani argues that the West should understand the political complexities of Iran and should fashion policies that "do not isolate it and revive the xenophobia of the past." Repswal observes that the US policy of destroying the self-evolved political system in Iran will only entrench the hardliners who are sure to get sidelined in Iranian politics.

Dr. Khaydarov seeks to isolate the causes that have led to weakening of the non-proliferation regime in the region and says that it is useful to have a functioning regional forum or organisation for greater interaction among the regional powers who can then take up issues concerning their security in a more meaningful way. Cameron Brown, defends the nuclear policy of Israel in his paper, and observes that nuclear disarmament will not come about until a comprehensive Arab-Israeli peace has been reached in the region. Nobumasa Akiyama from Japan probes the cause of Iranian sense of insecurity *vis-à-vis* Israel and argues that it is in Iran's interest to halt its enrichment programme and international community ought to find out a peaceful solution to the issue and force would not help.

The papers by Indian participants reflect the perceptions from India. In his paper, Hamid Ansari critiques the US policy in the region and cautions that if its policies have left the region worse off than before then it should evaluate its policy preferences. It should be sensitive to the feelings of the people of the region regarding Palestine, Iraq and other such issues. The importance of the region to India is well argued in the papers by Talmiz Ahmad, who focuses on the energy requirements of India, and Dr. Janardhan, who deals with the Indian workforce, their remittances and the conditions under which they work in the Gulf region.

The paper (opening address) by Hon'ble Minister of External Affairs, India, Mr. Pranab Mukherjee (who was Minister of Defence, when the conference was organised) and the paper by Ambassador S.K. Singh, Hon'ble Governor of Arunachal Pradesh, revisit history to underline the point that India has always had good relationship with West Asia and it is important to carry the relationship forward. By emphasising the need to seek non-violent solutions to the problems facing the world today, the chapter 1 in a way anticipates the overwhelming consensus among the authors that West Asia poses a unique case for global security and application of force may complicate rather than solve problems.

We would like to thank J.A.R. Murthy for helping us in the editing and proofreading and Savitri Malhotra for typesetting of the manuscript. Our thanks also go to Mr. Sanu Kapila of Academic Foundation for undertaking the publishing work and bringing it out on time. We owe a special debt of gratitude to our colleagues at the Institute, especially Shri Virendra Gupta, Vishal Chandra, Rushda Siddiqui and Tanya Mohan, who have worked towards the making of this book at different stages of preparation.

We hope that this book will help the readers in situating the problems in West Asia in the right context and contribute to a better understanding of its security dynamics.

N.S. Sisodia
Ashok K. Behuria

1

Security Situation in West Asia

Implications for India and the World

PRANAB MUKHERJEE

Throughout human history, trade and cultural linkages between the people of Indian subcontinent and the region which we now call West Asia, have ensured a lasting relationship. This relationship continues and in fact, vigorous efforts are being made to deepen and strengthen it.

Much is made out of the theory of "Clash of Civilizations", advanced by Prof. Samuel P. Huntington in his famous book. But such theories are predicated on division of the world in narrowly defined ideological terms. Our own experience as a functioning, multi-religious, pluralistic democracy completely debunks such divisive theories. Indeed, our interaction with the West Asian region has been traditionally warm and marked by cooperation and mutual enrichment, leading to cross-fertilisation of cultures. Ideas have been exchanged in the fields of mathematics, astronomy and medicine; and literary texts have been translated or transliterated for mutual benefit. The influence of Persian and Arabic learning in India, and the impact of the architectural styles and aesthetics of West Asia, are too obvious to be missed today. The birth of Urdu, the language spoken by a large number of people in India and its influence on other Indian languages symbolises the creative cross-fertilisation between India and West Asia.

The unceasing stream of mutual influences has contributed to the growth of an amazing interweave of cultures in the region. In the increasingly interdependent world today, there are easily discernible

---

Based on Inaugural Address by Mr. Pranab Mukherjee at 8th Asian Security Conference (ASC), organised by Institute for Defence Studies and Analyses (IDSA), January 30, 2006, New Delhi.

civilisational and cultural links among nations—stretching from the coast of Mediterranean to the Bay of Bengal. These remind us of the history that has gone into the making of the culture of interdependence that we emphasise so much today. We all live in turbulent and even dangerous times today. In this turbulent and tense phase of our times, it is appropriate for us to emphasise the unerasable commonalities between our cultures and peoples. We should work together to preserve and sustain this invaluable historical legacy.

Historians have pointed out that there were trade links between the cities of Mohenjodaro and Harappa and Mesopotamia nearly 4,000 years ago. The trade links between India and West Asia were seasonal and influenced by the monsoon winds. These links had developed between the Malabar Coast in India and the mouth of the Red Sea, as well as between other port cities in India and ports in the Persian Gulf and the Arabian Peninsula. There is evidence to show that there was trade in cotton cloth, spices, perfumed oils and handicrafts between India and the West Asian region from around 200 BC. The Arabs traded more than a 1,000 years ago with South West India for spices and carried them back to Europe.

It is also well-known that people-to-people contacts between West Asia and India continued over the centuries. In the post-colonial period, close political relations between India and other countries in the region gave us a common voice in shaping the global agenda through non-alignment, characterised by principles of equality, rejection of colonialism, and independence in decision making and action. These close relations also facilitated dialogue to resolve differences.

The unprecedented revolution in communication technology has activated cross-state linkages among non-state actors. Intra-State conflicts have the capacity now to snowball into international crises. This makes it necessary to isolate the factors causing international insecurity today and work in a cooperative manner to make our societies more safe and secure. As the traditional concept of militaristic state security is yielding ground to a more comprehensive notion of security, we now need to identify non-traditional threats to security.

Given the vast reserves of hydrocarbon in the region, West Asia has emerged as the principal source of energy in the world. Once described as a 'marginal crescent' in world politics, it has made its mark as a critical pivot in the international economy today. The entire world, thus, has a stake in the security and stability of the region.

India has an abiding interest in peace and prosperity of the region. There are about 3.5 million workers from India in the Gulf countries. The remittances they send home are an important source of foreign exchange for us. Our oil imports from the region account for nearly 60 per cent of our total crude oil imports. As regards our exports, the region is the third largest destination for Indian goods, behind only the EU and North America. Exports to the region account for nearly 16 per cent of our global exports.

India has always remained willing to share its experience with the countries in the Gulf and Middle East region. Economic reforms initiated in our country since the early nineties have led to a robust growth rate of over seven per cent in the GDP. This has been fuelled in a large measure by ever growing manufacturing base in the country, alongside the world leadership position achieved in the areas of cutting edge technology. If the world has already witnessed the phenomenal growth of IT and computer software industry in India, the coming years are expected to see India clocking similar, if not greater thrust in the biotechnology and other development defining sectors. I have dwelt on this issue at some length since in the present times, this aspect has increasingly begun to define relations between countries and peoples. I am happy to note that these relations are poised to grow even stronger in the coming years with new opportunities having been made available, for flow of investments, both ways.

Our relationship with the GCC has been growing. We have a mutual commitment to get engaged proactively. We have signed a Framework Agreement for Economic Cooperation with the GCC. We have also intensified our relations with all the key countries of the region. We have an ongoing dialogue mechanism with the Arab League and the possibility of further cooperation between India and Arab League Members, is being continuously explored.

In recent years, the menace of international terrorism has been causing increasing concern to the world at large. The depredations of this scourge are not restricted to this country or that country or any specific region alone. We are witness every day to the growth of ever new global linkages amongst terrorist and radical networks. While it took the tragic events of 9/11 to focus world's attention to the threat posed by international terrorism, India has been a victim of terrorism for much longer. The international community can no longer afford to find excuses for not dealing with this problem in a frontal and concerted manner. Our fight has to be directed not only against those who perpetrate the terrorist crimes but also those more powerful interests which harbour and support terrorist elements for their narrow and short term, but ill-conceived, gains.

Commitment to the Palestinian cause has been a bedrock of our foreign policy even before we gained independence. In a cable sent to Mufti of Jerusalem on 4 September 1937, Pandit Jawaharlal Nehru, who became our first Prime Minister after independence, had affirmed, "The Indian National Congress sends you greetings and assurance of full solidarity in the struggle for Palestine Independence." In the post-independence period we have remained fully supportive of the legitimate aspirations of the Palestinian people for a homeland of their own. We believe that the conflict can be resolved only through meaningful dialogue and not by force; particularly since Arabs and Jews have lived together in peace for long. A just, comprehensive and durable peace in the region can only be achieved through negotiations on the basis of relevant UN Security Council resolutions as well as the 'Land for Peace' principle. In this context, we support the 'Quartet Road Map' which has been accepted by both sides.

India welcomes the Gaza disengagement and hopes that the resurgence of violence will soon give way to peaceful negotiations. India has consistently urged an end to violence from all sides. We look forward to further progress in the peace process that would bring about a just and peaceful solution within a reasonable time frame, leading to a sovereign, independent State of Palestine with well-defined and secured borders, living at peace with the State of Israel. We also remain ready to join in the efforts of the international community for capacity building of the Palestinian people and for strengthening their administrative and democratic institutions.

Our people have traditionally enjoyed friendship and cordiality with the people of Iraq. We feel their pain and suffering and deeply empathise with them in these trying times. We are ready to respond to the needs of the Iraqi people for stability, security, political progress and economic reconstruction. We have committed US $20 million in assistance to the Iraqi people in response to the UN Secretary

General's appeal in 2004 and at the Donors' Conference held in Madrid in October 2004, we further committed an additional amount of US $10 million to the two Iraqi Trust Funds. We distributed milk powder worth US $1.2 million to Iraqi children through the World Food Programme in 2004 and have sent food grains to Iraq in 2005. We have also expressed our willingness to participate in the reconstruction of Iraqi petro-industry in future.

We believe that the UN has a crucial role to play in the process of political and economic reconstruction of Iraq. As such we welcomed the unanimous adoption of Resolution 1546 by the UN Security Council, on June 8, 2004. We also remain fully supportive of the right of the Iraqi people to freely determine their political future and to exercise control over their natural resources. We welcomed the Interim Government of Iraq as the first step towards transfer of full sovereignty to the people of Iraq. Only a fully inclusive political process leading to a truly representative Government in Iraq can effectively deal with challenges of fighting insurgency and rebuilding of the economy and it is in this context that we have welcomed the recent elections in Iraq. The priority before the international community now is to ensure gradual return to peace and normalcy in Iraq. A stable and peaceful Iraq is vital for the stability of the entire region, more so for India, since continued instability in Iraq can have dangerous consequences for our own peace and well-being.

West Asia stands at the crossroads of history today. The ongoing turbulence in West Asia will have a major impact on the global security system. With its energy potential and the rest of the world's critical dependence on its resources, the region will continue to attract international attention in years to come. Large and growing extra-regional military presence in West Asia, and the rising constituency of radical religious groups threatening to assert their subversive strength against the States, will continue to be of great concern to the international community.

We have all seen the disastrous consequences of an untamed tide of violence in the region. We will have to cooperate with one another to stem this tide and find more effective ways of addressing the root causes of such violence. The main task awaiting the international community is to bring lasting peace and security to the region.

Section I

West Asia and the International Security Environment

2

Changing Security Dynamics in West Asia

Implications for the Region and the World

HAMID ANSARI

The security dynamics of West Asia have changed repeatedly in the past 50 years. It is, therefore, interesting and instructive to take an initial reading of the situation in the region in the year 1956. It provides a perspective as 1956 was the year of the Suez War. What was the war about? An account written by an eminent statesman gives a graphic explanation:

> France and Britain had come to regard the Suez Canal almost as a part of their metropolitan territory. They believed, with exaggeration, that it was their jugular vein. How could they maintain their strength or prosperity without it? In Egyptian hands, it would be a weapon of extortion; it would also be beyond the Egyptian capacity to operate it. The atavistic habit of regarding the Canal as an extension of their own territory, as integral to them as Normandy, Kent and Sussex, was hard to break.

These sentences constitute the opening lines of Abba Eban's account of the Suez crisis in his *Autobiography*.[1] The Suez aggression was rolled back on account of the perception of such aggression as abject folly by the two superpowers.

Eleven years later, the region experienced a catastrophic war, initiated by Egypt on the basis of faulty intelligence provided by the Soviet Union.[2] The Arab world, and Egypt, paid dearly for it and its principal victims—the Palestinians—continue to pay a high price for it to this day.

Other wars and conflicts, in different parts of West Asia, followed. The Israeli-Egyptian Treaty of 1979, ruled out the possibility of another general Arab-Israeli war. The Iranian Revolution, also in 1979, changed the strategic balance in the region. Israel's experience in Lebanon demonstrated the limits of military superiority. The *Intifada* of December 1987 "marked the end of the twenty-year period in which Israel was able to control the West Bank and Gaza at a cost acceptable to Israeli society, allowing Israel to avoid painful and fundamental choices."[3] On the Arab side, Iraq's unwillingness to live with the *status quo* propelled it into an escalating series of follies; each, in turn, strengthened the military presence in the region of the United States in an era in which a competing external power was disappearing or had disappeared from the scene. Kuwait was followed by Madrid and Oslo, and by exercises in protraction at Wye River Plantation and Camp David. A glimmer of hope, at Taba, disappeared early enough and Palestinians and Israelis got driven by other, non-converging, imperatives. Iran remained isolated in a Thermidorian phase; defiant yet acutely in need of reconciliation and seeking openings for it sporadically.

The post-Cold War impulse drove the United States to the vision of a 'New American Century' of global leadership and, as Robert Kagan put it, to the awareness of "America's power and its willingness to exercise that power unilaterally if necessary."[4] The opportunity came with 9/11 and was avidly grasped. To quote Kagan again, "America did not change on September 11. It only became more itself."[5] The argument for it was summed up by the author of *The Pentagon's New Map* in an extended syllogism: "America is the initiator of globalisation; globalisation means growing global markets; stability is essential for markets and growth; growth requires resources, resources need infrastructure, infrastructure depends on money, money on rules, rules on stability, stability on the presence of a Leviathan having the will and the capacity to use force."[6] This was the thought process through which the Leviathan syndrome surfaced. Lest we forget, the description of Leviathan, given by Thomas Hobbes, still holds good, "that Mortal God, to which we owe under the Immortal God, our peace and defence."

This power was first exercised through the invasion of Iraq in 2003. The West Asian region, thus, became the critical pivot, and the principal ingredient, in the global security environment. Its intent was to be comprehensive on six counts:

- Deal with the WMDs in Iraq.
- Replace a dictatorship with a model democracy.
- Use the occasion to bring about modernisation of the Greater Middle East.
- Further the Global War on Terrorism.

- Settle the Israeli-Palestinian question on the basis of the proposed two-state solution
- Secure Israel.

By the spring of 2004, realisation was dawning, in the words of John Ikenberry, that the Iraq war was "a costly misadventure" based on misleading statements. His amplification of the point is noteworthy:

> Misleading the American public may not be inherent to the fundamentalist project, but is now part of the historical record. Hardliners were unwilling to go public with their private reasons for going to war in Iraq: transforming the Middle East and re-establishing fear of American power. Instead, the Bush Administration justified the war by citing the failure of the Iraqi regime to comply with past United Nations disarmament resolutions. Fundamentalists fancy themselves tough-minded thinkers. But they did not have the courage of their conviction to level with the American people on what this geopolitical adventure in Iraq was really about and what it would cost. Instead, they seem to have determined that American support for the war and post-war reconstruction can only be sustained if Iraq is linked to terrorism and 11 September attacks.[7]

Eighteen months later and by the end of 2005, stray opinions about the course of events in Iraq had matured into accepted theology:

> The United States needs a new strategy in Iraq and the Persian Gulf. The war is at best a stalemate; the large American presence now causes more trouble than it prevents. We must disengage from Iraq. We must do it by removing most American and allied military units within 18 months. Though disengagement has risks and costs, they can be managed.[8]

Thus, three years after the invasion of Iraq, it is difficult to quarrel with Anatol Lieven's judgement that, "US global power, as presently conceived by the overwhelming majority of the US establishment, is unsustainable" and need to be put "on a more limited footing."[9] Even right wing ideologues like Robert Kagan concede that the United States has "suffered a relative decline in that intangible but important commodity: legitimacy."[10]

The Implications of this Situation for West Asian Security

Enquiries into the dynamics of security in West Asia (as also elsewhere) unavoidably raise fundamental questions: security for whom, from what, and how? An analysis of the ingredients of global, regional, sub-regional and local security, as also of their points of intersection is therefore, essential to further the argument.

In conventional terms, security of West Asia is comprehensible in terms of the security of the national units within the region. Threats to their security could arise from aggression or subversion emanating from regional or extra-regional sources. Threats could also emanate from domestic sources within each national unit. The object of threat could be state units, sections of population within such units or, individual citizens. The sources of threat, in each case, could be states or non-state actors. Threats could have traditional or non-traditional dimensions.

The nature of the nation-state in the region, and the process of state formation in different decades of the 20th century, is also relevant to the analysis. The legitimacy of virtually all of these national units rests on secular foundations. This is relevant when considering the impact of religious fundamentalism on individual societies.

The traditional paradigm of security has remained riveted on the external threat. After the Kuwait War of 1991, however, resorting to outright war by a West Asian state against another in the region, seems remote. On the other hand, recent experience has shown that it is internal security that is the determinant of domestic stability. Here, the impact of four sets of variables has to be assessed: socio-economic changes, the failures of the state, the articulation of civil society and external factors. Each variable impacted differently in different societies. In this framework, 'the security dynamics of the region as seen from within, and from without, do not always converge.'

The regional scene in West Asia at the end of 2005, as seen from within the region, is characterised by:

- Immobility in the settlement of the Palestinian question in terms of a Permanent Status Agreement that had been visualised in the Quartet Road Map.
- Undiminished insurgency in Iraq.
- Uncertainty about the future of the regime in Syria.
- Renewed tensions between the West and Iran over the latter's nuclear programme.
- Initial steps by the Arab states in addressing the freedom, knowledge and gender deficits identified in the *Arab Human Development Report 2002* and in addressing, at an uneven pace, other imperatives of globalisation.

- Stabilisation of the domestic situation in Saudi Arabia and of Saudi relations with the US.
- Elections in Egypt, resulting in the emergence of the Muslim Brotherhood as a major political force.
- Identifiable gaps between the regime and public perceptions of the role of the US in the region.
- Unprecedented prosperity in the oil producing states.

The view from the outside, however, is somewhat different. It is summed up in *The National Defence Strategy of the United States of America,* published by the Department of Defence in March 2005. It mentions types of threats that could confront the US and its allies. These are classified into four categories: 'Traditional', 'Irregular' (insurgency), 'Catastrophic' (emanating from WMDs), and 'Disruptive' (coming from new technologies). These could be category-specific or overlapping. The document defines a clear cut strategy of global reach, through regional facilities, for dealing with these with the help of allies and friends.

Within this framework, three questions, relating to facts and perceptions, seem to arise:

- What is the impact of the Iraq experience on the US strategic perceptions relating to West Asia and the Gulf?
- How would changes in US perceptions influence the security dynamics of the region?
- What are the other centres of power in the region and how do they relate to the US power?

It is perhaps, too early to find official pronouncements on the lessons of the Iraq experience. The reason is obvious and understandable. Military strategists, however, are already drawing relevant lessons. One such study, emanating from the US Army War College in October 2005, highlighted the following:

- The US may have to scale back its expectation of Iraq's political future.
- The US needs to renounce interest in permanent bases in Iraq and concentrate instead on the small wealthy Arab states (Bahrain, Kuwait, Qatar and UAE) who view these bases as a source of protection.

> In planning future wars, the US must take care to maintain realistic expectations of what can be achieved through military intervention particularly in regard to the imposition of market economies and democracies on states whose inner dynamics are not fully known or understood.[11]

Would the Iraq experience induce a modulated approach? An option that has been explored as an alternative is that of broad-basing the responsibility and co-opting European members of NATO on the pattern of the arrangements in place for Afghanistan. The European Union's involvement in the Middle East Peace Process is of older vintage and has been useful though rarely decisive. In 2004, an exercise was undertaken to explore the role that NATO could 'play in enhancing security in the Gulf and the Greater Middle East'. By the end of 2005, the Organisation was asserting that it is interested in Gulf security because of: (a) the changing security environment, (b) the changing nature of NATO itself, and (c) the new dynamic in the Gulf region itself that has been unfolding over the past few years.

Three countries, on the western, northern and eastern periphery of the region, add a significant dimension to the security calculus of West Asia. These are: Israel, Turkey and Iran. All are non-Arab, each has a distinct identity, has a distinct history of relations with the United States and impacts on and is influenced by the security dynamics of West Asia.

The security of Israel, on Israel's terms—political and economic—has been for decades an article of faith in the US security doctrine. Israel is integrated politically, technologically, and culturally with the Western world. The terms of the debate over Israel and its policies are therefore, carefully structured and sustained. This accounts for the delay in resolving the Palestinian-Israeli problem, as also for the Palestinian rage.

Turkey emerged from being a faithful ally in the Cold War with twin impulses: on one side to weave a revamped Islamic identity into its secular state fabric and, on the other, to assert its European-ness by seeking to join the European Union. It remains to be seen if secular Europe, dedicated to the promotion of pluralism, would accept and accommodate this instance of diversity.

And finally, Iran. It has remained in the headlines for 27 years for having abandoned the Western alliance, disturbed the strategic balance in West Asia, questioned accepted theology on modernity, assisted the emergence of Hezbollah in Lebanon, benefited from the failed invasion of Iraq, insists on its right under the NPT and, in general, for being an obstacle to an unquestioned American hegemony in the region.

The discussion, thus, reverts to the basic question: in considering security dynamics of West Asia, do we view it from without, from within, or endeavour to develop a balanced perspective? The view from without is simple enough. If its premises are accepted, *The National Security Strategy of the United States of America,* supplemented by the new NATO doctrine, provides the answer: "that West Asia is too precious not to be dominated, not to be 'reformed' in terms of the Greater Middle East Initiative."

The perspective from within is altogether different. It has also evolved. Its infra-historical rhythm is now conceded, the implications of neo-patriarchy accepted, the three deficits of freedom, gender and knowledge, acknowledged. Helped by satellite television and the internet, a public opinion has surfaced and is sending signals of considerable intensity. Along with it, however, the anger over Palestine, Iraq and diverse other grievances are kept on the front burner of individual, group and national agendas and will not go away. Arab opinion surveys conducted by credible institutions in the United States bring this out unambiguously, and so do all sections of the media.[12] Effort to resolve the first set, while ignoring the second, will therefore, not succeed. In the absence of this success, and in the light of this failure to change public opinion, continuing US military presence in the region would only be seen in terms of might that, as Rousseau said long back, "does not make right."

The choice then is between a balanced and cooperative approach leading to harmony, and a hegemonic one with inbuilt potential for conflict. In times of change, international systems, in Kissinger's words, live "precariously" with their elements "in constant flux."[13] If, therefore, hegemony has not translated itself into primacy and is unlikely to do so after Iraq, then the obvious result would be a perpetuation of conflict. How would that promote security in the region? And, if the region is unstable, its impact on global security and economic well being of the world is not difficult to visualise.

International relations are about states in the global community. Such a community requires to be regulated by rules. Realists in statecraft in western societies (and eastern ones) need to be reminded of the founding principle of their own philosophy of state: that justice is the first virtue of social institutions, that rights secured by justice are not subject to political bargaining or to the calculus of social interests, and that the national interest of a just state are defined by the principle of justice.[14]

Wisdom in statecraft lies in anticipating disturbances and avoiding them. This would suggest a new security dynamics for West Asia. India, as a close neighbour, has an intimate interest in it.

Notes and References

1. Eban, Abba (1997). *An Autobiography,* Random House, New York, p.207.
2. Ibid., p.318.
3. Rabinovich, Itamar (2004). *Waging Peace: Israel and the Arabs 1948-2003,* Princeton University Press, New Jersey, p.308.
4. Kagan, Robert (2003). *Of Paradise and Power,* Random House, New York, p.61.
5. Ibid., p.85.
6. Barnett, Thomas P.M. (2004). *The Pentagon's New Map: War and Peace in the Twenty-First Century,* Putnam Publishing Group, pp.199-205.
7. Ikenberry, G. John. (2004). "The End of the Neo-Conservative Moment", *Survival,* Vol. 46 (1), Spring, p.19
8. Posen, Barry R. (2006). "Exit Strategy: How to Disengage from Iraq in 18 Months", *Boston Review,* January 7.
9. Lieven, Anatol (2005). "Decadent America must give up Imperial Ambitions", *Financial Times,* November 28.
10. Kagan, Robert (2006). "Still the Colossus", *Washington Post,* January 15.
11. Terill, W. Andrew and Conrad C. Crane (2005). *Precedents, Variables and Options in Planning a US Military Disengagement in Iraq,* Strategic Studies Institute, US Army War College, October, pp.43-47.
12. Arab American Institute (2005). "Arab Attitudes Poll 2005", *http://www.aaiusa.org/polls/Arab Attitudes2005.htm* and Aljazeera.Net Friday, December 2.
13. Kissinger, Henry (1994). *Diplomacy,* A Touchstone Book, New York, p.809.
14. Rawls, John (1999). *A Theory of Justice,* Revised Edition, The Belknap Press of Harvard University Press, pp.3-4 and 332-333.

3

Military Power

The Elusive Answer

SARI NUSSEIBEH

Reduced to its essential ingredient, the doctrine of military power and capability rests on the primary notion of fear, and specifically, fear to one's life. Even 'preemptive' military capability, including nuclear capability, is rooted in this primary notion of fear. Nuclear Powers do not tire from claiming that their capability is only preemptive, that it to say, only conceived as a defensive measure, or as a measure to address that primary notion of fear. Curiously, however, military capability, viewed in terms of the history of its own development, has constantly been found wanting. Renewal and regeneration of military capability seems to be an indispensable part of the military doctrine, conceived to address what turns out inevitably to be the insatiable need of fear. For every level of fear addressed through a military strategy or state of readiness, a further threat comes to be conceived for which further military capability is sought. Nine/Eleven, both as an occurrence and as a response, stands as a paradigm of this dilemma of the military doctrine: that cracks in the security strategy are dramatically revealed, which are then addressed through a reappraisal and refinement of the very same military doctrine that proved its failure to provide security in the first place. For any revealed failure in the security strategy, it is assumed, there must be a better strategy that could be found—hence, the pursuit of the elusive answer.

There is, besides the military doctrine's inherently paradoxical logic, and embedded deeply within its foundation, a cluster of assumptions concerning human nature which the doctrine simply takes for granted. Rather than seeking to find an answer to the dilemma of the military doctrine by pursuing its own line of logic (for example, by wondering what better security means can be found to address the basic 'fear to one's life' concern), this may be a good moment to revisit that cluster of assumptions, and to determine whether the 'original sin' of the doctrine is to be found there. By determining where the fault lies, perhaps an altogether different solution to security could be found. In considering that cluster, there are two interrelated dimensions to address—that of identity, and that of human association—which bear on our inquiry.

One may look upon the first dimension—that of identity—as a dimension of 'human ontology': briefly, separate and distinct human entities in this 'doctrinal' worldview are to all intents and purposes considered preset or predefined, as well as rigid or frozen. For example, 'being Palestinian' or 'being Israeli', or 'being Moslem' or 'being Hindu', are looked upon as hermetically-sealed, separate and distinct human entities possessing predefined and rigid identities on the world stage. Being an individual and possessing a singular identity is viewed in likewise fashion. Individual identities, in other words, are also viewed as being preset. But they are also viewed as being rigid or frozen. 'Being who you are' is a lifetime affair, so to speak. You are born into being a Hindu or being a Moslem, or into having your personal identity, after which you are determined to act in set ways which reflect that born identity. Being born Palestinian sets you in conflict with being born Israeli, simply insofar as being Palestinian is in conflict with being Israeli.

One may look upon the second dimension—that of human association—as a dimension of construction or of seeking to formulate relations between identities. I would like, at this juncture, to invoke Thomas Hobbes: assuming one has the human ontology one has as a given dimension, but further assuming that our preset and rigid units of discourse in this ontology are the individual human beings (rather than larger groups), political theorists have traditionally sought answers to explain, and therefore, find original justification for the existence (as we see them) of associations between those individuals. The answer Hobbes proposed was "fear to one's life." It is in search for 'securing' that life, that individuals are motivated to enter into associations with others. There are two interrelated but separate elements here, one having to do with what 'motivates' an association, and the other with what 'justifies' it, or is therefore the source of its legitimacy. An Army general who was also a historian and who lived long before Hobbes, by the name of Thucydides, already in the context of the war between Athens and Sparta identified 'fear,' besides self-interest and self-aggrandisement, as being the ultimate source of human political action. A human association in this view is, thus, predicated on security, both in terms of what brings it about as well as in terms of what justifies it.

I believe the conceptual source of the military doctrine (and the seed of its inherent dilemma) can be found in these two dimensions—the first relating to our understanding of human ontology, and the second to our understanding of how identities in this ontology relate to one another. I shall first try to explain in theoretical terms why this is so, and I will later try to relate this explanation to a practical case of conflict.

Starting first with a closer look at our human landscape it is immediately obvious that it is highly questionable that identities are set out in the way the military doctrine implicitly assumes. Let me here invoke another great thinker, a Nobel laureate from the Indian subcontinent, Amartya Sen, who has recently been addressing the issue of identities. Sen questions the fundamental notion of a uni-dimensional identity. Typically, he argues, an individual is by definition composed of multiple identity-layers, each one of which reflects a different role or association that individual happens to have in real life contexts. Sen uses the notion of this multiplicity to argue further in favour of enablement or freedom conditions which would allow the individual to 'choose' the strengthening and development of one or more specific layers rather than others. An Indian who is also a UK citizen should be free to choose whether to emphasise one or the other of these two aspects of her identity, as well as the extent of this emphasis. Sen applies this analysis to human culture or civilisation, essentially with a view to challenging the thesis of hermetically-sealed and rigidly predetermined identities. In other words, rather than viewing identities as being rigid and uni-dimensionally predefined, they should be viewed as being multidimensional, that is, already sharing commonalities, and malleable, or subject in their nature to the conscious choices of the human will. More generally, a cogent argument could be and has been made in favour of a human ontology whereby identities are neither rigid nor predefined, but are instead in constant processes of formation and change. As I shall argue below, changes in one identity can also cause changes in another, whether positively or negatively.

The second dimension in the conceptual source of the military doctrine relates to fear as a primary concern and hence, as a motivation for associations. While Thucydides, as we said, ascribed fear as a primary motivation to already existing political communities or groups, by the time political theorists of the Enlightenment began to consider the logical origins of the social contract, the emphasis had turned to single individuals, since the assumption was that the explanation needed was that of why such separate individuals are motivated to enter into associations in the first place. Various ideas were proposed, but one that is relevant to our discussion is that of fear, formulated by Hobbes. If fear to one's life from the aggression of others is what primarily motivates an association, then individual security for that life immediately becomes the *raison d'etre* of that association. Conceived primarily as a defensive mechanism, the notion of individual security naturally translated itself into a military capability, or into the capability of using force. However, while the notion of fear to one's life could be recognised as an important concern, it is not at all obvious why or whether it is indeed the primary concern. For a start, it is arguable, as the Moslem sociologist, Ibn Khaldun did in the 14th century, that the primary fear against aggression by others is that to the lives of one's blood, rather than to one's own life. But such instinct already presupposes compassion, that is, associations which are founded in altruistic rather than egotistic concerns. Of course, one could also propose primary concerns other than fear or altruistic compassion as sources of association, including, for example, a unified theory combining affect and calculative skill in the pursuit of a better life. Or one could propose other ideas proposed by various theorists, including, for example, Locke's property-legitimisation concern. That is to say, there is nothing in the fear-thesis that is more compelling than other possible accounts, and nothing therefore, to assure us that this thesis truly captures human nature at its pristine stage.

To return to the roots of the military doctrine one could now show that while these roots outline a human landscape where preset and rock-solid identities are destined to converge or clash in an almost independent law-governed movement, it is quite possibly the case that the only laws that are relevant in that landscape are those which Kant might have described as self-promulgated, affecting both one's identity as well as how one relates to others.

Applying the military doctrine to Israel's predicament, and given a rigid definition of 'what being an Israeli' means, the dilemma of the insatiability of fear implies an indefinite search for better security capabilities. For every security measure, there is a counter measure that breeds further insecurity. The acquisition by Pakistan or Iran of a nuclear capability neutralises the security achieved by Israel's own nuclear capability. Non-governmental terrorism neutralises governmental military superiority. Chemical weapons neutralise the security of cement walls. Demography itself can come to be viewed as a threat. In short, an ultimate answer to the military impera tive is simply never to be found, simply because there is none. However, in addition, and perhaps not so curiously now that we are aware of how identities can in fact change, what 'being a Palestinian' means comes to affect 'what being an Israeli' means. Often the effect is negative, in the sense of further reinforcing a sense of seclusion and retrenchment, and therefore, more fear. Now that the results of the Palestinian elections are out, 'being a Palestinian' for example has come to mean 'being Hamas', which already is beginning to affect what 'being an Israeli' means. Likewise, it is arguable that perhaps the very new face of Palestinians, judged negatively, is itself the product of how Israel behaved or behaves, that is, of 'what being Israeli' means. But just as there is verifiable negative influence on the other in identity-formation, surely a verifiable positive influence can also be hypothesised. To contemplate the powerful impact of this hypothesis (a hypothesis of positive change), imagine how it would be if, cognizant of the notion of identity malleability, either Palestinians or Israelis were to consciously work on themselves in the first place with a view to causing positive identity change in the other. Such an approach, based on a different point of departure, might well provide the answer to security which the military doctrine seems inherently incapable of providing.

Let me end my observations by elaborating a little on what this process is, and how it might work. Instead of its being a 'repellent' force, causing self-entrenchment in the other, 'being Israeli' could come to mean, through a radical internal process of self-reevaluation, being genuinely concerned for Palestinian rights and needs. Whether by word or by deed reflecting basic respect and recognition of responsibility, Palestinians could be 'enticed' or 'seduced' into reassessing how they see what 'being Israeli' is. A compassionate arm extended towards Palestinians, instead of the deployment of aggressive arms, could easily generate positive change in the Palestinian posture. Transforming oneself from being an enemy of the other into being a concerned friend, the other may similarly become transformed from being an enemy into becoming a friend. This thesis is predicated on the principle that, in this world, one doesn't only 'find' enemies or friends as rigidly predefined identities, but can equally and more importantly 'make' them, as well of course as lose them. But this is not a process that works only in one direction. It works in both directions. Palestinians, too, could similarly initiate an attraction dynamic, with a view to causing positive change in the other. But whichever direction it is, one could imagine the tool being used as some form of an Archimedean moral lever, by which one side, resting itself on the solid platform of human commonality, can so act as to raise the latent disposition to human compassion in the other side, which hitherto has been suppressed by layers of mistrust, fear, and the false sense of security through force. In the circumstances, one might call this a Gandhian lever.

Lest it be thought that I am proposing a totally idealistic theory, I should quickly add that egoistic self-interest is an integral part of the theory. In our present case, Israel's interest can best be served through an agreement with Palestinians—indeed through an agreement that provides us with acceptable terms for an independent State. Palestinian's best interests, likewise, are arguably served by the establishment of such a State. Therefore, rather than employ 'repellant' methods which verifiably make that agreement harder to achieve, it is more rational to employ an attraction dynamic which is more likely to bring its realisation about. However, from the point of view of the theory, even if a two-state solution is not in the best interest of either side, or is not in the mutual best interest of both sides, the important point is that a common interest, whatever it may be, is theoretically identifiable, but has to be identified by both sides together. It is not, like a military answer, elusive or inscrutable, or unilaterally determinable.

In conclusion, I believe it is tragic to give in to a human ontology of hermetically-sealed, discrete, and inherently rigid identities, and to proceed to construct international relations and military strategies on this basis. There is no inherent reason for why 'being Iranian' and 'being Western', or 'being Christian' and 'being Moslem', or 'being Hindu' and 'being Moslem', should be at odds with one another—any more than for why 'being Israeli' and 'being Palestinian' could not be complementary rather than mutually exclusive identities. I do not claim, of course, that acts of terrorism can automatically cease once the new approach is adopted, or that conventional security methods would all of a sudden become obsolete. But I do believe that much of the groundswell accompanying and encouraging terrorism can be defused, hence making targets more isolated and identifiable. And when they do become identifiable as abnormally extremist voices in a societal context, it is very likely the case that it is also within that context they could be contained.

4

The International Strategic Medial Path

West Asia in a Global Security Context

JAFFAR MIRGHANI AHMED

Ever since renaissance marked the modern era, Europe took the lead in shaping the contours of modern world affairs. The advent of colonialism led to globalisation of these essentially European ideas, ideologies, concepts and even the language of discourse, whereby such ideas flourished in all corners of the world. Thus, the entire world came to see itself through the European spectacles. It is a fact that the ideals and concepts that shape the world today continue to remain infected with Eurocentric idioms and vocabularies, which continually reinforce the hold of Europe on the 'mind' of the world.

Whether we admit it or not, the non-European world stands completely overtaken by the post-renaissance worldview of Europe at all levels, i.e., culturally, politically and economically. Colonisation started a process of phasing out of the native culture, politics and economy and sought to push them to archives and museums, duly recorded and displayed but unworthy of being replicated. The most popular way of dividing history of the colonised people into 'pre-colonial, colonial and post-colonial' signalled the unerasable influence of the European mode of thinking and 'inferioration' of the worldview of the colonised people.

Nothing can be more illustrative of aforesaid proposition than the following expression which imposes on the reader a European perspective of the area which we now call Middle East. For the strategic thinkers in the so-called 'West', the landmass at the conjunction of the three continents is important in as much as it can contribute to the strength of the predominant powers:

> The chief importance of the Middle East in world affairs is strategic and dates from the time the expansion of Europe turned the whole world into the theatre of power politics. The region constitutes a land bridge between three continents. Possession of it by a predominant sea power makes possible the limitation of the scope of a military conqueror by confining

> the area of his conquests to a single continent, his own. Neither group of powers can afford to see it become a vacuum into which the influence of the other flows unimpeded.[1]

This landmass was most suitably called the Middle East. Subjected to analysis such a vague expression would mean that part of the 'East' which bridges the three continents. The insertion of the word, 'East', points clearly to a Western perspective, which seeks to draw the line between the east and the west and create invisible borders in the mind. But what is exactly meant by 'Middle East' here? It means a geographical terrain that is more strategic for both the powers (from the West) than perhaps, it is for the people inhabiting this very terrain. And that is the reason why it has to be continually under the influence of the West. Influence of any other power over this area is an impermissible proposition, almost a theoretical impossibility.

A scholarly work published from USA in 1955, makes things even clearer. Right in the first line of its preface we read:

> The ghost of the Old Eastern Question, never properly interred...in the peace settlement of World War I, returned a quarter of a century later to stalk up and down the Middle East and haunt the great powers....[2]

Here again focus is on the 'East' as a problem which refuses to go away and demands critical attention and military engagement. The term, 'Eastern Question', designated, for the westerners, the problem of European territory controlled by the Ottomans since the 18th till the 20th century. Most of the European historians would argue that the First World War which saw the demise of the Ottoman Empire closed the old Eastern Question, but the problem of maintaining stability in the area continued to bother the succeeding powers in the region.

The 'Eastern Question' is, in fact, a European mint dating from the times of Ottoman dominance of the Eastern Europe. The lasting impact of the Ottoman forays perhaps, laid the foundation of the power politics divided between the East and West. Even considerable chunk of territory in Europe, which was under the Ottomans, was marked 'Eastern Europe'. The seemingly innocent categorisation of 'eastern' and 'western' Europe emphasised a cultural and civilisational and even religious discontinuity. The 'Eastern Question', as it was pejoratively used, implied that the 'East' was the eye of the storm. This notion ignores the fact that the two World Wars originated in the Western domain.

The location of the 'Middle East', if we agree to use this terminology, has made this strategic medial land bridge the theatre of power struggles throughout history. Actually that awareness of the strategic value of this medial zone dates back to the dawn of human civilisation. And there has been a perennial fight to access and control this terrain. Here, one may like to remember the confrontations between the early Nile-Euphrates powers, the Greeks-Romans-Persians and the Islamic-Persian-Turkish-European powers.

The Eurocentric notion of a world torn between two poles of East and West considerably blurred the memory of the people of this terrain and distorted their understanding of history. We are fast forgetting the days of undivided history and refusing to acknowledge the continuing intermeshing of the east and the west. In fact, there is a theory that the natural division of world into a 'Cold North' and a 'Warm South', necessitated regular exchanges between the two regions. This formed the basis of mercantilism, upon which, traditionally economy used to rest. The world has since those days experienced fair competition and jealous rivalry between these two blocks during the process of mercantile exchanges. Commodities from different parts flowed through the seas and overland from one end to the other, i.e., from the African hinterlands, West Asia, North and North East Africa to Europe. The flow of commodities was also accompanied by flow of knowledge and ideas as William D. Cooley, justifiably, stated in his treatise *Claudius Ptolemy and the Nile.*[3] Journeying all those quarters of warm climates and even some of the near reaches of cold ones, from Morocco to China, Ibn Battuta—the medieval Moroccan Muslim traveller who, 50 years before Marco Polo, visited the equivalent of 44 modern countries from West Africa to India, wherever he travelled or landed—was sure to find brothers and friends ready to welcome him with open arms. Despite this overwhelming sense of fraternity and interpolation and intermixing of cultures and civilisations, the human penchant for differentiating one from the other remained. Thus, we see new barriers, new borders of the mind seeking to divide the world according to perception of the dominant power of the world. Thus, we see European terms and terminologies crowding the international political landscape. The nomenclatures they set upon others have so far survived and with them, the perceptions that led them to name a place or people one way or other have stayed alive too. For example, they called the Afro-Asian world lying in-between, or within the near reaches of the tropics as Ethiopia in Greek or Sudan in Arabic. And the names have survived.

In his account of the reasons behind the Greco-Persian wars, Herodotus, the father of History, relates that the Persians declared war on the Greeks when the latter attacked the country of the Phoenicians; because the Persians claimed that Western Asia was legitimately under the sphere of their influence. But the Greeks considered west Asian roads lying under Persian control, denied them access to trade in the Indian Ocean region. Some classical writings dating back to the Greek and Roman eras attest to this. Alexander the Great endeavoured to bring under his sway an empire which would comprise this terrain. History is, thus, witness to the fact that the medial landmass which we call Middle East has been one of the most sought after corridors in the history of mankind and this has attracted the attention of the world powers in different periods of history. The Moslems were the only people who could bring the intermediate world, the Middle East under their effective control.

In old Islamic literature, the notion of a whole integrated Afro-Asian world of warm climates is to be found. In his book titled: *The Glory of the Sudan*, Gahiz, a renowned Arab writer who lived 12 centuries ago, entertained the idea of an integrated Afro-Asian world of warm climates and elaborated on it. Under the generic name (Sudan), he counts the Chinese, Indians, Abyssinians, Nubians, Copts, Berbers and all other nations living in-between.[4] This world of in-between has also been identified by western authors as predominantly the world of Islam. They would say that it has been afflicted with a persisting sense of being in the middle, between the north and south, east and west. The problem of this intermediate landmass is well-characterised by Fernau in his book titled: *Moslems on the March*:[5]

> The intermediate position to the burdens of which the world of Islam is exposed makes its appearance abruptly, without softening gradations, in Persia. In Persia the means of escape from pressure alike from the ocean and from the massive block of the Eurasian continent are of the frailest. Since Russians and Anglo-Saxons have faced each other threateningly and with suspicion, the territory between Turkistan, the Caucasus, and the Persian Gulf has been one of the weakest parts of the girdle separating the two great groups of powers. Thus the main trends of modern international policy have relentlessly dictated developments in Persia. Any gains to be

> had from successfully playing off the powers against one another are more than counterbalanced by the instability produced within the country in its weakness by the changing policy of the Great Powers.[6]

The author goes on to characterise in the early days of the Cold War that the instability in the Moslem world which is so much in discussion in the world today:

> Lack of stability characterizes the world of Islam. That is the outward sign of a historical process. Its causes lie in the great shiftings of power that have come in the outer world, and in the profound changes that have taken place, in the course of decades, in the peoples of Islamic civilization.[7]

The allusions in the passage as in many such observations by western scholars would make us alive to their perceptions that the politics of this intermediate world is dependent more on external power plays than on internal transformation, reform or revolution. In view of the developments in the entire region straddling North Africa and Western Asia, it is apt to say that the analysis stays relevant even today even if the nature of world politics has changed rapidly since the end of the Cold War.

It is of great significance to note that this intermediate position, as defined in Fernau's work, is nothing but the Middle East of today in its widest geopolitical definition. In referring alternately, once to the peoples of Asia and Africa, thence the peoples of Islam, Fernau collates the Afro-Asian-Islamic worlds; coupled with the phrase, intermediate position chosen by the peoples of Islam, and implies that developments in this intermediate region has to be continuously monitored and for the benefit of US, the region has to be kept under its influence.

Analysis of history reveals recurrence of analogous situations. Post-Cold War situation is not much unlike that in the aftermath of Second World War. The victorious campaign of the allied powers in the 1950s, as well as the so-called success of liberal democracy over communism in the 1990s, was confronted with severe economic crises and the threat of new pattern of rivalries. The political and economic world situations in the world in 1950s and the 1990s resemble each other.

Through all these vicissitudes of time, Middle East has remained quite important in the calculus of the world powers. In today's world, with the threat of terror looming large in the horizon, there are many who indicate that the roots of terror lie within the Middle East region. This has again made the region immensely important from security point of view.

It is unfortunate, however, that the political dynamics of the region is still determined by the forces outside the region, who have a vested interest in the hydrocarbon resources of the region. The security dynamic of the region has received a big jolt with the Second Iraq War, which has conveyed a strong message to the ruling regimes in the region that any country seeking to alter the security dynamic of the region will be sternly dealt with by the US and its allies.

As far as the international community is concerned, it must put in place more equitable and just measures to ensure that the people of the region enjoy greater political freedom to decide their future. The region demands greater attention and the politics within states in the region merits careful analysis.

Notes and References

1. Royal Institute of International Relations (1955). *The Middle East: A Political and Economic Survey*, London, (Reprint).
2. Hurewitz, J.C. (1953). *Middle East Dilemmas: The Background of United States Policy*, Harper, New York.
3. Cooley, William Desborough (1854). *Claudius Ptolemy and the Nile or an inquiry into that geographer's real merits and speculative errors, his knowledge of Eastern Africa and the authenticity of the Mountains of the Moon*, J.W. Parker and Son, London .
4. Hence a natural sense of unity among them which extended to the colonised Afro-Asian countries of the region. The Bandung spirit of the 1955, which sought to carve out a middle path between the two power blocks during the cold war days, some analysts have argued, was the extension of this feeling of being in the middle.
5. Fernau, F. W. (1954). *Moslems on the March: People and Politics in the World of Islam*, trans. E.W. Dickes, Knopf, New York.
6. Ibid., p.160.
7. Ibid., p.277.

5

Security and Development in West Asia

Need for New Thinking

MOSTAFA ZAHRANI

In its classic definition, state, as Max Weber argues in the general framework of sociology of state, has two major functions: first, to provide internal order and second, to defend the nation from external invasion. Based on this definition, in the recent released literature in the field, the concepts of security and development have also been considered as the main functions of states. While this may be true, the concept of security as it is defined in the west does not have universal appeal. Every region, religion, culture and civilisation has its own approach to security and this has to be factored into any discourse on security or security dynamics.

Coming from Iran, an Islamic country in the Middle East, I thought it better to share my own understanding of the trend of politics in the region and the way the things are moving there. My central hypothesis is that the countries in West Asia lack a security framework in the region. There is also an absence of any common understanding of security, broadly because there is no culture of security. It is partially because the term 'security', as we have tried to define and understand it, is of Anglo-Saxon origin and based on a different set of realities. It has become fashionable today to focus on the third world while talking about security primarily because majority of conflicts are happening in the third world countries in last 50 years and not in Europe or US. Hence, if one wants to understand the meaning of security in today's world one has to fathom into the social, political and economic reality of the third world. We have to have a new conception of security. The traditional as well as the realist school of thinking (primarily Anglo-Saxon in origin) may not quite grasp and explain the security problems, we, in the third world, experience.

These two schools of thought are state-centric in nature. They cannot explain the problems we encounter, which do not necessarily flow from the state. The traditional school assumes that the threat to state comes from another state in a form of military operation in an organised manner. However, many of the security problems we face in our midst are different; they come from intra-state and non-state actors and sometimes they may not be quite visible. They are quite nebulous. So, state-centric approaches will not help. The strategy of combating these kinds of threats will have to be different. Since the threats that the states encounter today are not necessarily military threats, the solution may not be sought through military means alone. There has to be a developed concept of security.

When we talk about our region, the economic security of our region assumes enormous importance. When I talk about economic aspect of security, I do not mean geo-economy alone. Those who use the concept of geo-economy assume that since the end of the Cold War, economic instrument can be used as a coercive force instead of military force. So when I say economy, I do not mean economy as a source for military buildup or as a source of coercive diplomacy. Instead, I am talking about economic development. That is a very important instrument in our region.

Security and development complement each other. At the same time, as far as states are influenced by international and regional environments, both security and development are constrained by these two factors. To find out this effect, one should refer to what West Asian states make of the world surrounding them. What is important for observation is that the Asian states are incorporating ideational factors into their interests. As a result, no single state in Asia perceives security; for example, in the Middle East the way Washington does. Regionalism being an alternative to a polar international system, some regional big powers shaping Asian Identity are on the rise, getting more powerful due to the US policies in the Middle East, notably the invasion of Iraq. Pressurising Iran on nuclear issue by the West which costs deeper rift between Asia and the West would also be another example.

Another thing which is important, when we talk about a new conception of security, is the security dilemma. Nowadays, especially in our region, we are faced with a security dilemma. It is within the States, mainly in the context of economy, that this dilemma appears in many forms. One of them is how to ensure economic security along with physical security. I will give you some examples. In Afghanistan, those who grow opium are being overlooked because the external forces want to stabilise that country first and address the opium issue later. Since the economy is in ruins, opium production also gives the people some hope of recovery and they grow it even if it is illegal. However, this creates insecurity for the countries outside, for example countries like Iran, Russia and countries in the West. It is not military insecurity but social insecurity. This is a big problem that we have. This is a big dilemma for the external powers in Afghanistan. If you take away opium entirely then you cause economic instability and provoke the local warlords who depend on drug production for their survival. If you do not interfere, then you create social instability in the opium or drugs receiving areas. You are trying to create stability and security for some people through something which is creating non-military insecurity for the others.

When I talk about the security in terms of development, I subscribe to the view that a government should be able to establish an acceptable level of welfare for its citizens. That brings security to its people and the state. However, sometimes this also creates a security dilemma. When the government undertakes measures to bring development to its people, it gives a boost to the expectation of the people and more often than not it becomes quite difficult to fulfil these expectations. This brings in popular apathy and breeds popular disaffection and in a multicultural setting, where some communities may feel neglected, this may lead to internal security problems. Iran is a case in point.

As you know in Iran, a country with a population about 70 million, the economic expectation of the people is the highest in the region and this has grown in last 20-25 years, after the Islamic revolution. Those who have been in Iran before the revolution and after it, they can compare it. Before the revolution, I very much remember that if you had a big family with five brothers, two sisters, all these five brothers lived in a single house with different rooms; all of them would marry and continue to live together. However, at the moment, this is impossible even for a father and a son to live together, let alone a big family. All of them need houses; all of them need cars; all of them need all kinds of things.

Thus, the economic expectation is very high. As a result, to give this level of welfare to the people the state has to adopt a policy of sustainable development. A country like Iran has the oil, has the gas, but can we rely on it and for how long? Nuclear technology is an important instrument for the sustainable development. This search for nuclear energy has been there from the Shah's time, when some of the plants were given by the West, by Americans. If the regime wants to be secure and ensure that there will be no further revolution, no unrest and instability, they have to address the economic needs of the people and so they have to develop and they have to have sustainable development. For this we consider nuclear power plants absolutely essential for sustainable development. To be self-sufficient in this direction, we have gone for enrichment. This has led to lot of misunderstanding in the world today. The Iranian intentions are very clear. Our nuclear efforts are peaceful and aimed at sustainable development. This is, what I call, a new type of security dilemma. Regimes or governments in Iran, whosoever is in power, regardless of the type of ideology, will have to address the needs of these people. They need sustainable economic development. Another crisis in the region is the difference between the people and governmental elite. This is likely to grow. The governments will have to give emphasis on individual civil right, which is a big problem in our region.

Now, I will try to explain some of the trends which I see in the region. The security analysts from the region should come to terms with the specificity of their situation and evolve a new concept of security. This is very important. Identity is very important; ideology is very important. Many countries in the region including Iran do suffer from an identity crisis, especially in their foreign policy. Two specific periods of this crisis, I must identify. One was during our revolution. At that time, the decision from within created that kind of an identity crisis in our foreign policy. We were coming to terms with the same and evolving a new identity when another change in external environment brought about another crisis in our foreign policy. This second period was after the Cold War. This time we did have this crisis because Iran had been in interaction with other players of international system and changes in their foreign policies compelled us to readjust our relationship with them. I think other countries in the region also suffer from similar identity crises. It is the root of the problem.

The sources of identity are: territory, culture and religion. Some people have come to a conclusion that if you try to establish your identity based on citizenship, you will not have any problem. How many countries in our region have been able to establish such an identity based on citizenship? But it is a good solution I think. However, culture or religion in the region is very important to understand the security concerns of the states in the region.

In the 20th century, there were two experiences specific to the region. One the fall of the Turkish Ottoman empire and the other was the decline of the Iranian Persian empire. The collapse of these two brought three different types of Muslim movements. One was radicalism; the other was rationalism or modern understanding of religion and the third was Islamic secularism. The modernist or rationalist approach will continue but Islamic secularism will not prevail. Radical Islamic movements will, however, be a recurrent theme in the region, as a typical response to the external stimuli. The struggle between radicalism and rationalism for the soul of Islam will characterise the nature of Islamic politics in the region in future.

Look at the last Palestine elections which brought Hamas to power. You will have both radicalism and modern understanding of Islam. Modern understanding of Islam not only accepts some of the dimensions of modernism but also tries to justify its position on Islam based on the Islamic texts. But no one can say whether radicalism or modern Islamism will prevail. I think both of them will be there, depending on the behaviour of the actors, both international and domestic. Look at Iraq for example. You have those elements of rational understanding of religion and radical understanding. You do have the same thing in Egypt. You do have the same thing in other countries as well.

Globalisation has given a new dimension to the understanding of these trends. If globalisation is a prevailing trend, it ought to bring together different religions by emphasising on the universal values in every religion. The only aspect which is universal among all the religions is spirituality. So many institutionalised religions will have to forsake their traditional roles like they had in the past. If they want to work in the age of globalisation they should concentrate on the 'universal' in every religion and something which is acceptable to everybody. As I said, spirituality is one such value common to all religions and this will help us in overcoming so many problems that we have. The metaphysics of religion which we had emphasised in the past is not prevailing in this new era of globalisation.

As I have said earlier, you have two trends within Islam in the region and the world—one is radical and the other is reformist. Iran has experienced both of them. In this context ethnicity also becomes important because of the penetration of the concept of nation-state in the region. Ethnic assertions at the periphery weaken the capacity of the government. Usually our experience in Iran showed that the political and economic expectations of different ethnic groups are different and such demands go up all the time creating unusual tension in the governmental machinery. But our experience at the time of Fatami government showed that if the Government is really strong then it can seek to address the expectations of the different ethnic groups by emphasising on human development. Every state and region will have to evolve their own paradigm to address such issues which will continue to problematise their security thinking in the coming days.

In the changed setting, what will then be the elements of development and security of the region especially in an ideational context? Assuming Japanese protectionism and not extreme Western individualism as a model for this region, reducing the threat to Asian way of life may be taken up as an important element of this security framework. Secondly, neither an anarchical structure, nor a hierarchical one should replace the bipolar order we left behind. It should also be clear that American hegemony cannot provide security to Asia. The international community should also not try to deprive nation of its right to peaceful usage of nuclear technology as a mean to sustainable development is not acceptable. Having said that, I must emphasise that international security can be assured if the Asian approach to pluralism forms the basis of security thinking. It is not only a step towards a more globalised world, but also a guarantee for the global order based on cooperation and understanding of each other. Any attempt to impose one's worldview on others has to be forsaken.

6

The Post-9/11 Security Predicament in West Asia

The Imperatives for Ending External Intervention

AKMAL HUSSAIN

The security situation in West Asia presents some new characteristics in the post-9/11 era. The perceived insecurity, emanating from within the region, has engaged the US militarily in a manner which was never seen in the Cold War time. The previous American strategy in the region was to build up its allies militarily instead of taking the onus into its own hands. The idea of containment of former Soviet Union was extended to the region by tying up the pro-western countries in military alliances like Central Treaty Organization (CENTO) and bilateral defence treaties.

The terrorist attack on the World Trade Centre and Pentagon on September 11, 2001, has sensitised the international community about terrorism in such a way that has not been noted before. Currently, many states have come together, under the leadership of the US, to launch an international campaign against terrorism, primarily focused on prevention of these activities. The Americans have extended their war against terrorism from Afghanistan to Iraq and kept the option open to extend it further.

Although fighting terrorism has been the basic thrust of new policy shift in the US, the justifications for an assertive American posture after Cold War was provided by the neoconservatives in the American foreign policy establishment even before the George Bush Jr. assumed the Presidency. The brutal attack on the World Trade Centre came as a boon for the 'neocons' to put their theoretical tools into practice. After Afghanistan, the occupation of Iraq was conditioned by this school of thinking that advocated a central role for the US in the world today. The US has to usher the world into the "New American Century," some of the neocons held.

Regrettably, the military occupation and subsequent democratisation process in the West Asian region have neither turned these countries peaceful nor secured a stable region for the US. The threat of terrorism has not been reduced regionally and globally. Religion-based terrorism, once fuelled by the US to counter communist influence in Afghanistan, has turned as a Frankenstein's monster against the US itself. The military challenges and or resistance to US occupation and political guidance in Iraq contribute to security predicament of the region. The socioeconomic stability has become very fragile in the given context.

There is no denying the fact that external intervention in military terms has proved counterproductive for the region. There is no let up in the spilling of blood in Iraq since the first US bullet was fired in 2003.[1] The democratic experiments under foreign guidance too have not brought the desired success in establishing a tolerant society. The clear majority of the Shi'a and Kurdish sections in the election held on 15 December 2005, in Iraq cannot ensure participation of the Sunnis in the democratic process. The alliance among the Sunni and secularist Arabs has already rejected the election results. By mounting mass protests they have demanded a new round of elections and threatened a boycott of parliament session.[2]

The instability in Iraq, resulting from the ongoing militant attacks, a backlash of foreign occupation, weakens the very foundation of these experiments. Iraq has not become anymore a peaceful country than it was during Saddam era. The continuing military hostilities and the shaky election results may push the country to more chaos, leading to an ineffectual central government.

Therefore, the thrust of this paper is to argue that the situation may change with the end of all physical presence of external entities paving the way for the indigenous forces to handle their own affairs. Although there have been announcements by the US and the British governments about withdrawal of some portion of their forces from Iraq, this token withdrawal would not solve the problem. It should not be construed that Iraqis will fail to govern themselves once the foreign forces will be withdrawn.

While writing his book *The Persian Gulf and American Policy* (1982), Emile A. Nakhleh commented that although the US was able to project its power in West Asia, it was unable to decide the region's destiny. Nakhleh may have sounded true in those days in view of weak American physical presence in the region. However, his statement is no longer appropriate in the present context. The military and diplomatic steps during the last one and a half decade in this region has established the United States in the arbiter's seat. Its contemporary position could be explained as the realisation of its cherished vision of being the protector of free and democratic world in the post-World War II era from the 'totalitarian', meaning the socialists forces, and, after the Soviet disintegration, from the terrorists of different Muslim dispensations.

The American strength, prestige and influence in the early 50s urged its leadership to assume the guardian's role of the Western world and unleash the Cold War competition with the Soviet Union. The values, hitherto propagated by European powers, had then been forcefully pursued by the US. After Winston Churchill coined the term "iron curtain" to explain the divide between the so-called 'free' and the 'socialist world', the Americans took over with terms like 'peril of communism', 'containment', 'Alliance for Progress', etc. In giving a perspective to these new concepts, they had pointed to the 'menace of totalitarian communism' and its contribution to de-stabilisation of the international order.

Because of its strategic location, emanating especially from its hydrocarbon resources, West Asia attracted US attention for mainly industrial and military reasons. And its physical proximity to the Soviet Union had always been the most important factor guiding US foreign policy towards this region. President Truman enunciated the Truman Doctrine (1947) to underscore the imperative of protecting the region from 'communist subversion'. Subsequently, President Eisenhower (1957) and President Carter (1980), in their foreign policy goals, underscored the importance of this region for US security.

Like any big player in international scene, the US treats its own security and survival as the prime goal of its foreign policy. However, a critical perspective of American ideology states that American national interest was pursued by ingenious use of moral values, organised Christian religion, abstract principles and clichés such as 'fighting aggression', 'practising non-intervention', 'respecting self-determination', 'helping poorer people through benevolently conceived economic aid', 'providing military aid to those threatened by communist totalitarianism', 'working for collective security through effective international organisations', 'striving for four freedoms', and 'making the world safe for democracy'.

One may say that the American worldview was conditioned by a feeling among its ruling elites that the US is the 'number one' power and therefore, it must have allegiance of other nations. The psychology of powers like the US had been aptly explained by Senator J. William Fulbright, once the Chairman of the Senate Foreign Relations Committee.

> The more I puzzle over the great wars of history, the more I am inclined to the view that the causes attributed to them—territory, markets, resources, the defence or perpetuation of great principles—were not the root causes at all but rather explanations or excuses for certain unfathomable drives of human nature. For lack of a clear and precise understanding of exactly what these motives are, I refer to them as the 'arrogance of power'—as a psychological need that nations seem to have in order to prove that they are bigger, better, or stronger than other nations. Implicit in this assumption...that force is the ultimate proof of superiority—that when a nation shows that it has the stronger army it is also proving that it has better people, better institutions, better principles, and, in general, a better civilization.[3]

The US strategy and the tactics employed in post-world war days swivelled around the idea of 'containment'. Accordingly, they had to paint the adversary as black as possible and understand that a high-risk foreign policy required selling the threat.

It is interesting to note that the threat of Islamic militancy seized the Western attention even before the 9/11 incident. After the advent of an Islamic regime in Iran and the resultant strained relations between Iran and the Western countries, there had begun profound academic and political interests in Islam in the West. The triumph of Islamic Salvation Front in Algerian elections, the activities of Egyptian and Sudanese Islamic forces and Al Qaeda attack on US missions were regarded as threats to liberal values and interests. President Bush Junior very clearly underlined the US goals when he spoke in the joint session of Congress immediately after 9/11 incidents:

> Our war on terror begins with Al Qaeda, but does not end there. It will not end until every terrorist group of global reach has been found, stopped and defeated...They want to drive Israel out of the Middle East. They want to drive Christians and Jews out of the vast regions of Asia and Africa.[4]

It may be argued that the current formulation of US policy for the West Asian region is connected with the ideas and thoughts of the 'neocons' in the administration of President Bush Junior. The neocons have opted for an iron-fist policy for the region. The advent of these neocons should be seen in the backdrop of competing ideas around American foreign policy in the post-Cold War period.

In the initial days after the end of the Cold War, there had been an euphoric vision about a new world in which the arms race will be reduced, war will be less frequent and peace will pave the way for more economic progress. It was also believed that regional powers would also cut down their military spending. Even within the American administration, there were people who argued for reduced military spending and smaller size of armed forces. Neither this new vision was realised nor the arguments on reduced defence spending were acceptable to the neoconservatives in the American administration. Persons like Donald Rumsfeld, Dick Cheney, Paul Wolfowitz, Jr. Bush, Francis Fukuyama, Zalmay Khalilzad and others grouped together and formulated what came to be known as Project For New American Century (June 3, 1997). In their words:

> As the 20th Century draws to a close, the United States stands as the World's preeminent power. Having led the West to victory in the Cold War, America faces an opportunity and challenge: Does the United States have the vision to build upon the achievements of past decades? Does the United States have the resolve to shape a new century favorable to American principles and interests?
>
> We are in danger of squandering the opportunity and failing the challenge. We are living off the capital-both the military investments and foreign policy achievements-built up by the past administration. Cuts in foreign affairs and defence spending, inattention to the tools of statecraft, and inconstant leadership are making it increasingly difficult to sustain American influence around the world...But we cannot safely avoid the responsibilities of global leadership or the costs that are associated with its exercise.

The project planners have laid importance on the following goals in future American foreign policy: to increase defence spending significantly and modernise the armed forces; to strengthen ties with democratic allies and to challenge regimes hostile to US interests and values; to promote the causes of political and economic freedom abroad and to accept the responsibility of America's unique role in preserving and extending an international order to "our security, our prosperity, and our principles."

As the defence budget of the Clinton administration was unacceptable, the project planners unashamedly opted for an enhanced military budget that, in their view, should be commensurate with the stronger role of the country. Secondly, their standard of treating an ally and a foe is based on a simple principle of assessing who is for and against the American interests and values, not on the nature of the regime. Judging by this criterion, a dictatorial or despotic regime may be treated as an US ally if it supports its interventionist role in world affairs. The third goal could be defined as an attempt to underscore the need for a globalised world where attempts to protect one's own economy from the encroachment by big economies would be regarded as an offence. And lastly, the US should play a unique role in creating a global order, which offers the US enough scope for intervention in any part of the world in defence of its interests.

One could discern the neoconservative ideas surrounding US foreign policy in the academic writings of Condoleezza Rice, Robert B. Zoellick and Kenneth M. Pollack in *Foreign Affairs* journal before and after the election of Bush Junior for the first term. Rice who later became National Security Advisor and now the Secretary of State held that the next administration's foreign policies must help attain the American values that the Clinton administration could not attempt. Writing on the eve of presidential election in 2000, Rice argued for a strong military so that it can deter war, project power and fight in defence of American interests if deterrence fails. Having criticised Clinton's defence policy about military expenses and deployment of forces, she argued in favour of building the military of the 21st century rather than continuing to build the same on the structures of Cold War. Secondly, the US military, she held, must be able to meet decisively the emergence of any hostile power in Asia-Pacific region, the Middle East, the Persian Gulf and Europe. Thirdly, the US foreign policy should be geared towards 'humanitarian intervention', if there was a need for that in any corner of the world. Rice was unambiguous about her views on how to deal with the 'rogue regimes' like Saddam's by pointing that he (Saddam) has no "useful role in international politics, nothing will change until Saddam is gone, so the United States must mobilize whatever resources it can, including support from his opposition to remove him."

Zoellick started his article stressing on the next President's primary task as building public support for the strategy that could shape the world so as "to protect and promote American interests and values for the next 50 years." Like Rice, he made critical comments about Clinton's foreign policy and opined that the latter failed to define a new internationalism for the US and let historical opportunities slip away. The five principles he suggested for the future Republican foreign policy were: matching America's power to its interests; building and sustaining coalitions and alliances; judging international agreements and institutions as means to achieve ends, not as forms of political therapy; linking the US to communications, technology, commerce and finance sectors around the world through new networks of free trade, information and investment; and, finally, facing enemies who developed the WMDs and missiles.

Pollack was more outspoken than Rice and Zoellick. He advocated outright invasion of Iraq and replacement of Saddam regime with a successor prepared to abide by its international commitments and live in peace with regional neighbours. In his opinion, the strategy of containing Saddam regime had started to unravel and Saddam was out to destabilise the region again. Even the strategy of deterrence relying on the threat of US intervention to dissuade Saddam from future aggression would not be a viable alternative, he strongly argued.

On 28 June, 2004, the US occupation authority handed over power to an interim government in Iraq, two days before the deadline. This interim government was composed of men from different sections of Iraqi society, handpicked by the US. Before that, the UN Security Council, due to intense diplomatic negotiations with the US and British governments, finally adopted Resolution 1546 terming it a legal framework of transferring sovereignty to the Iraqi people. To the US and British governments, these steps were significant for the transitional period that would usher in a new Iraq. But ironically, there was no mention of when and how the occupation forces would be withdrawn. There was rather a consensus among them that security was an essential element for a successful completion of their mission.

In spite of the interim administration and the parliamentary election held in January 2005, the situation remained tense, as the divide along Shi'a-Sunni, Arabs-Kurds, Kurds-Kurds continued to offer a frustrating picture for an integrated Iraq. In the given situation, with the inter-sectarian and inter-ethnic relations going berserk, the task of returning peace to Iraq through an elected government in Baghdad has turned into a nightmare.

The insurgent operations that followed Saddam's overthrow were initially weak but they grew in strength overtime. In 2003, General

John Abizaid, the commander of US command was boasting about their capability in these words: "I want to emphasize to the people that there is no military threat in Iraq that can drive us out. We have the best equipped, best-trained army in the world—positioned in the most difficult areas we have to deal with."[5] But the situation worsened soon afterwards. The US troops faced increasing attacks from the insurgents and the casualties of troops and the burden of economic costs climbed up day by day. It was reported by the mid-2006 that the US administration was considering a phased withdrawal from Iraq. Support for intervention from the American public has waned overtime too, placing pressure on the Bush administration to begin to withdraw troops.

However, there are some implications that once the troops are withdrawn, it may lead to the following consequences. Firstly, it would serve as an example of Washington's inability to deal with guerrilla conflicts. Considering this point, an observer has commented that any such withdrawal would embolden future insurgencies. Secondly, there will be civil war in Iraq, involving all sectarian and ethnic divisions of the country, causing spill over in the rest of the region and threaten the world's energy supply.[6] Moreover, any unfettered American access to Iraqi energy source will be threatened in a situation of weak governance in Iraq.

On the other hand, it may be politically and economically costlier to sustain American presence. Not only that, there is an anti-intervention constituency growing steadily among American people in direct proportion to the growth in US casualties; the fear of terrorist attack is also growing day by day. This fear, in fact, has led to the formulation of some stringent rules and regulations by the administration, affecting the freedom of American people.

It is evident from the way things are going in Iraq that a decisive military victory for the US is unattainable. Moreover, the political and economic costs are also growing. The American public opinion is growing against the continuation of the occupation. Considering this perspective, a quick military withdrawal may remove the primary cause for militant activities in Iraq and rest of the region. It may also encourage the local populations to find out how best they can accommodate their political and sectarian differences. Iraq's Arab neighbours may play an effective role in mediating among various Iraqi groups.

From the beginning of its offensive in Afghanistan, the administration advocated that this was a war against terrorism. For Iraq, though a different pretext of WMD was used, the hostile forces were neither subdued in Afghanistan nor in Iraq. Terrorism has now become a global phenomenon. In fact, the tentacles of terrorism cannot be cut by military actions but by political, economic and cultural measures. A sustained fight against it should first begin by finding out the political and cultural causes that breed terrorism and by taking appropriate corrective measures.

Notes

1. Agence France Press, compiling its figures from Pentagon News releases, mentioned that as of January 8, 2006, over 2,200 U.S. troops had lost their lives since the start of the invasion on March 19, 2003. In addition, from October 7, 2001, the official beginning of the Afghan war, until January 1, 2006, some 187 people from all branches of U.S. Armed Forces, plus 4 from the CIA, have died in the fighting there. While the casualties on the Iraqi side estimated in October 2004 by the prestigious British medical journal, Lancet was 100,000. Based on the Lancet figures, the current casualty toll for the Iraqis has been estimated as between 130,000 and 150,000, including combatants and civilians. See, Dunkel, G. (2006). "US occupation's problems increase in Iraq", January 14, *www.workers.org*.
2. "Iraq's Election Aftermath Reveals a Failed State," 5 January 2006, PINR, *www.pinr.com/report.php?ac=view_report&report_id=420&language_id=1*.
3. Fulbright, J. William (1967). *The Arrogance of Power*, Penguin Middlesex and Random House, New York, p.17.
4. Address to joint session of US Congress and the American people
5. "U.S. Faces Pressure to Pull Troops from Iraq," PINR, *op. cit*.
6. *Ibid*.

References

Freedman, Lawrence (2004). "War in Iraq: Selling the Threat", *Survival*, Vol.46(2), Summer.

Gupta, M.G. (1986). *Foreign Policies of Major World Powers*, Agra.

Nakhleh, Emile E. (1982). *The Persian Gulf and American Policy*, Praeger, New York.

Pollack, Kenneth M. (2002). "Next Stop Baghdad", *Foreign Affairs*, Vol.81(2), March-April.

Rice, Condoleezza (2000). "Promoting the National Interest", *Foreign Affairs*, Vol.79(1), January-February.

Zoellick, Robert B. (2000). "A Republican Foreign Policy," *Foreign Affairs*, Vol.79(1), January-February.

Section II

Regional Security: Different Perceptions

7

West Asia Security Policies after 9/11

A Case Study of GCC Countries

FAISAL O. AL-RFOUH

Paradigms of national security as well as global security have undergone tremendous changes in the aftermath of terrorist attack on New York based World Trade Centre and the Pentagon in Arlington, Virginia, on 11 September 2001, also known as 9/11. It has warranted the need for redefining not only the American national security albeit, the international security as well. Vulnerable regions, especially West Asia, in general, and Arab or Arabian Gulf region, in particular, have become the focal point of serious consideration in the changing security scenario in view of the region's geoeconomic, strategic and geopolitical significance. Internal developments, dynamics of regional developments and machinations of extra-regional actors entail the potential of adversely impacting upon the security of West Asia, in general, and Gulf region, in particular, which has already witnessed four wars. The fluid politico-strategic situation currently pervading the region does not augur well for peace and stability of the region, in particular, and West Asia, in general, in the long run. It is in the backdrop of this scenario that an attempt has been made in this paper to briefly appraise the security policies of West Asian states with specific emphasis on member-states of the Gulf Cooperation Council in the aftermath of 9/11.

Geostrategic Importance of Gulf Region

The geographic contours of the region called West Asia, also popularly known as Middle East, usually extend from Morocco and Mauritania on the Atlantic Ocean to the Gulf countries on the Arabian Gulf, including Arab members of the League of Arab States and three non-Arab countries: Turkey, Iran and Israel.[1] In the post-Cold War period, the United States, the European Union (EU) and NATO, while maintaining no direct geographical boundaries with the countries of the region, have come to influence the security of West Asia to a large extent. The decade of the 1990s has witnessed security boundaries of the West Asian region being managed through several security instrumentalities like bilateral cooperation with the United States, security dialogue and confidence building measures via the NATO Mediterranean Dialogue, and the EU Security Charter within the framework of the Barcelona process.

The Gulf region derives its importance from its geostrategic location and availability of oil and natural gas. Following British withdrawal from East Suez in 1970, the United States (US) had rushed in to fill the security vacuum and to deny the former Soviet Union access to the warm waters of the Gulf and threaten the oil resources in the region. Until the close of 1970s, the US depended on Iran and Saudi Arabia for the security and stability of Gulf region. Advent of the 1979 Islamic revolution in Iran, Soviet invasion of Afghanistan in end-December 1979 and the eruption of the First Gulf War between Iraq and Iran in 1980, the concept of 'Gulf security' came visibly to the forefront in US policy towards the region. Following Iraqi invasion of Kuwait in 1990, the US deployment of troops in the Gulf region doubled. The size of US military presence reached its zenith during the Third Gulf War against Iraq in 2003, which is frequently referred to as 'the War for the liberation of Iraq'.

Geographical location of the Red Sea and the Arabian Gulf, in the Gulf region also adds to the geostrategic significance of the region. The Red Sea serves as a pontoon bridge between the western seas mainly the Atlantic Ocean and the Mediterranean Sea to the Eastern ones, as the Arabian Sea, the Indian Ocean and the Arabian Gulf. The Arabian Gulf serves as the hub of oil production and loading, whereas the Red Sea is the 'main artery' through which oil is transported to the West.[2]

Availability of oil and natural gas in the Gulf region and the industrialised world as well as developing countries' heavy dependence on these resources adds to the economic importance of this region, frequently referred to as 'geoeconomic' significance of the Gulf region. Dependence of the US, Western Europe, Japan and other countries on import of oil from the Gulf region is shown in the Table 7.1.

Table 7.1

Net Oil Imports from the Arabian Gulf Region

	As Per cent of Demand			*As Per cent of Total Net Oil Imports*		
	US	*W. Europe*	*Japan*	*US*	*W. Europe*	*Japan*
1990	12	29	66	27	45	65
1991	11	27	65	28	41	64
1992	10	26	67	26	42	66
1993	10	29	70	23	47	69
1994	10	25	70	21	45	69
1995	9	23	71	20	44	70
1996	9	22	69	19	41	70
1997	9	23	76	19	44	75
1998	11	26	76	22	47	77
1999	13	22	73	25	43	74
2000	13	22	75	24	42	75
2001	14	18	76	25	33	76
2002	12	16	74	22	29	74
2003	12	17	78	22	30	76

Source: *'Persian' Gulf Oil and Gas Fact Sheet*, September 2004, available at *http://www.eia.gov/emeu/cabs/pgulf/html*

It is observed from Table 7.1 that the percentage of total net oil imports by the United States from the Gulf region, which was 28 per cent in 1991, stood at 22 per cent in 2003. Though the West European countries have shown a declining trend in their share of oil imports from the Gulf region from 45 per cent in 1990 to 42 per cent in 2000 and 30 per cent in 2003, yet Western Europe meets about one-third of its energy requirements by imports from the Gulf region. It is also discerned from Table 7.1 that Japan is fast becoming heavily dependent on oil imports from the Gulf region, from an average of 65 per cent per annum during early 1990s, to 70 per cent in 1995 and 76 per cent in 2003.

According to broad estimates, the volume of Gulf oil exports will more than double by 2020. As bulk of these exports from the Gulf ports to their respective destinations in Asia and elsewhere is transported through sea-lanes of communications (SLOCs) of the Indian Ocean like Strait of Hormuz, Malacca Straits and other waterways, presence of a powerful Navy is necessary for ensuring safe and uninterrupted flow of oil through the strategic waterways of the

Indian Ocean. The United States, having massive naval presence in the Indian Ocean, enjoys strategic advantage over other great powers *vis-à-vis* the Gulf region. In other words, the geostrategic and geo-economic importance of the Gulf region has provided a greater security role for the United States in this region.

9/11 and the Gulf Region

Smitten by these sudden terrorist attacks, the United States launched a worldwide campaign to combat terrorism by eliciting the cooperation of its allies and the international community. In the beginning of October 2001, the Bush Administration embarked on a three-pronged strategy. The first stage of this strategy entailed gathering irrefutable evidence of the involvement of Al Qaeda group in 9/11 events and maintenance of high-level alert, with particular focus on domestic defence. The second stage involved building up of a multinational political, social and economic coalition against Afghanistan-based Taliban-Al Qaeda combine. Finally, the US was to sustain and coordinate this coalition over an indefinite period against, what President Bush termed as, "a conflict without battlefields or beachheads." While launching its offensive against Taliban-Al Qaeda combine in Afghanistan, the Bush Administration took measures to garner domestic political support as well as launched diplomatic offensive to foster global coalition in its war against terrorism. Washington also rallied support of NATO members by invoking collective defence obligations.

The United States called upon all countries, including the Gulf Cooperation Council (GCC)[3] countries, to combat international terrorism by employing a wide array of measures ranging from diplomacy, international cooperation and constructive engagement to economic sanctions, protective security measures, covert operations and military force. Apprehending the adverse impact of international terrorism on the security of the Gulf region, the GCC member countries lost no time in immediately and publicly condemning the September 11 attacks[4] and adhered to quick steps to stem the flow of terrorism financing, augment security measures and to comply with the provisions of the UN Security Council Resolution 1373. The decade of the 1990s' witnessed Iran and Iraq being projected in regional media and conferences as the main sources of threat to the GCC member countries. There seemed to be unanimity of views between the United States and its Arab Gulf allies with regard to precipitins about threats from Iran or Iraq. Given the conflict over weapons inspections and doubts over whether the Iraqi regime was complying with the UN resolutions on disarmament, most threats were perceived as emanating from Iraq.

However, the developments obtaining in the aftermath of 9/11 wielded tremendous impact on the pattern of US-GCC relations, particularly in view of the fact that all the 19 hijackers involved in the 9/11 were from GCC countries and American President George W. Bush's speech talking about America's war on terrorism: "You're either with us or against us," which left no middle ground for the GCC countries to safeguard their interests.[5] GCC states felt they were being dragged into a conflict with their own people, who were outraged by the Western media's campaign against Islam, growing American influence in the region and American apathy to what was going on in the occupied territories of the West Bank and Gaza Strip had outraged the domestic public opinion in the GCC. The ruling elite in the GCC countries found themselves on the horns of a dilemma. On the one hand, they were being dragged into a conflict with their own people, and on the other, they had to choose between the options of 'either with us or against us'.

Weighing all options under the prevailing circumstances, the GCC countries chose to align with the American alliance in the war against terrorism and rendered substantial support, most of it behind the scenes, to the US military campaigns in Afghanistan and Iraq. Bahrain dispatched a frigate to participate in maritime operations in the Indian Ocean during the US campaign in Afghanistan.[6] Besides, the UAE and Oman opened their airbases for US warplanes conducting operations in Afghanistan; and Saudi Arabia's Prince Sultan airbase served as the command and control centre for war operations. The GCC countries even tailored their weapons procurement programmes to deal with the threat of terrorism.[7] Despite differences over the definition of terrorism and the issue of Arab responsibility for the 9/11 attacks, military cooperation between the US and GCC countries has continued particularly in training and equipping of the Special Forces.

GCC and War on Iraq

The GCC countries had disapproved of the US strategy of waging a war against Iraq. Until 2002, most of the Arab officials nurtured the impression that Washington's intention of waging a war against Iraq was directed against Iraq's weapons of mass destruction and at that juncture the pronouncements articulated by Arabian Gulf officials laid emphasis on the fact that Iraq no longer portended a threat to the existence of any of them.[8] Having total disregard for the views of the officials of the Arabian Gulf countries, the United States went ahead with its invasion of Iraq. By early 2003, it had become apparently clear to the GCC countries that, unlike the 1991 Gulf War, the American war against Iraq was no longer an Arab war but "America's war for purely American interests."[9] In view of the United States having gone for war in Iraq, the GCC countries have extended limited support to Washington. The American failure to find WMD in Iraq and growing opposition to the coalition forces in Iraq have added to the credibility crisis for the US in the Arab world, in general, including the GCC.

The US reluctance to withdraw from Iraq has given rise to various speculative theories about American long-term interests in the region, including Washington's desire to exert control over Arab oil wealth. According to one opinion, "All other US rhetoric about threat assessments, national security and long term goals has been discarded as mere window dressing for patently imperialist goals."[10]

GCC and Weapons of Mass Destruction (WMDs)

Use of chemical weapons, namely poisonous gases, during the First World War represented one of the most dangerous developments in human history. During the Second World War, the use of such weaponry continued at a larger and far more destructive scale. The use of atom bombs by the United States on the two Japanese cities of Hiroshima and Nagasaki, which claimed hundreds of thousands of innocent civilians lives, demonstrated the genocidal capacity of this new warfare. It is noteworthy here, that ever since the emergence of such weapons as a new element of warfare, sane voices have called for the prohibition on the use of such weaponry precisely because of their capacity to inflict indiscriminate lethal harm and damage on a massive scale.[11] During the Iran-Iraq War (1980-1988), Iraq became the region's first testing ground for the use of chemical weapons when Iraqi troops launched a chemical attack against the city of Halabja, in northern Iraq, in August 1988, in which thousands were reported to have been killed. Media reports indicated that weapons of mass destruction (WMDs) were indeed used, though on a limited scale, during the Iraq-Iran war. The Gulf region, thus, became one of the few regions in contemporary history in which WMDs are known to have been deployed.

Extensive use of chemical weapons in the Iran-Iraq War was an alarming development for almost all countries of the Gulf region and added a new dimension to the security of the Arabian Gulf countries. Another worrisome question was the use of ballistic missiles in the Iran-Iraq War, particularly the Scud missiles that provided a serious threat to the security of Gulf region and proved instrumental in adding to the worries of the Gulf countries. It was in the wake of these developments that the countries of the region had come together under the banner of the Gulf Cooperation Council. During the Iran-Iraq War, the GCC extended support to Iraq in its war with Iran, and so did the United States. This common emphasis was galvanised by the declared policy of Iran to 'export its brand of religious revolution' to neighbouring Gulf countries and to provide support to the Shiite communities living in those countries.

However, by 1990, the scenario in the Gulf region underwent a drastic change in the wake of Iraqi occupation of Kuwait and Iraqi Scud missiles targeting Bahrain and Saudi Arabia. The United States provided a defence shield to the GCC member countries against Iraqi Scuds. This helped Washington to entrench itself deeply in the Gulf region. The GCC-US military cooperation had actually commenced in the aftermath of the outbreak of the Iran-Iraq War when the Gulf countries had sought Washington's help to protect their oil tankers and keep shipping lines open in the Arabian Sea and Gulf basin. Most of the GCC militaries were less than a decade old on the eve of the outbreak of the Iran-Iraq War. Under the changed scenario the GCC countries embarked on ambitious and costly armament programmes in the early 1980s to boost their defences. The defence spending of GCC countries in the five-year period 1976-1980 which stood at $71.73 billion, went up to $146.44 billion, during 1981-1985 period. The defence expenditure of the GCC countries was about $101.63 billion during 1985-1990.[12] The subsequent period has witnessed massive hike in the defence budgets of the GCC countries.

The UAE-based Gulf Research Center (GRC) has initiated a new research project entitled "The Gulf Region as a Weapons of Mass Destruction Free Zone," that calls for the prohibition of all WMDs, whether they are nuclear, chemical or biological. The key motivation driving the GRC to promote this concept stems from the conviction that if a WMD-free zone (WMDFZ) were achieved in the Gulf region, it would be the first phase of a more comprehensive plan to declare the entire Middle East region as WMDFZ. In this context, the Gulf WMDFZ project represents a partial yet critical step in promoting awareness of the issue and reaching the core objective. It is generally believed that success in declaring the Gulf region a WMDFZ will certainly help create an ambience conducive to applying political and moral pressure on the remaining countries to expand the initiative to cover the whole Middle East region.[13]

Presently, there are five such geopolitical regions that collectively include 110 countries that are members of the United Nations. It is worth mentioning here that in the majority of these five regions, the concept of declaring a WMDFZ evolved over a long period of time. In most cases, the inception of the project was modest in scope in that each of the WMDFZs initially only included a limited number of nations interested in the issue. However, these regions were able to push forward and consolidate their experiences, thus, leading to the burgeoning of their projects to involve all of the countries in their respective regions.

Iranian Factor

Iran has been a potent factor in influencing the security of the GCC member countries. Prior to the advent of Islamic revolution in Iran in 1979, Iran, under the regime of Shah possessed a large and best-equipped army and its friendly relations with Israel, was seen as a threat representing age-old Persian desire to dominate the region. Owing to Cold War geopolitics, Iran had got a special status in American foreign and defence policies. In the post-Revolutionary Iran, the United States lost a trusted ally and for the Arab states it represented a 'radical Shi'ism' determined to export 'religious revolution' in the predominantly Sunni Arab countries, including the GCC. It was in this backdrop that the GCC countries had supported Iraq against Iran in the Iran-Iraq War.[14]

Uneasy relations between Washington and Teheran in the post-Revolution period resulted in the adoption of the 'containment' strategy by the United States toward Iran in the form of economic sanctions and forward military deployments in the region. In the aftermath of 1991 Gulf War, the 'dual containment' of Iran and Iraq became the main bulwark of the US policy towards the region. Recent years have seen further hardening of American attitude towards Iran in the wake of latter's dubious nuclear policy. However, at the same time there has been some 'incremental' improvement in Iran-GCC relations in recent years.

There have been exchange of high-level visits of officials from Iran to individual Arabian Gulf countries on a bilateral basis, including the signing of many trade and security pacts. Saudi Arabia moved ahead in 2000 with a series of trade and mutual-cooperation pacts with Iran. Both the countries signed a security agreement in April 2001 and a judicial-cooperation memorandum of understanding in July 2003. Another security-cooperation agreement was signed between Iran and Qatar in October 2002, encompassing cooperation on various aspects of border security, including measures to counteract drug running and money laundering.[15] Undoubtedly, the historic dispute between the UAE and Iran over the Abu Musa and Tunb islands has continued to impact upon relations between the GCC and Iran and UAE and Kuwait have been critical of Teheran's military buildup there, yet both sides have emphasised on the peaceful solution of the outstanding issues. Historical factors continue to dominate the overall pattern of Iran-GCC relationship which is marked by suspicion and distrust towards Iran as well as the attitude of the United States towards Teheran.

The US Factor

The United States is a significant factor in the overall security framework of the Gulf region. Many countries in the region still favour a strong American military and diplomatic presence owing to persistent inability of Gulf countries to evolve their own balance of power. Washington's physical presence in the region, global political dominance and growing role as facilitator of regional-security frameworks makes the United States as a *de facto* major regional partner in the Gulf. There prevails an opinion that there is no alternative to the continuous presence of large-scale American troops in order to address the inherent regional imbalance of power. According to one opinion: "The imbalance of power creates a vacuum, and [this] invites aggressors to make adventures to destabilise the region. [Therefore], the United States should be part of any security arrangement."[16]

The Gulf countries have been historically relying on external powers. Until 1970, the region was dependent on United Kingdom and after the British withdrawal the United States stepped in to fill the power vacuum and since then American involvement has further deepened, from Kuwait to Saudi Arabia and now to Iraq. Iran's potential threat to the region and US strategy to 'contain' Iran strategically suits Washington to strengthen its military grip over the region. Besides, the Gulf countries are heavily dependent on American weapons, which provides additional leverage for the US to continue to dominate the region.

There is no denying the fact that the GCC has reached a point where member states cannot convene a meeting to discuss security structures without the presence of the US officials. The United States is a major player, and its approval of any security structure is a basic reality in the region. The GCC countries could not succeed in proposing a genuine regional-security formula, either among themselves or with Iran or Iraq because of the 'extreme sensitivity of size,' that is, the differences in national population and financial and military resources available to individual states. The balance of power is a major problem in the region and it is still heavily relying on external powers. The Gulf region suffers from a severe crisis of trust. There seems lack of trust between and among GCC countries and they do not trust each other, but they do trust foreigners and foreign powers. According to one argument, "Foreigners have the best jobs, the best salaries and, supposedly, the best formula for security arrangements."[17]

Undoubtedly, the US action in Iraq has been instrumental in eroding its credibility in the region for the time being and there is growing disenchantment with the United States' so-called war on terrorism among the people of the Gulf region in particular and Arab world in general but dependence of the monarchical regimes of the Gulf countries on the United States is likely to continue in the near future for their survival and it is the interests of the ruling elites that sustains American interests in the region.

Conclusion

Despite the existence of Gulf Cooperation Council for about 25 years now, it has not been able to evolve a security framework for the region. The varied interests of the members, lack of common threat perceptions and predominance of the United States as well as Gulf Countries' excessive dependence on Washington has served as a deterrent against evolving a common security framework. Iran is seen as a potential threat to the security of the region but the Gulf countries of there are unable to confront Teheran and they are dependent on the US. Besides, the US perceptions of Iran as a threat to American interests in the region gets jumbled with region's member countries' perceptions and that goes to serve the American strategic interests, and not the interests of the GCC countries. The US which has managed to acquire a strong foothold in the region's geopolitics can not be expected to cede that position to any other outside power. At the same time, the Gulf countries should maintain a semblance of balance between domestic opinion and their foreign policies, particularly towards the United States in order to contain the domestic 'dissidence' and incremental dose of political reforms should be allowed to take place.

It will be good augury if Gulf countries involve the European Union in any future security dialogue for the region as a counterbalance to American predominance. The EU-Mediterranean partnership which initially could not make much headway owing to some unknown reasons should be given a fresh consideration in order to revive the 'multilateral' character of 'security framework' for the GCC region instead of relying exclusively on the 'unilateral' aspect pioneered by the United States.

Notes and References

1. Kadry, Mohamed (2004). "Potential Egyptian Contribution to A Security Framework in the Gulf", *Middle East Policy*, Vol. XI(3), Fall, p.69.
2. For further details on the importance of the Red Sea for the Gulf Region see, "GCC-Red Sea Security", available at *http://www.grc.ae/gcc-redsea/htm*, accessed on 12 November 2005.
3. The Gulf Cooperation Council was established in 1981 by the Gulf countries- Bahrain, Kuwait, Oman, Qatar, Saudi Arabia and the United Arab Emirates (UAE) as its founding members in response to the outbreak of Iran-Iraq war. The main objective of the GCC is to coordinate resistance to outside intervention in the Gulf region and to strengthen cooperation in areas such as agriculture, industry, investment, security and trade among its members. See "Gulf Cooperation Council", available at *http://www.grc.ae/gulf cooperation council/html*, accessed on 18 November 2005.
4. The Ministerial Council of the GCC at its 25th special session held on 23 September 2001 in Jeddah, Saudi Arabia, considered with 'deep concern the grievous consequences' of the devastating attacks on economic and governmental establishments in the United States of America. It voiced its censure and condemnation of these savage acts of terrorism and conveyed its condolences and sympathy to the Government and people of the United States of America. The meeting affirmed that the GCC States would support and would cooperate fully in international efforts to identify the perpetrators

of the terrorist acts and bring them to justice. See "Ministerial Meeting of the Gulf Cooperation Council, Jeddah, Saudi Arabia, 23 September, 2001", available at *http://www.grc.ae/gulf cooperation council/html*, accessed on 18 November 2005.

5. Kahwaji, Riad (2004). "US-Arab Cooperation in the Gulf: Are Both Sides Working from the Same Script?", *Middle East Policy*, Vol.XI (3), Fall, p.54
6. Kahwaji, Riad (2004). "Interview with Bahraini Defense Minister General Sheikh Khalifah Bin Ahmad Al-Khalifah," *Defense News*, April 26.
7. Kahwaji, Riad (2004). "For Arabs, Spec-Ops are the Front Lines,"*Defense News*, April 26.
8. Kahwaji, Riad (2002). "Qatar Remains Wary of Campaign Against Iraq," *Defense News*, July 8.
9. For details see, Kahwaji, Riad (2003). "Gulf Support for Iraq War Rise", *Defense News*, January 6.
10. Kahwaji n.5, p.56.
11. For more details in this regard see, Goldixat, Josef (2005). "Introduction: Why is the GWMDFZ Project Important?", *Security and Terrorism*, Issue. 1(1), pp.5-6.
12. Kahawji n.5.
13. For further details on this aspect see, Alami, Mustafa (2005). "The Case for GWMDFZ", *Security and Terrorism*, Issue. 1(1), pp.7-9.
14. For more details on Iran-GCC relations see, Al-Suwaidi, Jamal (ed.) (1996). *Iran and the Gulf: A Search for Stability*, The Emirates Center for Strategic Studies and Research, Abu Dhabi.
15. *US-Qatar Journal*, No.33, December 6, 2002, online at: *www.qatarbusinesscouncil.org*.
16. Cited in "Conference Report", in *Middle East Policy*, Vol.XI(3), Fall 2004, p.6.
17. Cited in Ibid.

8

The Security Challenges in the Mediterranean Area and their Links with Western Asia

SMAIL BENAMARA

The Political Strategic and Military Context

In spite of the end of the Cold War and the collapse of the Soviet Bloc, the Mediterranean region has not been able to benefit from the advantages of peace so far. Indeed, if the end of the East-West rivalry has been able to reduce the tension in many parts of the World and quieten down some centres of conflicts, the Mediterranean region has known nothing of the sort. No doubt, it remains the most disturbed region where the stakes and the struggle for influence go beyond the framework of the neighbouring countries. The changes occurring on the international scene have also certainly led to decline in interest in the Mediterranean, but the region remains as important as ever. Analysts have cited other reasons and justifications to make this region retain its strategic value.

Among these reasons, one has to distinguish the old from the new ones, linked to the politico-strategic context in the Mediterranean. One has to seek convincing answer to the following question: What makes the region so strategic? It is a truism to say that the importance of the Mediterranean region is linked inevitably with the vital strategic interests of the West in general and those of the United States in particular. Firstly, the Mediterranean region remains close to the region which is the main source of hydrocarbons for West. Secondly, the specific geographical configuration of the Mediterranean Sea makes it the most important maritime waterway in the world and the most sensitive as well. The Mediterranean Sea, situated at the crossroads of the three continents (Europe, Africa and Asia), links the Atlantic Ocean with the Indian Ocean through narrow passages (the Gibraltar Strait, the Suez Canal) and gives access to several internal seas (the Adriatic Sea, the Black Sea, the Red Sea, the Aegean Sea, etc.).

The disturbances in the neighbouring region, the Middle East, make the Mediterranean a zone of the utmost importance. The Mediterranean also remains one of the most highly militarised regions in the world today. It has not benefited, as it were, from the reduction of the armament and disarmament measures following the improvement of the East-West relations. On the contrary, disarmament in Europe shifted the attention of the West towards the Mediterranean. The war fleets are still there, much more powerful, sophisticated and dedicated to their missions. Even American and Russian fleets are present as part of their strategies to retain their influence and power over the maritime waterways. The countries on the southern shores of the Mediterranean are confronted with the challenge of more and more sophisticated, more and more mobile and extrovert military presence of the Northern Powers.

Besides the geographical, historical and political factors mentioned above, the Mediterranean has been witness to some important events that have contributed to increase of interest in the region. Thus, the nature of the threat, its roots and the way to counter it are perceived in different ways now by the external powers.

A feeling of threat coming mainly from the countries on the southern shores of the Mediterranean has replaced the threat of USSR/Soviet Union of Cold War days. The threats are no longer military but rather trans-border in nature. This threat of terrorism is rooted in the prevailing socioeconomic conditions. The threat of terror is not directed at states lone but at individuals, business concerns, cultures or groups who are perceived to be enemies in a theo cultural sense.

The Mediterranean has also witnessed during the last few years, the emergence of different types of conflicts worldwide. Unlike the earlier days when conflict was perceived essentially in inter-state terms, now conflicts like the ones in ex-Yugoslavia have become more frequent and posed serious challenges for states. The Mediterranean region has been some sort of a laboratory for the adaptation of defence policies and joint force exercises, to respond to the changes occurring in the strategic landscape. Most of the concerned Powers have created a Rapid Deployment Force to intervene in Mediterranean zone, i.e., the French Rapid Action Force, the Italian FIR, the Spanish

FAR, the NATO Rapid Reaction Force and the Eighteenth American Corps. These Rapid Action Forces do not seek to defend national territories against external attacks but to intervene in external theatres of conflict, for performing special tasks. This change has been encouraged by the development of peacekeeping forces (Bosnia, Somalia, Rwanda), and this has been woven into national military strategies.

At the military level, the threat, still felt from the South bank, concerns the proliferation of weapons of mass destruction (nuclear, chemical and biological) as well as the development of ballistic missiles. In case of armed conflicts, the Europeans know that they have an advantage and an assured level of superiority, that is, the "ability to move forces and the mastery of the space." What they rather fear is the possibility of this kind of weapons falling into the hands of hostile countries or regimes.

Various Frameworks of Dialogue

There is no doubt that since the mid-1990s, we have witnessed a profusion of initiatives aiming at setting up security and cooperation frameworks in the Mediterranean. The change in international and regional environment was considered favourable for the setting up of these frameworks of dialogue. However, there was no collective framework in Mediterranean, like the OCSE (Organization for Cooperative Security in Europe) for security in Europe. The Mediterranean states are seeking to benefit from the changes that have occurred in the international relations following the disappearance of the Soviet Union to try to launch initiatives aiming at stability and security in the Mediterranean. Some of these frameworks are:

- The Euro-Arab dialogue;[1]
- The Mediterranean Forum;[2]
- The Mediterranean dialogue with the CSCE which has changed into OCSE;[3]
- The Euro-Mediterranean dialogue or the Euro-Mediterranean process of Barcelona (started in November 1995);[4]
- The 5+5 dialogue in the Western Mediterranean;[5]
- The NATO-Mediterranean Dialogue (MD);[6]
- Multilateral negotiations on the Middle East;

- Conferences on the economic cooperation in the Middle East in North Africa;
- The Great Middle East Initiative (GME).

Before examining some of these frameworks, it is necessary to underline what constitutes the fundamental elements that distinguish them from one another and to judge their effectiveness. For this a critical analysis of the following points is needed:

- The number and group dynamics (homogeneity or heterogeneity) of the participating countries;
- Fields of competencies: general or specialised frameworks;
- Formal or informal character of works and recommendations;
- The role of external actors.

The issue of the composition and the number of participating countries relates of course, to the process of obtaining a consensus while fashioning out action plans. When the number is too high or the composition is heterogeneous, it usually becomes difficult to arrive at any consensus. As regards the Euro-Mediterranean framework of Barcelona is concerned, the problem of working out a consensus is very real and it hinders the advance of a collective security framework or a pact of stability in the Mediterranean.

It is also noticed that if the field of competence is precise and specialised, it becomes easier to concentrate the collective efforts and more likely to obtain results. From this point of view, the NATO-Mediterranean Dialogue has a comparative advantage since it concentrates on the issues relating to security alone, even if it defines security in the broad sense of the term.

Finally, the question of formalism also has its importance and would have to start from the following postulate: to avoid not only any excess of rigid and procedural 'formalism' which slows down considerably the pace of work, but also the excess of 'informalism' which leads to sterile discussions, without consequences and especially making operational decisions-making difficult.

As an example, the process of Barcelona can be regarded as too formal and complex in the decision making process while the Mediterranean Forum that includes 11 countries (and is not hostage to the Israeli-Palestinian conflict, since Israel and Palestine are not members) is a victim of too much 'informalism'. The meetings generate useful discussions but without concrete decision making.

The Promises of the 5+5 Framework

A less known framework, but which presents interesting characteristics, is that of the 5+5 framework, which consists of countries of the Western Mediterranean region—the Maghreb countries and the European countries on the Western shore of the Mediterranean. This framework is characterised by comparatively less number of participants (10), and relatively more homogenous member countries. Additionally, there is a commonality of interests because of shared problems as well as geographical, historical and human proximity that makes it necessary, even obligatory, for the countries to look for solutions to the problems arising within the region, even if they are complex. Whether it is because of their colonial links in the past, the well established economic and commercial relations, or the common problems of terrorism or immigration, there is constant interaction between these countries and this reinforces their willingness to deal with the problems through dialogue through a forum like the 5+5.

It is true, of course, that the potential of this 5+5 framework is far from being optimised and it is affected by the side effects of unfavourable economic considerations by some countries and in particular the blocking of the Maghreb Union. But it represents, nonetheless, a promising sub-regional framework that will be able to contribute to the security and stability of the Mediterranean region.

During these last several years, there has been a real will to reactivate the dialogue between the Maghreb and the South European countries. The starting point was the first summit of the heads of state of the Western Mediterranean countries in December 2003. As far as the security issues are concerned, the heads of state have reiterated the link between peace, security and development, and their commitment to endeavour together to fight terrorism in all its forms, i.e., organised transnational crime, drug trafficking, arms dealing and money laundering.

It should be noted that within the 5+5 framework, the security issues are taken up only in the regular and annual meetings of the Ministers of Defence, Ministers of Interior and those in charge of immigration. For example, on 12 December 2005, a meeting of the

Ministers of Defence was held in Algiers and an action programme was adopted, which dwelt on aerial and maritime monitoring and assistance in the event of natural disasters.

There is also another framework of Mediterranean Dialogue with the OSCE, which is similar to the NATO-Mediterranean Dialogue (MD), in its form and agenda. These two for a will have to coordinate their activities and avoid duplication to address the issues of security and stability in the Mediterranean more meaningfully.

Specificities of the NATO-MD

It is necessary to discuss the particular case of the NATO-MD, in order to see what are its specificities and if it has any advantage compared to the other existing frameworks. The dialogue is, first of all, restricted to a particular field of competence and concentrates primarily on the issues of stability and security in the Mediterranean. Thus, the NATO-MD has a distinct advantage over other dialogic tracts and this helps this framework to concentrate its efforts on concrete and operational measures.

The NATO-MD has also another specific arrangement to engage several countries on individual basis and there is an emphasis on free and voluntary association by each Mediterranean country, independent of others. Such flexible arrangement helps the NATO-MD framework to adapt to the rhythm of each Mediterranean partner separately and makes it difficult on the part of any country to block the entry of any other country into the NATO-MD arrangement. Such security-centric arrangement also remains immune to the developments in the Middle East, because by emphasising on addressing the security concerns of each individual country it makes the arrangement beneficial to each one of them separately.

The setting up of the bilateral and multilateral dialogue in the 19+1 and 19+7 format constitutes, from this point of view, a simulating innovation. In fact, NATO's plans to establish and strengthen political dialogue with the Mediterranean countries include its innovative offer to take advantage of NATO's already very flexible modus operandi, under which meetings can be held between all the 19 alliance members and one partner country (known in NATO parlance as "the 19+1" formula), or between alliance members and all Mediterranean dialogue countries (19+7 formula), or between the alliance and any number of partners (the 19+ "n" formula). It is necessary here to mention the specificity of the NATO itself, which is a well-structured and well-organised military organisation. The strategic advantage of the NATO-MD may be that of gathering in its midst the USA, and influential European states who find in such an arrangement to stay engaged in the issues of stability and security in the Mediterranean.

The nature of the activities initiated by the NATO within this framework also benefits the partner states from the region. They provide an opportunity to them for training their military and civilian workforce and for knowledge-sharing as well. Whether it is training, military exercises, security evacuation, the organisation of relief works or the numerous seminars and workshops that gather senior officers and even members of Parliament, academics and representatives of the civil society, all these activities make it possible to bring the elite from all these countries together and help them better understand one another and provide a forum to discuss their points of view.

In the case of my country, i.e., Algeria, in spite of its late association with this dialogue, it invested itself in the arrangement with determination. This allowed the NATO countries to better understand the decade old Algerian struggle against terrorism. This dialogue has also made it possible to bring many Arab countries closer to the NATO and helped them in appreciating this organisation better, even if this impetus were somewhat hindered by the war in Iraq. The Arab populations—whether rightly or wrongly—considered the involvement of NATO improper, against the tenets of international law and hence, illegitimate.

Links between Security in the Mediterranean and Western-Asia

It is necessary to recognise that the phenomenon of globalisation has reduced regional specificities as regards security and stability considerably. In both these regions—Mediterranean and West Asian—we can notice that the nature and the types of security threats have changed during the last few years. They have become more numerous and more complex. New threats have emerged in the post-Cold War period. Most of them have arisen because of internal political transitions and socioeconomic disparities. The threats from non-state groups have increased manifold, i.e., terrorism; trans-national crime; drug trafficking; arms dealing, etc. If we analyse the document prepared by the Institute for Defence Studies and Analyses of New

Delhi for our meeting, we note that many of issues mentioned there remain same for the Mediterranean area. The issues like: (i) the growing international interest in the region; (ii) the emergence of non-state actors as threats to state security; (iii) the rising levels of popular disaffection; (iv) the gradual rise of religion extremism; and (v) the increasing presence of external actors.

For all these reasons, we can say without any doubt, that there is a link between the security of these two regions. Thus, the situation in Afghanistan has a real-time impact on Algeria, for example. During the last decade, most of the terrorists operating in Algeria came via Afghanistan which acted as the nerve centre for training ground for terrorists around the world. Algeria has suffered directly because of this export of fanaticism, violence and terrorism from this part of Asia. On the other hand, the negative impact of non-resolution of Arab-Israeli conflict also directly affects the West Asian as well as Mediterranean public opinion and contributes to the disaffection of the people against their governments.

They must consider the question of stability and cooperation on priority basis. The two regions will have to evolve answer to new threats to stability and security from non-state groups such as terrorism or organised crime. It is necessary on the part of the governments in these two regions to work hard towards establishing useful and operational security frameworks, where dialogue, consultation and cooperation will be the main principles governing the interaction between the governments at one level and between them and the external actors on the other. There will have to be a multiplicity of frameworks of dialogue because these two regions have suffered for a long time from a deficit of communication and security dialogue among themselves. There is a need to establish good working relations with external actors, and in particular with US superpower, based on mutual trust and understanding. Both the regions will also need to work seriously, in cooperation with international community, towards the settlement of the Arab-Israeli conflict, with the creation of an independent Palestinian state, meeting the legitimate aspirations of the Palestinian people as well as the peaceful acceptance of Israel as a sovereign state in the region.

Lastly, the challenge in the future will be to set up a harmonious framework of security and cooperation which can bring these two regions together and include at the same time subregional and regional frameworks. Such dynamic security framework will seek to deal with both the issues of security and stability and contribute to the idea of creating shared prosperity.

Notes and References

1. After the Yom Kippur War and the oil crisis in 1973, the then European Community (EC) put in place a structure of Cooperation and Dialogue with the Arab League. The Euro-Arab Dialogue (EAD) began as a French initiative composed of representatives from the EC and Arab League countries.
2. The Mediterranean Forum came into existence before the Barcelona Declaration in 1995 and continues to be active. It is an informal cooperation among 11 Mediterranean countries (Greece, Portugal, Spain, France, Italy, Morocco, Algeria, Tunisia, Egypt, Malta and Turkey), without financial support from the EU budget for these programmes.
3. This is an initiative by the Organization for Security and Cooperation in Europe (OSCE) to establish close cooperation with six Mediterranean countries (Algeria, Egypt, Israel, Jordan, Morocco, Tunisia) in security matters.
4. The Euro-Mediterranean partnership was first defined in the Declaration of Barcelona, adopted on November 28, 1995 by the 15 Ministers of Foreign Affairs of the European Union and by the Foreign Ministers of the 12 Mediterranean Partner Countries, Algeria, Cyprus, Egypt, Israel, Jordan, Lebanon, Malta, Morocco, Syria, Tunisia, Gaza and Cis-Jordan territories. It is a political dialogue framework between the European Union and the Countries of the Mediterranean Basin, aimed at addressing southward the relationships that the Union has been developing with Eastern Europe Countries since 1989.
5. Algeria, Spain, France, Italy, Libya, Malta, Morocco, Mauritania, Portugal and Tunisia came together in the Conference of the Western Mediterranean Countries in 1990 to set up the "5+5 Dialogue" framework.
6. NATO's involvement in the Mediterranean goes back to the Cold War. The Mediterranean was important to NATO primarily in military terms, a fact reflected in it being identified as the Alliance's "Southern Flank." The NATO-Mediterranean Dialogue was launched in 1994 as a forum of cooperation between NATO and seven countries of the Mediterranean (Algeria, Egypt, Israel, Jordan, Mauritania, Morocco, and Tunisia). The overall aim of NATO's Mediterranean Dialogue is to contribute to regional security and stability, and achieve better mutual understanding between NATO and its Mediterranean Partners.

9

Maritime Terrorism Off the Arabian Peninsula

A Post-Limburg Security Assessment

GURPREET S. KHURANA

The 21st century has ushered in immense prospects and opportunities for economic globalisation. However, it is also witness to a retardant in the form of terrorism. The USA initiated "Global War on terrorism," but its only 'achievement' thus far has been the dilation in scope of terrorism. A year after the September 11, 2001 attacks, Al Qaeda operations diversified to energy denial. The bombing of the French oil tanker *Limburg* was followed by Bin Laden's ominous warning to the West, "By God, the youths of God are preparing for you things that would fill your hearts with terror and target your *economic lifeline* until you stop your oppression and aggression."[1] The strategic choice was apt. The hunger of the industrialised countries for energy resources is driven by the need to nourish their economies, and in turn, their military might. The USA alone accounts for a quarter of the world's oil consumption of over 82 million barrels a day (mbd),[2] half of which, it imports. With global consumption rising steeply, production barely exceeds demand and any disruption in supply would cause oil prices to skyrocket, hitting the American economy hard. The US natural gas demand also constitutes a quarter of the world's total, which is growing even faster. However, oil imports are critical for the US economy considering that these constitute 89 per cent of its energy imports and its domestic oil production is declining.[3]

Southwest Asia has acquired a significant place in the era of globalisation on many counts. However, its attributes have also long provided the key ingredients for terrorism to thrive, *inter alia* religious fundamentalism and means of funding. The *jehadis* now have two more motives to pursue their ominous agenda; the region's substantial energy exports to US (and its allies)[4] and the instability in Iraq, where insurgents (or the 'resistance', depending on the viewpoint) perceive the 'occupation' to be motivated by the quest to control Iraqi oil. By 'economic lifelines', Bin Laden may not have implied sea-lanes *per se* but commercial ships that carry two-thirds of the world's oil trade and increasingly, Liquefied Natural Gas (LNG). With 90 per cent oil exports from Persian Gulf being seaborne[5] and Bin Laden having specifically expressed his 'penchant' for "...oil facilities in the Persian Gulf...the main artery of the American economy,"[6] the threat to maritime transportation is evident. The peril can also be extrapolated to the region's energy infrastructure, much of which is located in the littoral region and susceptible to strikes from the sea.

As elsewhere around the globe, the seas adjoining the Arabian Peninsula are also plagued by seaborne crimes. Conventional military threats due to intra- and extra-regional misgivings also persist. It is however, feared that sooner or later, terrorism would emerge as a major cause for insecurity in the region's maritime environs. The

Figure 9.1

The Arabian Peninsula and its Energy Sea Routes

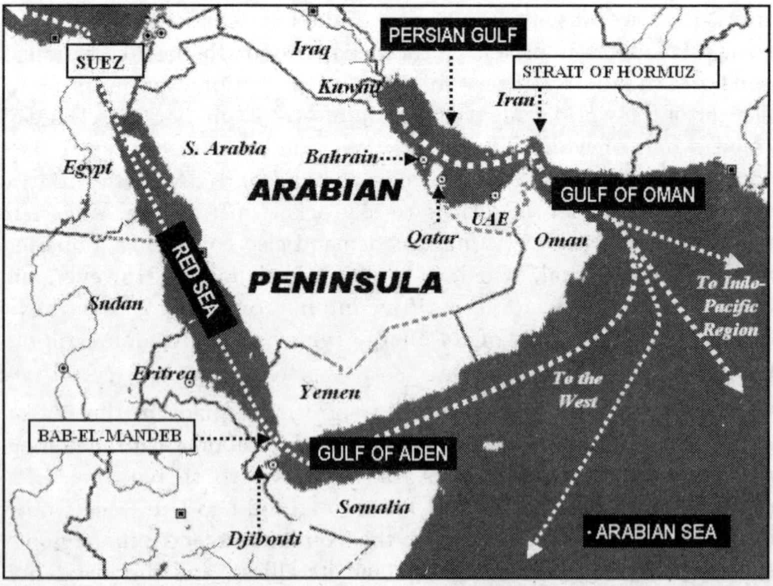

imminence of such anarchy at sea also causes extra-regional anxieties, which transcend economic dimensions of security into geopolitical, humanitarian and environmental domains. In its first section, the paper appraises the probabilities of maritime terror attacks and terrorists' link with seaborne crime. The second section examines the existing security mechanisms and makes some recommendations.

THE SECURITY CHALLENGE

Trajectory of Maritime Terror

Attacks at or from the sea are tactically demanding. These require not only skill, but also intricate planning through scientific computation, while accounting for a host of interlinked factors like the stability of own/target vessel and elements of nature, which often baffle mariners despite thorough reckoning. Al Qaeda had begun attacking US warships as early as in January 2000, using bomb-laden boats, but it was a learning process. The first one on USS *The Sullivans* failed because the boat sank under the weight of explosives. The next one in October 2000, on USS *Cole* at Aden harbour was a partial success—it did not sink the ship, but nonetheless succeeded in crippling it seriously and killing 17 of its sailors. The aim of these attacks was to mangle the symbols of US military power, but these have also helped Al Qaeda acquire a degree of nautical expertise.

Energy Terminals: Eligible Targets

Stationary targets may be easier to strike, but the offshore oil terminals in the Gulf are far from being easy prey. In particular, the Iraqi *Al Basra* and *Khor Al Amaya* platforms are heavily defended by the coalition since these are responsible for 95 per cent of Iraq's oil production and 80 per cent of its GDP. However, in April 2004, two suicide attacks occurred 20 minutes apart on each terminal. The attacks were thwarted when the bombs exploded short of their targets, but three US servicemen were killed and four were injured.[7] The aim was to cause maximum destruction to the platforms. A sufficient damage or fire to one of the two supertankers berthed alongside *Al Basra* platform could also have placed the terminal out of operation for months. It did not succeed beyond damaging an electric generator, but certainly demonstrated the potential for success since even the ensuing 24-hour shutting down of the terminals was sufficient to cause significant economic loss amounting to US $28 million. A successful strike would have led to a serious effect on global oil prices, along with stalling the Iraqi economy. A 20-minute gap between the attacks is significant. It indicates that the second one was planned as a 'backup' in case the first failed—to follow, notwithstanding the security alert. Strikes on these vital nodes against all odds only prove their criticality for Al Qaeda to further its agenda. More such suicide attacks cannot be ruled out, particularly on Gulf's key offshore terminals, loading berths, oil tanks and pipelines leading to mooring buoys. Saudi Arabia's *Ras Tanura* is one such eligible target.

'Moving Ducks'

In October 2002, the French supertanker *Limburg* was bombed as it approached Yemen's *Ash Shihr* oil terminal. In the attack, both the inner and outer skins of the double-hulled ship were penetrated, rupturing a cargo hold and spilling 90,000 barrels of burning oil into the sea. Striking ships underway is extremely difficult even for seasoned experts in naval warfare. However, the successful attack on *Limburg* while cruising was not surprising. While the global shipping industry is inherently vulnerable due to various factors,[8] the modern oil tankers are 'moving ducks'. An Ultra-Large-Crude-Carrier (ULCC) is a modern marvel carrying three million barrels of oil, but it displaces over 320,000 tonnes. Driven by economic prudence, the tonnage of commercial vessels is rapidly increasing, making them unwieldy and attractive targets. All merchantmen have only high-pressure fire hoses for self-defence, but unlike other vessels, the tankers cannot even install 'electric fences' (to prevent boarding) due to their inflammable cargo. No wonder that the 4,000-odd tankers plying the world's oceans are victims in a quarter of the pirate attacks, whereas these constitute a mere 10 per cent of the world shipping fleet.[9] Many of the modern ships, though cumbersome, are capable of sustained cruising at more than 20 knots. Nonetheless, tankers would remain easy targets while cruising at slow speed in restricted waters.

Who Thought of Environs!

An attack on a crude oil tanker could severely affect the maritime environment due to the ensuing oil spill, particularly a major one over 700 tonnes that necessitates a Tier-3 response. The aftermath of the

Limburg attack was witness to a massive slick in Gulf of Aden caused by 12,000 tonnes of spilled oil. A survey conducted by the French government revealed tar balls and oil cakes over vast carbonated stretches of land caused by burning oil along 70 km of the Yemeni coastline. The survey found that even "ghost crabs from long sandy beaches were seriously affected." The symptoms of the tanker attack could be observed miles inland from the coast.[10]

While it is difficult to quantify the effect of such disasters on marine life, it is nonetheless annihilating. Besides the immediate economic implications due to the depletion of fish-stocks over a vast area, major oil spills have grave long-term ecological effects. These could be particularly severe in narrow waterways and areas with fragile marine ecosystem such as in Gulf of Aden and Red Sea due to their coral reefs. A major oil spill off the Arabian Sea coast could also have adverse effects as far as the Indian subcontinent due to ocean-currents.

Appraising Other Probabilities

Will the Lull Last Long?

In November 2002, a top Al Qaeda man Abd al-Rahim al-Nashiri was arrested. It was a major breakthrough since he was the operational commander in the region and an expert in maritime terrorism who had planned all the attacks.[11] It may also be significant because there have been no major maritime strikes since the *Limburg* incident in 2002. Even in case of warships, June 2002 can be considered as the 'cut-off', when *jehadis* in Morocco were nabbed with plans to target US and British warships transiting the Strait of Gibraltar.[12] However, the lull may be a brief one, considering reports that Hamid al-Utaibi, another expert in maritime terror, has taken his place.[13]

Sub-Surface Terrorism

The revelations by Al-Nashiri make the lull more precarious. He talked about existence of a terrorist naval manual—a compilation of tactical tips such as vulnerable areas in different types of vessels, how much of explosive is required and handling of weapons and boats. The manual mentions the use of underwater scooters and limpet mines, which indicates the likelihood of attacks by Scuba divers, at least on fixed installations. Al Qaeda's frequently used bomb-ridden-boat tactics is believed to be motivated by the *Sea Tigers* of the Sri Lankan terrorist group, Tamil Tigers (LTTE),[14] who are also well versed with Scuba diving. Besides, the mean depth of Gulf waters is only 40 metres and currents are slow, which make diving operations easy. Hence, the probability of 'sub-surface' attacks by Al Qaeda is very high, particularly considering the tightened security around the offshore energy facilities since April 2004.

LNG Carriers as Bombs

The manual also refers to use of hijacked LNG carriers as floating bombs to devastate vital ports and littoral energy infrastructure. However, sufficient grounds exist to discount the threat since the gas is inert during transportation as a liquid. In case the tank is breached, it would vaporise into its inflammable gaseous form in contact with warm sea. But it will also slowly dissipate into the atmosphere and if ignited, the gas will not explode. An LNG carrier is therefore, an unattractive option for terrorists, relative to vessels loaded with other dangerous cargo such as ammonium nitrate. Nevertheless, attacks could be carried out on LNG facilities and carriers to disrupt energy trade.

Can Choke Points be Choked?

Fears are often expressed that terrorists could scuttle a huge hijacked vessel in a choke point, thus blocking the route for other ships with grave consequences for global economy, including spiking insurance rates. Three of the world's six principal geographical choke points lie off the Arabian Peninsula, *viz.* Suez Canal, Straits of Hormuz, and the Bab-el-Mandeb passage. Around 80 per cent of the West Asian oil is shipped through at least one of these waterways.[15]

The Suez Canal is used by about 25,000 ships every year, carrying about 3.8 mbd[16] oil to the West. Besides, it is vital for mobility of the coalition naval forces operating in the region. Being on sovereign territory, a tighter security cover is possible for the Canal relative to the latter two waterways, which are international straits. However, considering its navigational width of only about 200 metres and a uniform depth of only 21 metres, the Canal is highly vulnerable and thus more insecure. Terror threats and ensuing security-alerts are frequent, as in fall of both 2004 and 2005.[17]

3.5 mbd oil transits through the eastern channel of Bab-el-Mandeb used by northbound vessels.[18] The waterway is 3 km wide and its least depth is less than 30 metres due to coral growth. While it is theoretically possible to sink a vessel here, it may be difficult to settle it at the critical spot to block passage for other ships. However, it can be easier done using seamines, Al Qaeda's access to which cannot be discounted in the future. The only hurdle would be laying bottom mines accurately, which can be overcome by using divers to guide their positioning.

The Strait of Hormuz is a critical waterway since 17.4 mbd oil passes through it.[19] However, the fear of it being choked by sinking a ship is unfounded since each of its two, 3 km wide navigable channels (for entering and exiting the Gulf) are more than 80 metres deep. Nevertheless, if Al Qaeda could procure moored mines, it would not be difficult to lay them in the southern outbound channel used by laden oil tankers and LNG carriers. An activation of even one mine by any vessel would be sufficient to disrupt energy exports for many days due to fear among the shippers.

Cruise Shipping at Risk

Nashiri also spoke about Al Qaeda strategy of targeting "tourist ships...if warships became too difficult to approach."[20] The probability of its occurrence is high and increasing rapidly. The largest of the cruise ships today displace 150,000 tonnes, and carry more than 4,000 passengers and crew. An attack on a single vessel could cause mass casualties and panic, with attendant publicity for the terrorists. Considering the steady growth of cruise tourism in West Asia in recent past,[21] it is cause for a major insecurity. Besides, it also bears the potential for severe trans-regional ramifications.

Piracy & Terrorism: Sinister Allies

Piracy is common in northern Persian Gulf and Red Sea. The crime is particularly rampant off Somalia and in the Gulf of Aden, where ships are normally hijacked for ransom. In November 2005, the cruise liner *Seabourn Spirit* cruising off Somalia was fired upon using rocket-launchers and machine guns.[22] Although reported as an incident of 'piracy', it was certainly 'terrorism' for the victim vessel. In this part of the world, pirates often use sophisticated weapons, making the dividing line between piracy & maritime terrorism hazy. It is a common fear that the 'skills' of pirates could prove useful for maritime terrorism; after all, boarding ships has been a matter of 'livelihood' for pirates around the world. Al Qaeda may either mimic the pirates or even employ them to hijack ships. Another possibility goes further; some among the *jehadis* may be 'super-specialised' in piracy as a source of generating funds, considering the ongoing international efforts to freeze their finances. Piracy *per se* constitutes a threat to shipping, but the attacks in these waters were not taken seriously hitherto and considered as a 'problem' of the ship-owners. However, following the high probability of the link between terrorism and piracy, a new dimension of the threat has emerged.

Terror's Maritime Logistics

Terrorism too relies heavily on sea transport for its nourishment and activities. This is 'good news' for the civilised world because these lifelines could be severed. The hurdle however is that the maritime realm is inherently unregulated and with a little ingenuity, it can easily conceal the transactions. It was known that the Persian Gulf is a major transit route for the heroin produced in Afghanistan bound for markets in some Gulf States and Europe. However, hard evidence that Al Qaeda fosters drug trafficking (to finance its operations) was only obtained in December 2003, when its three dhows were seized near Hormuz Strait.[23]

For long distance transportation, Al Qaeda owns more than 15 sea-going ships flying 'Flags of Convenience' (FoC).[24] Besides being used to sustain terrorist operations, these vessels may be used to make money by benign commercial transactions. Considering the virtual impossibility to track ownership of FoC vessels, these are ideal to ferry instruments of terror and even Weapons of Mass Destruction (WMD). The US-led *Proliferation Security Initiative* (PSI) aims to address this insecurity. The increasing use of shipping containers for trade has compounded the peril since, even with modern non-intrusive detection technology—an essential element of the US *Container Security Initiative* (CSI)—it is feasible to inspect only a miniscule portion of the global container traffic. Recent findings also suggest that crude-oil tankers could be used to transport nuclear material, since the hydrocarbon atoms in oil would prevent detection of gamma rays. Experiments are ongoing to find the means to detect them (such as through Muon-detection).[25] The possibility of some tankers in the Al Qaeda 'fleet' cannot be discounted. Along with threat of WMD proliferation, it also has grave implications for the security of ports being visited by tankers. Is a Tanker Security Initiative (TSI) in the offing?

A new dimension of Al Qaeda-LTTE link has also emerged. Besides its other sources, Al Qaeda also gets the weapons from LTTE that are sourced from Cambodia. The gunrunning network in India's maritime east is well-known. Its hubs are Cambodia and the illegal arms market in Thailand and its spokes extend to all regional insurgencies—in Sri Lanka, India's north-eastern states, Myanmar and Southeast Asia. It is believed that Al Qaeda may be paying for these weapons in drugs. It is known that at times, the payment is made directly in Thailand. [26] This means that Al Qaeda may even be dealing directly with the Thai arms market, facilitated by LTTE. Although plenty of arms are available in Afghanistan and surrounding regions, Al Qaeda may have the need to obtain other types of weapons like shoulder-launched missiles, and may be even seamines that the LTTE are known to possess.

THE SECURITY RESPONSE

Maritime Forces in the Region

The Coalition: Omnipotent, or just Omnipresent?

Maritime security in the region is largely being handled by multinational forces led by the Commander, US Naval Forces Central Command (NAVCENT). Designated Combined Forces Maritime Component Commander (CFMCC), he is headquartered in Manama (Bahrain) and has about 45 warships (and 15 maritime patrol aircraft) under his command, 25 of which are American and the rest from countries like the UK, France, Australia, Germany, Italy, Japan and Pakistan. Some others navies join occasionally. The assets are presently divided among three Task Forces—TF-58, TF-150 and TF-152. Commanded by Commodores (normally) from all coalition members in rotation of about five months each, their respective areas of responsibility are northern Gulf, central/southern Gulf, and beyond Gulf (western Indian Ocean/Arabian Sea, Gulf of Aden and Red Sea).[27] The Destroyer Squadron 50 (DESRON-50) is the US Navy's resident command in Bahrain, which is the also the base for US Naval and Coast Guard patrol vessels, *Mobile Security Detachments* (MSD) and *Law Enforcement Detachments* (LEDET).

The mission of naval forces is to deny the use of the sea to terrorists. It involves interdicting smuggling of drugs and weapons, anti-piracy and maintaining the vessel traffic database. TF-58 is also entrusted with security of Iraq's waters (and oil terminals), except riverine areas that are presently being patrolled by the Marines. The UN Security Council Resolution (UNSCR) 1546 of June 2004 provides the mandate to coalition naval forces to board and inspect all vessels *en route* to and from Iraqi ports, and detain those carrying prohibited cargo. Since the Navy does not wield legal powers, the TF-58 units carry Coast Guard LEDETs. The security of oil terminals involves a layered defence. The 35 MSD personnel manning each platform constitute the inner layer. Up to 2 km around the terminals are exclusion zones, within which only authorised vessels can enter. At an outer perimeter of 3 km from each terminal are three naval vessels on patrol with their boarding teams and integral helicopters, which clear the merchant ships to enter the exclusion zone. The coalition navies also assist Kuwait and Saudi Arabia to enhance security of their oil terminals.

The UN mandate for boarding vessels in international waters is explicit on approaches to Iraq. However, the authority has been extrapolated by the coalition to rest of the Gulf and even beyond Hormuz Strait. These interdictions also serve the aim of the US-led Proliferation Security Initiative (PSI) that seeks to counter the spread of WMDs.[28] The photographic information of the interdictions is passed on to the headquarters through digital data-link called CENTRIXS (Common Enterprise Regional Information Exchange System).

Despite the overwhelming coalition naval presence, all is not well in the region. Even piracy has not subsided in the northern Gulf. These waters are among the busiest in the world and it is difficult to identify the miscreants. Another handicap is the dearth of reliable statistics. The US naval intelligence has its own record-keeping for trend-analysis, but the many unreported incidents of piracy lead to an inaccurate assessment. Alike their counterparts ashore, combating terrorism and crime at sea is not what the coalition naval forces are particularly 'cut-out' for. Although the naval helicopter pilots looking out for suspicious boats would not admit, they are certainly trained and equipped for locating submarines rather than an amorphous enemy. The MSD personnel guarding oil terminals too may not have acquired the necessary expertise, since they began to be deployed for such tasks only after 9/11. The naval forces are now being pushed to share the onerous burden of land forces. The new US Naval Expeditionary Combat Command, established on January 13, 2006, has been directed to take over the Iraqi river patrols from the Marines by early-2007.[29] It remains to be seen how successful the coalition naval forces would be under such increasing pressures.

Iraq's Nascent Navy

The Iraqi Coastal Defence Force was build from scratch in mid-2004 through a roadmap prepared by the coalition. After about six months, the force assumed the Navy title. Currently, its personnel strength is about 700 (including naval infantry) and its assets are five 27-meter Predator-class patrol boats, three Al Uboor–class patrol boats (both armed with heavy machine-guns), 24 Fast Assault Boats (FAB) and 10 rigid inflatable boats. Based at Umm Qasr and under Iraqi Navy's own command and control and logistics support, these vessels maintain a 24-hour patrol in the territorial seas.[30] There are plans to acquire two 1000-tonne Assad-class corvettes and two offshore support vessels, but exact schedules have not been announced since these may be linked to achieving necessary training standards. The 500-strong Iraqi Riverine Patrol Service (IRPS) uses the British Combat Support Boats to patrol inland waterways along with the US and Royal Marines. The naval infantry assist the US sailors to guard oil terminals.

Although the Iraqi maritime security establishment has achieved significant milestones in a short time since its inception, it is still very distant from acquiring the capacity to safeguard the nation's vital interests. Despite the 400-strong naval infantry, the coalition goal to transfer security of platforms to Iraqi personnel by November 2005 has not been met. In fact, as on mid-2005, no Iraqis were deployed in Khor Al Amaya platform since its MSD also had women sailors.[31] Even with the planned addition of three *Al Uboor* boats by September 2006, the total patrol assets may not be sufficient to combat the prevailing threats out to the Exclusive Economic Zone (EEZ) considering their minimal weapon fit, the Predators' dire need for new radars, and the fact that there are no plans to back surface patrols with airborne surveillance. By the time the Iraqis are proficient to operate the 1980s-built Assad-class corvettes, these would be old with problems of maintenance & spares. Even for the existing vessels, there is a severe dearth of equipment, spares and even fuel. Sailors of the engineering branch complain that, "spares shortages prohibit the patrol boats from staying underway for more than 12 hours (a day)" and that "requests...go unheeded by the Iraqi Ministry of Defense."[32]

The land forces are no better off—a US General serving in Iraq has stated that the forces are not being provided adequate logistics in terms of weapons, ammunition, equipment and even salaries.[33] This raises doubts whether the Iraqi government can sustain the Navy financially, let alone its acquisition plans.

Whither Go Other Navies?

The Gulf States have acquired substantial naval hardware since the 1991 Gulf War. The focus has been on sea-denial assets like fast attack craft, anti-ship missiles, midgets and mines, which highlights the insecurities due to conventional threats. Submarines are vital for Iran to deter the US control of Hormuz Strait. Saudi Arabia and Oman are inducting modern stealth frigates and the UAE has plans for high-speed corvettes. States involved in disputes over islands (including in Red Sea) have also invested in amphibious capabilities.[34]

In the process however, the acquisition of patrol vessels for constabulary duties has suffered. While warships are versatile to be used for peacetime roles, the larger vessels are not optimised for low-end tasks and are limited in numbers. Besides, their employment for policing causes material degradation. Only Saudi Arabia, Oman and Kuwait have adequate patrol vessels, but these cannot operate helicopters. In general, all vessels are qualitatively unsuitable in other respects too. Either these cannot operate in shallow waters of the Gulf; or the ones that can, are inferior in terms of speed and firepower.

In the Aden Gulf–Red Sea area, the presence of regional navies is sparse. While Egypt's focus is on its Mediterranean coast, Saudi Arabia's assets are also divided between its two coasts. Other than Yemen, which is acquiring new patrol boats, states can ill afford adequate policing assets. Hence, these waters are highly unregulated. Somalia has asked for external naval assistance to combat piracy, but its neighbours do not have the capabilities to take on even these pirates who are well armed.[35]

Multilateral Cooperation

Cooperative efforts to counter non-traditional maritime threats have been rare in the region. There has been a reticence among some states to accede to global treaty-based conventions like the *Convention for Suppression of Unlawful Acts against the Safety of Maritime Navigation,*

1988 (SUA Convention). This vital legal instrument provides a framework for cooperation to counter non-state violence at sea. Within it, states have obligations to establish their jurisdiction over the listed offences, punish the offenders themselves or extradite them and exchange information needed for criminal proceedings. While 126 countries all over the world have acceded to the convention, key regional states like Saudi Arabia and Iran (and also Iraq) have not yet done so. Saudi Arabia objects to its provisions on extradition and arbitration. Iran's resistance to join it probably stems from the fact that the Convention's 2005 Protocol has now expanded its ambit to transportation of WMD[36] at the behest of the USA, and Iran perceives itself to be the target.

Specific to the regional seas, the only existing arrangements deal with non-contentious issues like disaster relief and environmental preservation. The Gulf Cooperation Council (GCC) has been discussing terrorism and measures to strengthen security coordination and intelligence exchange, but there has been little emphasis on maritime security, probably due to the persisting maritime contentions among states. Besides, the forum does not include all relevant powers. In 2004, the initiation of a wider and more representative forum—the IISS-sponsored Regional Security Summit (*The Gulf Dialogue*)—augurs well for the region. It has brought together security establishments of regional states and key extra-regional ones—the USA, France, Russia, UK, Australia, Japan and Singapore. China, Germany and India have also joined since the second Summit in December 2005. It was however, aptly observed during the maritime security deliberations of the 2005 Summit that "inter-state tensions seemed likely to rule out an over-arching multilateral regional security arrangement in the foreseeable future."[37]

The Way Forward

The region's security response to maritime terrorism and the increasing intensity of other non-traditional threats at sea would essentially comprise a three-pronged approach as follows:

Individual Measures by States

- While planning, regional navies would need to lay adequate stress on constabulary tasks. This is endorsed by Geoffrey Till, who recommends for world navies, "...the diversifying maritime threats...may well require a shift in emphasis from the military to the civil aspects of sea power."[38]

- Furthermore, it may be necessary for maritime forces to tailor their equipment fit, tactics and Rules of Engagement (RoE) suitably to counter prevailing threats. Vessels would have to be faster, capable of operating in shallow waters and armed with the right weapons. A good example of such 'tactical realignment' is the use of high-rate-of-fire Close-in-Weapon-System (essentially designed for warship defence against sea-skimming missiles) to counter terrorist bomb-laden speedboats.
- At government level, states may consider joining the global legal instruments like the SUA Convention, and to counter WMD proliferation, its 2005 Protocol and UNSCR 1540 (2004). Appropriate adjustments in domestic laws may then follow.
- Compliance of International Ship and Port Security (ISPS) Code would mitigate inherent vulnerabilities of ports and commercial ships. Also, experts recommend deployment of underwater surveillance technology to secure their ports.[39] Such technology could also be used to counter under-sea terror attacks on offshore terminals and naval assets.

At Regional Level

- Sincere efforts to resolve maritime disputes among neighbours would serve to build confidence and remove a major hurdle towards cooperative security against non-traditional threats.
- To enhance security in Hormuz Strait, maritime forces of bordering states may need to maintain a continuous presence, coordinate their patrols and cooperate in compiling the surface-picture of shipping traffic.
- Notwithstanding participation of GCC states in the biennial 'Arabian Gauntlet' naval exercises with coalition forces,[40] they may be wary of joint operations. However, these states could still play an important role in ongoing coalition efforts through tactical coordination and information sharing.
- It is significant that Iraqi Navy participated in 'Arabian Gauntlet' exercises of March 2005, as an observer. Despite Iraq's mere 58 km coastline and two ports, strengthening of its maritime forces holds the key to the region's security and

stability. States would thus need to contribute towards their capacity building.

- A cooperative arrangement is necessary to combat piracy in the region. The Asian *Regional Cooperation Agreement on Anti-Piracy* (ReCAAP) of November 2004 may be worthy of emulation. It provides for information-exchange among members through an Information Sharing Centre, aimed at generating better statistics and reports to facilitate policy decisions.

Beyond the Region

The inclusion of extra-regional states in the *Gulf Forum* is significant and an imperative. These states not only have stakes in the region, but can also usefully contribute towards its security with their areas of expertise and experiences in combating terrorism. Besides, states of Southern Asia could play a significant role to disrupt the terrorists' logistic-lifelines to the East.

The regional states are now seeking to involve states beyond the Gulf in their security matrix and have realised the rising eminence of India as a potential stabilising factor. The states have also noted the reach and wherewithal of its Navy to assist the Indian Ocean littorals whenever required, the latest being the December 2004 Tsunami disaster. India too has high stakes in the stability of the region and sea-line security. Hence, it ought not to dither in delivering in the event of a contingency in regional waters, notwithstanding the presence of coalition naval forces. It is however certain that it would act only when requested for, and as far as possible, in conjunction with regional maritime forces. Thus far, India has been having regular naval exercises only with Oman and, with other Gulf States, naval/ coast guard interactions are limited to port visits and training. Hence, at the outset, there may be a need to expand the scope of joint exercises to facilitate interoperability.

The regional states form a significant subset of the international community. Intra-regional deliberations on maritime security could be followed by raising issues at global fora like cooperation on capacity building and modification of current legal mechanisms to combat new threats. Specific issues could include technological cooperation for non-intrusive inspection of shipping containers and abolition of outdated norms of the shipping industry such as FoC registrations and self-defence for merchantmen.

Conclusion

While energy transportation of the USA and its allies may be at a greater risk due to terrorism, it cannot be 'business-as-usual' for other countries. In next 20 years, the energy dependence of industrialised countries on the Gulf would increase incrementally, but it would be an exponential surge for the developing states. Industrialisation may not be a panacea to solve all socioeconomic ills prevailing in these countries, but it is an imperative for economic growth, which may empower governments to resolve many such issues. Besides, port infrastructure, choke points and cruise shipping are also vitals of their economies. So far, the year 2002 and *Limburg* may be the last that we heard about maritime terror in the region, but it is purely due to the complexities involved in carrying out maritime attacks. Al Qaeda has strong motivations to overcome the hurdles, its tendency for patient and deliberate planning is well known and its ingenuity is beyond doubt. If the *jehadis* have their way, among the worst hit would be the growing economies of Indo-Pacific region that have embarked upon the path of development. Not the least affected would be the energy-producing states of the Gulf, for many of whom energy resources account for nearly all export earnings.

It is undeniable that crime and disorder in waters off the Arabian Peninsula have received a fillip in the past few years. Maritime insecurities may grow in future when the presence of multinational forces is eventually scaled down. The regional states thus, need to reorient their security mechanisms to pre-empt illegitimate use of the seas. The role of the key extra-regional states is also necessary. The energy-deficient ones are being compelled to resort to diversification of energy sources and building strategic oil reserves. These may be pragmatic imperatives, but the transnational security imbroglio deserves a comprehensive solution.

Notes and References

1. "Station Airs Alleged Osama Tape", *CBS News*, October 6, 2002, *http://www.cbsnews.com/stories/2002/09/13/attack/main522015.html.*
2. *World Energy Outlook 2005*, OECD/IEA, p.83.
3. Report of US National Energy Policy Development Group, Washington DC, May 2001, p.1-11, *http://gcrio.org/OnLnDoc/pdf/nep.pdf.*
4. Oil imports from Gulf constitute 22 per cent, 30 per cent and 76 per cent of the demand for the USA, Western Europe and Japan respectively, *EIA Fact Sheet*, September 2004, *http://www.eia.doe.gov/emeu/cabs/pgulf.pdf.*

5. *EIA* Fact Sheet, September 2004 at *http://www.eia.doe.gov/emeu/cabs/pgulf.html*
6. "Map of Future Al-Qaeda Operations", *SITE Institute*, February 14, 2005 at *http://www.siteinstitute.org/bin articles.cgi?ID=publications19805&Category=publications& Subcategory=0*
7. "Suicide Boat Bombers Attack Key Iraqi Oil Terminal", *Jihad Watch*, April 24, 2004 at *http://www.jihadwatch.org/archives/001704.php*
8. Khurana, G.S. (2005). "Maritime Terrorism in Southern Asia: Addressing the <Precursors>", *Indian Defence Review*, Vol. 20(1), Jan-Mar 2005, p.60
9. "Piracy dips but Indonesia remains violent black spot", *The New Zealand Herald*, February 9, 2005, at *http://www.nzherald.co.nz/index.cfm?c_id=3&ObjectID=10010024*
10. "Pollution due to the explosion of oil tanker Limburg", *Le Cedre* website, May 2004 at *http://www.le-cedre.fr/uk/spill/limburg/limburg_pollution.htm*
11. "Suspected Qaeda Chief Cooperating", *CBS News*, November 22, 2002 at *http://www.cbsnews.com/stories/2002/11/17/attack/main529656.shtml*
12. While August 2005 was witness to rocket attack on US warships *Kearsarge* and *Ashland* berthed in Aqaba (Jordan), it cannot be considered maritime terror since the rockets were fired from a warehouse.
13. "The Threat of Maritime Terrorism", *Defence Update News Commentry*, December 10, 2005 at *http://www.defense-update.com/2005/12/threat-of-maritime-terrorism.html*
14. *Al Qaeda* is also reported to have sent a team to LTTE to sharpen its skills after the not-too-successful USS *Cole* bombing.
15. No. 2, p.265
16. This includes 2.5 mbd through Sumed pipeline to reduce the draft of vessels transiting the Canal.
17. *"Suez Canal at ISPS Level 2?", Reuters, December 09, 2005 at http://sentinelmaritime.com/*
18. No. 2, p.263
19. Ibid.
20. Christopher Dickey, "High-seas terrorism", *Newsweek*, January 27, 2003, at *http://www.iissm.com/newsletter/february-03/terror.htm*
21. Jamila Qadir, "Mideast cruise industry set to boom", *Khaleej Times*, February 02, 2003 at *http://www.menafn.com/qn_news_story_s.asp?storyid=13103*
22. "Piracy on the rise off Somalia", *BBC News*, November 8, 2005 at *http://news.bbc.co.uk/2/hi/africa/4415196.stm*
23. Bruce Hoffman, "Navy Makes Persian Gulf Drug Bust", *CBS News*, December 19, 2003 at *http://www.cbsnews.com/stories/2003/12/19/terror/main589500.shtml*
24. More than half of the global commercial fleet today flies the flags of states like Panama, Honduras and Liberia, which offer easy vessel registration and demand low taxes and no restrictions on standards/ employment of crews.
25. Jonathan Medalia, "Potential for Terrorist Nuclear Attack Using Oil Tankers", CRS Report for US Congress, December 07, 2004 at http://www.fas.org/irp/crs/RS21997.pdf
26. "Armed and dangerous: Taliban gear up", *Asia Times* online, December 22, 2005 at *http://www.atimes.com/atimes/South_Asia/GL22Df01.html*
27. "News Briefing with Vice Adm. David Nichols", US Department of Defense, September 22, 2005 at *http://www.defenselink.mil/transcripts/2005/tr20050922-3963.html*
28. Harold Kennedy, "U.S.-led coalition seeks to block weapon shipments", *National Defence*, January 2004 at *http://www.nationaldefensemagazine.org/issues/2004/Jan/US-Led.htm*
29. Louis Hansen, "Command will give sailors a larger role on the ground", *The Virginian-Pilot*, January 14, 2006 at *http://home.hamptonroads.com/stories/story.cfm?story=98139&ran=183893*

30. Cordesman, Anthony (2005). "Iraqi Force Development", *CSIS*, November 21, *http://www.mafhoum.com/press9/257P7.pdf*

31. Tiron, Roxana (2005). "Defense of Iraqi Oil Depots Tests Mettle Of U.S. Sailors", *National Defence*, June, *http://www.nationaldefensemagazine.org/issues/2005/Jun/defense_of_iraqi.htm*

32. Axe, David (2005). "Resurrection", (US) *Navy League*, November, *http://www.navyleague.org/sea_power/nov05-30.php.*

33. Schmitt, Eric (2006). "Iraqi Forces May Turn into Rebels Due Lack of funds: General", *Times of India*, January 7, p.25.

34. *Jane's Fighting Ships: 2004-05*, Jane's Information Group, Woodbridge Company.

35. Brown, Nick (2005). "Somalia Calls for Piracy Help", *JDW*, November 02, p.17.

36. Boese, Wade (2005). "Treaty Amended to Outlaw WMD at Sea", *Arms Control Today*, December, *http://www.armscontrol.org/act/2005_12/Dec-WMDsea.asp.*

37 "The Gulf Dialogue", *IISS News*, Winter 2005, pp.3-4, *http://www.iiss.org/showdocument.php?docID=712.*

38. Till, Geoffrey (2005). "Navies and the New World Order", *USNIP*, March, p.60.

39. McKelvie, Duncan (2005). "Gulf Ports Warning", quoted in *CCTV Info*, February 21, at *http://www.cctvinfo.com/news_article.aspx?news_id=2106*

40. Mahdi, Mazen (2005). "Running the Gauntlet of Mines in Sea", *Arab News*, March 29, *http://www.arabnews.com/?page=9&section=0&article=61268&d= 29&m=3&y=2005&pix=community.jpg& category=Features%22*

10

The Human Aspect of Regional Security

A View from North Africa

AHMED A. AL-ATRASH

A great number of theorists and practitioners in the field of international conflict analysis and management concur that most inter-state conflicts are internal in origin. Some contend that, "since the end of the Cold War, armed conflicts have increasingly taken place within, and not between, states."[1] This is not to deny that domestic unrest (which is usually labeled as civil wars) is, somehow, vulnerable to internationalisation, especially because there is an international consensus taking shape over the need to strengthen human security around the world. This also validates the view that the human dimension of international security should be seen and dealt with as a priority in the process of maintaining peace and security, as well as attaining prosperity for all human beings.

Unlike the conventional interpretation of security that heavily relies on military capabilities as a driving force, the contemporary understating of the term (security) is becoming multidimensional. According to the new explanation, "traditional security policy emphasises military means for reducing the risks of war and for prevailing if deterrence fails while the proponents of Human security, without eschewing the use of force, have focused to a much greater degree on non-coercive approaches. These range from preventive diplomacy, conflict management and post–conflict peacebuilding, to addressing the root causes of conflict by building state capacity and promoting equitable economic development."[2]

Due to the circumstantial variables of the post-Cold War era, the United Nations Development Program (UNDP), in 1994, was the first to introduce, in detail, a new multidimensional look at the concept. These include the economic, environmental, personal, societal, political, as well as the human aspect of security.[3]

The paper examines the human aspect of regional security within the framework of North Africa-Southern Europe states (or 5 + 5), while taking into account the Euro-Libya cooperation in tackling the problem of illegal immigration. The core question here is what are the underlying factors that cause human exodus from Africa to Europe? In other words, should the parties concerned (especially the Europeans) concentrate their efforts on its roots or its manifestations to put an end to this phenomenon?

The West Mediterranean Region: A Case Study

One of the problems of undertaking research of this nature is the decision that has to be made concerning the number of cases to be used. In other words, can a single case, particularly on security, contribute to the development of theory? Does the particular apply to the general?

The answer may depend primarily on the characteristics of each situation and the method that is to be utilised in presenting it as a case study. Single case studies can be useful to the extent that they throw light on general and theoretical issues. According to Bercovitch, "single cases provide a powerful impetus to the development of a general explanation as long as they are historically grounded and their description is not couched in purely idiosyncratic terms."[4] The traditional assumption that "no two cases are the same" just as no "twins" are the same might be misleading, if not invalid. The fact that could not be overlooked is that crises generally share a number of common patterns.[5] According to Webb, all cases are both general and unique. Webb contends, "uniqueness is composed of constellation of particular characteristics each of which may be general, but which have never before occurred in that combination."[6]

The core of the above arguments applies, by and large, to most of the concerned areas of today's world. The origins and main characteristics of major regional security issues in the West Mediterranean area are regarded in this study as applicable to other regions, including West Asia. In order to illuminate the relationship between the general and the particular, it is of importance to throw light on the most common features of regional security from a different perspective.

As stated above, the term 'security' can no longer be perceived as a taboo. It is being subjected to different interpretations. One explanation is heavily based on human needs and rights as a prerequisite for a more secure world. The world community, through its institutional level, namely through the UN, had recently recognised the significance of looking at the question of international security from a different angle. That was due to the rising threats to the world security at large. Following the announcement of UNDP report in mid-1990s, a number of concerned governments had to re-think their views on the issue. The most striking example comes from Canada and Japan. In 1996, for instance, the human security approach was adopted and skillfully developed by a great number of Canadian scholars and practitioners.[7]

In the context of Afro-Europe cooperation, it is becoming evident that human security constitutes a precondition for any security arrangement on larger scales (national, regional, and so on). The Europeans engaged in the 5 + 5 gathering (North Africa-Southern Europe) need to realise and confess that the setbacks deter the process of cooperation between them and the Arab Maghreb states,[8] which are heavily associated with human rights and economic needs. European officials are keen to maintain and materialise their interests in the area and are usually eager to have the lion's share of the exploits.

When thousands of sub-Saharan Africans attempt to illegally and hazardously cross the Mediterranean by small private boats seeking 'better' life in Europe, the blame usually is put on Libya and other North African states for not taking serious and practical steps to tackle the problem. In the process of Libya rapprochement with the West, European officials (at various levels) raise the question of how best to stem the tide of wave of Europe-bound African illegal immigrants without taking into account the more pressing question of what are the underlying factors that cause this human exodus. In order to put an end to this problem, the parties concerned should devote their efforts to cooperate in dealing with its roots.

In order to clarify the official European self-interest, handling of the increasing waves of illegal immigrants to Europe, it is of importance to throw lights on the main elements and causes that largely hinder any effort to stop such crises. It should be apparent here that there is an overwhelming agreement, as records of concerned international bodies indicate, that the African continent, for a long period of time, remains the least developing part of the contemporary world.

The UN Human Development Index (HDI) of 2005 (based on 2003 figures) that "ranks nations according to their citizens' quality of life..."[9] clearly shows that all the 20 "Least Livable" countries of world are in Africa, with Niger on the top of the list.[10] On the contrary, the 20 "Most Livable" countries of the world are from the West.[11] It is interesting to notice here that the former group constitutes a source of 'illegal immigrants' while a great number of the latter become their ultimate target(s).

Though little progress has been noticed, Africa, according to a recent statistical survey prepared by the Canada Fund for Africa, is facing the following challenges:[12]

- 1 in 2 Africans lives on less than US $ 1 a day.
- 45 million African children are not in school.
- More than 26.6 million people are HIV positive.
- 1 in 5 Africans is affected by conflict.
- 14 countries suffer from water scarcity; 11 more will join them by 2025.
- About 500 million hectares of land, 65 per cent of it agricultural, has suffered degradation since 1950.
- Africa has 13 per cent of the world's population, but about 2 per cent of global trade and 1 per cent of global investments.

The above figures clearly indicate serious evils pertinent to conflicts and wars, hunger, poverty, illiteracy, human rights, corruption, diseases, natural disasters, lack of democracy and other major developmental obstacles. These, in our viewpoint, are the core sources of the problem that need to be dealt with in any security arrangement; and without which any effort may fail. Wars and conflicts are major stumbling blocks to development. Human beings are generally the scapegoats in such cases. For example, in the last decade, 80 per cent of the victims of wars were civilians.

The ongoing and intensive shuttles of Western diplomats and high officials (including a number of heads of state and government, as well as distinguished delegates from the EU) to Arab Maghreb states (notably from Libya) have focused on tackling specific issues that constitute a threat to world security. But it is becoming evident that their aims and efforts are based on their interests to maintain high level of security for their countries without addressing the others' needs and ambitions. An exception to this conduct that deserves to be mentioned and can be a model that required to be emulated is that of Canada. In 2005, the Canadian Prime Minister visited Libya and expressed his country's willingness to cooperate with African states (including the host state, Libya) in putting an end to the Darfur human disaster.[13]

Terrorism and WMD: West *vs.* Rest

The key argument of the paper also applies to other vital but conflicting issues. In the eyes of the West, exclusively after 9/11, the most serious problems that negatively affect their security (but not necessarily the others' security) come under the rubric of terrorism and/or weapons of mass destruction (WMD). Consequently, it is worth briefly throwing light on these questions from different perspectives, namely the 'West' (including its strong non-Western allies) and the 'Rest' (including its Western 'sympathisers or advocates').

For the West, terror must be globally fought and eliminated root and branch, and usually dealt with as being connected to Islamism. For the Rest, the term 'terrorism' must be clearly defined, bearing in mind the distinction between it (terrorism) and legitimate resistance against colonialism and foreign occupation. This view contends that terror is not necessarily a product of Islamic movements, but can be associated with and derived from radicalism, even within the Western societies.

For the West, since the late 1950s, the target has been to create nuclear-weapons-free zones on continental level; namely in Latin America and the Caribbean, South-East Asia and Africa. This is based on the explicit belief that, "the establishment of a nuclear-weapons-free zone in one region can encourage similar initiatives in other regions."[14] According to this theory, the African Treaty of 1996 (ANWFZ) may enhance similar initiatives in the 'Middle East'. In this context, it has also been argued, that "...settlement of the Arab-Israeli conflict is not a necessary precondition for such an initiative. Measures to establish a nuclear-weapon-free zone would automatically alleviate tension and advance negotiations. Such an initiative should not also be dependent on Israel joining the Nuclear Non-Proliferation Treaty (NPT). There should, however, be a treaty banning the acquisition, possession, production, and stipulating the dismantling and destruction of any nuclear weapons or installations established prior to its signing."[15]

For the Rest, this question must be dealt with comprehensively and on a global scale, without ambiguity, exclusion or selectivity. In relation to the 'Middle East', some disagree with the above belief by stating that "(t)he fact some countries in the Middle East might have nuclear capabilities makes it more difficult to establish nuclear-weapon-free zones in Africa, Central Asia or even Central Europe or South Asia, as all these regions lie within the range of weapons in the Middle East."[16] It has been also suggested "...Arabs should work towards establishing nuclear programmes for peaceful purposes in order to attain balance with Israel. In this case, transparency would be vital in order to avoid misconceptions regarding their aims and capabilities."[17]

Conclusion

Though the UNDP was first to address the importance of human security, the awareness that the human dimension of security as a precondition for state security can go back to the Cold War period. That was clearly stipulated in the UN constitutional foundations (noticeably, the Charter), as well as in declarations and decrees of the various international and regional arrangements. This is at the theoretical level. In practice, however, there is a pressing need, at the grassroots level, to translate such ambitions into reality. This is based on the fact that peace and security are correlative and complementary, and there will be no development without them.

The international conflicts and crises that occupy the agendas of politicians and decision makers in the developed world are frequently perceived as being transmitted from the developing and underdeveloped countries. They also stress on how to put an end to these trends without taking into account the significance of looking at the reasons and diagnosing their origins, thus finding proper remedies in accordance with the existing circumstances, at all levels. Contrarily, in the developing regions, including the Arab world, Africa and West Asia, wars, conflicts, starvation, poverty, illiteracy, human rights and needs, natural disasters, corruption, dependency, debt, diseases and other major developmental obstacles are the core issues that need to be addressed because they are interrelated and are the mainsprings of conflicts and crises in the world. Any security arrangement that either ignores or refuses to deal with these issues is doomed to failure.

Notes and References

1. *www.humansecuritycentre.org.*
2. Ibid.
3. For details, see "Chapter 2: New Dimensions of Human Security", *Human Development Report 1994, www.hdr.undp.org.*
4. Bercovitch, J. (1986). "A Case Study of Mediation as a Method of International Conflict Resolution: The Camp David Experience", *Review of International Studies,* Vol.12(1), p.43.
5. Groom, A.J.R. (1988). "Problem Solving in International Relations" in E.E. Azar, & J.W. Burton (eds.), *International Conflict Resolution: Theory and Practice,* Wheatsheaf Books, Brighton, p.85.
6. Webb, K. (1995). *An Introduction to Problems in the Philosophy of Social Sciences,* Pinter, London; New York, pp.129-130.
7. For more details visit for example: *www.dfait-maeci.gc.ca* (official) & *www.humansecurity.info/CCHS* (non-official).
8. The Arab Maghreb region is comprised of Libya, Tunisia, Algeria, Morocco and Mauritania.
9. *www.infoplease.com.* The criteria used in the HDI for ranking the most and least livable states in the world includes life expectancy, educational attainment, and adjusted real income.
10. As put in order in the HDI, these are: Niger, Sierra Leone, Burkina Faso, Mali, Chad, Guinea-Bissau, Central African Republic, Ethiopia, Burundi, Mozambique, Congo. Dem. Rep of, Zambia, Malawi, Tanzania, Cote d'Ivoire, Benin, Eritrea, Angola, Rwanda and Nigeria. For more pertinent information on a particular country, visit for example, www.infoplease.com.
11. These are: Norway, Iceland, Australia, Luxembourg, Canada, Sweden, Switzerland, Ireland, Belgium, United States, Japan, Netherlands, Finland, Denmark, United Kingdome, France, Austria, Italy, New Zealand and Germany. For more relevant details on any of these counties, visit for example, www.infoplease.com.
12. The Canada Fund for Africa (2004). "New Vision, New Partnership", *www.acdi-cida.gc.ca.*
13. For more details on Canada's efforts in the African continents in general, visit, *www.acdi-cida.gc.ca.* See also, *Canada World View,* No 21, Winter-Spring 2004, Department of Foreign Affairs and International Trade, Ottawa, Canada.
14. Hammad, F and A. Ahmed (2001). "Comparative Study of Nuclear-Weapon-Free Zones", *Al-Siyassa Al-Dawliya(World Politics),* No.144, p.9 (English Text).
15. Ibid.
16. Ibid.
17. Ibid. For more details on the 'Middle East' case see the Arabic text of this article in *Al-Siyassa Al-Dawliya* (*World Politics*), No.144, April 2001, pp.35-38.

11

The Palestine-Israeli Issue and the Regional Security in West Asia

CHEN SHUANGQING

With its strategic position and abundance of natural resources, West Asia was the region, which both the colonists and the imperialists scrambled for since the end of 18th century. Until the end of the Second World War, the external powers ensured their presence in West Asia, maintained their influence and intervened aggressively, upon any threat to their interests. After the Second World War, along with retreat of the British and French powers from the region, and the interference of the two superpowers, and particularly with the establishment of the state of Israel and the revival of the Islam movement, the dynamics of security in West Asia underwent a paradigmatic change. While the big powers contended for hegemony behind the curtain, the intra-regional dynamics revolved around the Arab-Israeli conflict which also assumed the shape of Islam *versus* Zionism at times. The Palestine-Israeli conflict was indeed at the core of the regional insecurity.

Ever since the end of the Cold War, the security situation in the region assumed critical shape after Saddam Hussein attacked Kuwait. However, after the September 11, 2001, non-traditional security challenges emerged on the horizon, for example terrorist attacks, proliferation of WMD as well as the combination of both. Although the Palestine-Israeli conflict no longer occupies as much space and importance as it did in the past, it is still regarded as a thermometer of the regional security. The unresolved issue of Palestine has been at the root of Arab rage against the West and the locomotive of Islamist terrorism.

Regional Security

The regional security in West Asia transformed from traditional security to non-traditional security ever since the Cold War, and can be divided into four distinct phases: The first phase extends from the establishment of the state of Israel to the collapse of former Soviet Union. During that period, there erupted five wars between the Arab states and Israel, resulting in large number of territories of the Arab states being occupied by Israel. As a result, the confrontation between the Arabs and Jews were so intense that the Palestine-Israeli conflict became a key factor of the regional security. On the other hand, the war between Iran and Iraq that broke out in 1980 lasted eight years, leading to the deterioration of the security situation in the Persian Gulf region.

The second phase began from the collapse of former Soviet Union and lasted till 9/11. This phase witnessed three milestones, i.e., the Gulf War, the dismantling of Soviet Union and the Madrid Peace Conference. All of these happened in the year 1991. The content of the regional security in West Asia changed in two aspects since then. Firstly, under the "Madrid Framework", both Palestine Liberation Organization (PLO) and Jordan signed peace agreements with Israel that demonstrated a breakthrough and initiated the peace process. The peace process, however, did not go on smoothly as people expected, but suffered a setback by the assassination of Israel's former Prime Minister Yitzhak Rabin as well as the rise of the right wing in Israel. Along with the failure of the Camp David Summit which was held on July 2000, the Palestine-Israeli conflict worsened. After assuming power as Israeli prime minister on March 2001, Ariel Sharon began to carry out "active defence policy", as well as "targeted killing policy" against the Palestinian radical groups, retaliating severely to the terrorist attacks. Sharon's military policy stimulated the Palestinian radicals to launch more attacks, and the attacks in turn met with worse retaliation by the Israelis setting off a vicious circle. Secondly, during this phase the American hegemony in West Asia was strengthened after the break up of the Soviet Union. The US was seen to be putting greater pressures on Iraq and Iran, aggravating the security situation in Persian Gulf.

The evolution of the security situation in West Asia entered the third phase after 9/11. During this period, the US has taken up West Asia as a critical pivot for global war on terror. This has suited Israeli agenda in West Asia and it has hitchhiked the chariot of counter-terrorism driven by the US by military strike against Palestinian radical groups. As a result, there were 29 suicide bombings against Israeli targets between December 2001 and June 2002. This prompted retaliation from Israeli side.[1] On the other hand, in order to topple Saddam's regime in Iraq, the US launched the Second Iraq War in 2003, on the pretext that Iraq had WMD. The US army took over Iraq with surprising speed and almost zero resistance and Saddam was ousted from power only to be captured alive later. But before the US could celebrate the fall of the Saddam Hussein regime and claim total victory, the insurgency in Iraq picked up and the US was embroiled in an unending counter-insurgency/terrorism operation.

Thus, the security situation in West Asia entered the fourth phase after the Second Gulf War or the Second Iraq War. During this period, the terrorist attacks increased in West Asia spreading from Iraq to Turkey, Israel, Saudi Arabia and Egypt, etc. At the same time, the possibility of WMDs falling into terrorist hands became another key issue for the security practitioners worldwide.

In order to deal with the new challenges, the US is pursuing a strategy of democratic reform in West Asia. Accordingly, promoting the peace process, stabilising Iraq situation, as well as containing Iran and Syria are being considered integral parts of its strategy for integrating the whole region. One of the American's major objectives has been to resolve the Palestine-Israeli problem through promoting reform in the Palestinian Authority.

The Palestine-Israeli Issue and its Impacts on the Regional Security in West Asia

After 9/11, the issue of countering terrorism and prevention of proliferation of WMD has engaged the attention of the American policy makers. Nevertheless, the Palestine-Israeli issue is still regarded as one of the key factors which has been contributing to lot of turbulence in the region.

Firstly, the terrorist attacks against Israel have increased since 9/11. In spite of that, it was interesting to see the tension between Israel and Palestine coming down for a short period after the Iraq war. In March 2005, 13 Palestinian groups consented to stopping violence against Israeli targets within a year. But on December 5, 2005, a suicide attack in a marketplace in Netaya, a town in Israel, caused more than 40 casualties and which was indicative of the way things were moving in Palestine.

Secondly, the Palestinian Authority is plagued with severe internal crisis ever since Mahmoud Abbas succeeded Arafat in January 2005.

On the one hand, the power struggle between the reformist faction and the 'old guard' intensified, as a result of which a former Fatah leader Balgouti, who was under Israeli detention, established a new party named 'Party of Future'. Some Fatah leaders such as Muhammad Dahlan, former chief of local Preventive Security Unit in Gaza joined this party. Fatah, as the mainstream faction of the Palestinian Liberation Organization (PLO) suffered a serious crisis through this division. At the same time, the prestige of Fatah declined in the eyes of some Palestinians due to the electoral loses it suffered during the municipal elections. On the other hand, the authority of Fatah has been eroded in the face of the rise of Hamas.

The rise of Hamas in the municipal elections in the West Bank, which took place on December 2005, is a clear sign of the political trends in Palestine and a warning for moderate forces. Many analysts have been saying that the rise of Hamas is a reflection of the declining popularity of the corrupt and inefficient Palestinian Authority and Fatah. The Fatah-led Palestinian authority had, in fact, failed terribly in providing effective public services in the areas of health, education and welfare. Furthermore, Hamas has definitively claimed ownership of the narrative of the Israeli disengagement from Gaza which was brought about, in the opinion of the public, as a result of the continued armed struggle and 'resistance.'[2] Hamas is contrasted with Fatah as a cleaner and more sincere outfit by the Palestinian public.

Finally, the Jewish radical groups in Israel and the violent methods they employ against the Palestinian Arabs should also be regarded as a dangerous potential threat to the regional security. For a long time, the Palestinian radical groups have been the focus of international attention whereas the Jewish radical groups and their influence have largely been neglected. Nevertheless, the 'Hebron massacre', which occurred in February 1994, as well as the assassination of the former Israeli Prime Minister Yitzhak Rabin, on November 1995, surprised and shocked the whole world. International condemnation of the attack also put pressure on the Israeli government. However, the Jewish radical groups have not ceased their activities after this incident. They strongly oppose the 'disengagement plan' which Sharon had agreed to implement. They threatened to assassinate him if he would carry out the plan. Furthermore, it has been reported in the media that the Jewish religious extremists were also planning to bomb the Al-Aqsa mosque which is located on the Temple Mount in the old city of Jerusalem.

Moreover, the Palestine-Israeli issue also impacts other regional issues. First, the issue became an excuse for Islamic radical groups to engage in terrorist activities. It is not only regarded as one of the most important causes of Muslim hatred towards both Israel and the US, but also as the chief motivation for Islamist terror. It is even regarded as one of the causes of the 9/11 attacks. The Arab Muslims have been hostile to Israel ever since the establishment of the state of Israel, and have treated the US as the protector of Zionists. In the eyes of some Arabs, the sophisticated weapons supplied by US to Israel—such as the missiles and F-16 fighters—are mainly used by the Israeli army to attack the Palestinians in the occupied territories.[3] According to a report issued by the General Staff of Israel Defense Forces, on the eve of the September 11 event, the Palestine-Israeli conflict was aggravated by the terrorist attacks and military retaliation from both sides, which almost pushed the region to the brink of a war.[4] Therefore, the September 11 is regarded as a direct reaction to the Israeli aggression and the US was selected as a target because it is a common perception that the US has been taking sides with Israel for a long time and without the help of the US, Israel would be less assertive and aggressive. The address by Mr. Amre Mousa, the secretary general of the League of Arab States, to the American-Arab Anti-Discrimination League on October 2001, also hinted at this. He claimed that the origin of wrath and despair of the Arab world is from the inequitable treatment to which the Arabs are subjected, as well as their disappointment about the way the peace process was moving.[5] A series of terrorist attacks which occurred either in or outside the region after the September 11 were also linked to the 'Israeli factor', such as the unclaimed missile attack against the Israeli airplane in Kenya in November 2002, three explosions in Taba in Sinai peninsula in October 2004, as well as the series of explosions on July 23, 2005 in Sharm el-Sheikh in Egypt, etc.

Secondly, the Palestine-Israeli issue has an adverse impact on the issue of proliferation of WMD in the region. Opposed by the Arab countries, Israel apprehends military attacks from its neighbours at any moment, which has conditioned the Israeli thinking on military preparedness. There is a consensus among Israeli leaders that Israel must hold a continuous military advantage *vis-à-vis* its Arab neighbours to protect itself, and the Israeli quest for nuclear weapon is partially born out of this thinking. Thus, research and development of nuclear weapons has become an integral part of the basic defence strategy of Israel. In order to deal with the pressure from the international community, Israel has always adopted a "nuclear illegibility policy," which neither recognises nor denies if it has nuclear weapon. According to this principle, Israel not only refuses to sign Non-Proliferation Treaty (NPT), but also it never allows the officials of IAEA to supervise or inspect its nuclear facilities. Furthermore, Israel also reached understanding with the US in 1996 that the US would not force Israel to sign NPT if Israel does not claim openly that it was in possession of nuclear weapon. At the same time, the US also adopts a policy of 'double standards' in this regard by conniving with Israel in its nuclear weapon venture at one level and trying to restrict Iran and other 'rogue countries' in the region from developing nuclear weapons, on the other.

Being the only country in West Asia that possesses nuclear weapon, Israel's stiff stance toward the nuclear issue and its unique "nuclear illegibility policy," are likely to reinforce the motivation of other countries in the region—especially Syria and Iran who are hostile both to Israel and the US—to develop nuclear weapons. In addition, even Egypt, who has diplomatic relations with Israel, is also worried about Israel's possession of nuclear weapon. It has already warned that Israeli possession of nuclear weapons would accelerate arm race in the region. Other Arab countries also view this with lot of concern, The GCC countries in their summit meeting on December 19, 2005, unanimously called for the establishment a non-nuclear area in West Asia, and considered Israel, not Iran, as a potential nuclear threat in the region.[6]

Ever since the Islamic revolution in 1979, Iran has regarded Israel as an enemy, and supported Islamic radical groups such as Hezbollah in Lebanon and Hamas in Palestine to take on Israel. Hence, Iran has been regarded as a serious threat to the security of Israel. In recent years, the tension between Iran and Israel aggravated the security situation in West Asia. Iran has adopted a stiff stance on the nuclear issue and the new leadership in Iran has been quite acerbic in his remarks about Israel. For instance, the new Iranian president, Mahmoud Ahmadinejad, delivered a series of radical speeches denying the Holocaust, suggesting to 'transplant' Israel in Europe, and even remove Israel from the world map. Reacting to that, Israel threatened to launch military strike against Iran's secret nuclear facilities. Responding to Israel's threat, Iran also pledged to destroy Israel's secret nuclear base. The escalation of the hostility between Iran and

Israel could greatly affect the proliferation dynamics in the region, as well as threaten regional security.

Finally, the Palestine-Israeli conflict is also regarded as the fundamental contradiction determining West Asian security after the Second World War. The conflict is not only related to the Palestinian sovereignty over the occupied territories, but also the dignity of the Arabs and their religion, i.e., Islam. In the Arab world, no other issue can stimulate the ethnic sentiment like this issue.[7] The Palestine-Israeli conflict, as well as the hostility which it generated towards the US in Arab and Muslim countries, also poses a range of domestic problems. There is a surge in political activities against most of the ruling regimes in most of the countries in the region because these regimes are perceived to be pro-American, anti-Arab and anti-people. The pro-American regimes in West Asia are, thus, in a dilemma over how to deal with the Palestine-Israeli problem. On the one hand, they fear intense domestic opposition if they do not support the Palestinian struggle, which could endanger the solidarity of the regime. On the other, they are not willing to damage their relations with the US by taking tough or radical measures against Israel.[8] For instance, in order to maintain its hold over the people, the ruling regime in Saudi Arabia takes the Palestine-Israeli issue as priority in foreign policymaking.[9] On March 2002, Abdullah, the crown prince of Saudi Arabia submitted a "New Middle East Peace Plan" with a view to realising comprehensive peace between the Arab countries and Israel. It was formally passed on the 14th summit of League of Arab States. Soon after, the president of the US, George Bush, however, worked out another "New Peace Plan of Middle East", calling for Palestinians to reject Yasser Arafat through democratic reform and election. Thereby, as Arafat's supporter, Saudi Arabia, was heavily embarrassed.

It is the top priority for any government to maintain the survival of the regime. Some of the West Asian regimes, therefore, have in the past used anti-Israel hostility as a safety valve for political expression, allowing demonstrations against and criticism of Israel as one of the few forms of political expression. Nevertheless, such demonstrations have often spilled over into anti-regime protests, either because the regime is perceived as insufficiently hostile to Israel or because the organisation for the protests was captured by anti-regime forces. Both the Arab nationalists and Islamists have long exploited Arab-Israeli tension to criticise the pro-western regimes and their cooperation with the United States.[10] After September 11, some radical groups also exploited the anti-Israel as well as anti-American sentiments of the people to confront the pro-American regimes, causing instability in the political situation in the region. To deal with the challenge, as well as to avoid political crisis, the government of Saudi Arabia openly criticised the United States who took sides with Israel for a long time, especially the Middle East policy of inaction which president Bush adopted, which it held was responsible for the prolongation of the Palestine-Israeli conflict. This situation is likely to persist in the coming days if the US does not take into account the popular sentiments and aspirations in the region.

Notes and References

1. "Israel's History of Bomb Blast", *http://www.bbc.com.*
2. Baskin, Gershon (2005). "What to do about the rise of Amas", *Jerusalem Post*, December 19.
3. "Why the Americans are Hated?" *The Economist*, Weekly, Japan, November 6, 2001.
4. China News Network, August 6, 2001.
5. *http://jczs.sina.com.cn,* October 23, 2001.
6. By Reuters, "Gulf Summit Raps Israel, Not Iran on Nuclear Weapons Issue", Haaretz, December 19.
7. Tiezheng, Wang and Lin Songyie (2000). *The General History of The Middle East Countries-Saudi Arabia*, The Commercial Publishing Company, Beijing, November, p.151.
8. Ottaway, David B. and Robbert G. Kaiser (2002). "After Sept. 11, Severe Tests Loom for Relationship", *The Washington Post*, February 12.
9. Cordesman, Anthony H. (2002). "Saudi Arabia Enters the 21st Century", Revised on July 25, *http://www.csis.org/burke/saudi21/index.htm.*
10. Nora, Anthony H. and Daniel L.Byman (2004). *The Future Security Environment in the Middle East: Conflict, Stability, and Political Change*, Rand Corporation, p.311.

12

The New Government in Iran

Implications for International Security

PREMARANI SOMASUNDRAM

President Mahmoud Ahmadinejad's anti-Israeli outbursts and chilling millenarian visions[1] have raised concerns that a throwback to the militant foreign policies of the early days of the Islamic Republic is imminent. His statements could have been easily dismissed since the Iranian presidency has never had a monopoly on policymaking and Ahmadinejad is a political novice, but viewed against the backdrop of Iran's continuing intransigence on the nuclear issue, they merit attention. Besides, Ahmadinejad's rise to the presidency came on the heels of what is widely considered a 'conservative' consolidation of power through the parliamentary elections of February 2004 and was evidently supported by Supreme Leader Ali Khamene'i. In addition, Ahmadinejad is believed to have the support of the elite paramilitary forces, whose political influence appears to be growing.

This paper examines the labyrinthine power structure in Iran and the ideological inclinations of the various political factions in order to better understand the decisionmaking process and the foreign policy course that Iran is likely to adopt. It holds that, contrary to the perception of a consolidation of power by conservatives, there is, in fact, an intense power struggle going on within the ranks of the so-called conservatives themselves. It also contends that notwithstanding the ideological divides, Iran's leadership is bound together by common threat perceptions and, on balance, seeks fundamentally similar foreign policy goals and that any differences are often in approach to critical issues. This paper suggests that the international community recognise these complexities and continue to engage the Islamic Republic instead of isolating it.

Power Configuration

Iran's political structure is characterised by competing formal and informal power centres that act as checks and balances. The formal power centres are the representative institutions such as the Presidency and the Majlis (Parliament) and other constitutionally-mandated Islamic oversight institutions such as the Office of the Supreme Leader and Guardians' Council while the informal power centres include various politico-religious associations, the huge government-linked charities or *bonyads*, and the paramilitary groups aligned with various factions of Iran's clerical leadership. Thus, policy formulation is the result of a process of negotiation between competing formal and informal power centres, and gridlock is often common.

At the apex of Iran's unique political system is the Supreme Leader. It is the Supreme Leader who acts as commander-in-chief of the armed forces and has the power to appoint or dismiss the various commanders as well as the heads of the judiciary and of state radio and television as well as half the members of the powerful Guardians' Council, the body that vets candidates for election. While the president, as chief executive, has significant policy input, the Supreme Leader exercises oversight through a sort of shadow government involving what Buchta calls "clerical commissars" positioned in significant ministries and government institutions.[2] Likewise, the Majlis has significant policy input but the aforementioned Guardians' Council, whose role is to ensure that all legislation passed is Shari'ah-compliant, can check the Majlis. Another body that has grown to become one of the most powerful political institutions in Iran is the Expediency Council, which was created in 1989 to break the legislative gridlock between the Majlis and the Guardians' Council. Last, but not least, is the Assembly of Experts, a body of 86 clerics popularly elected to eight-year terms, who in turn elect the Supreme Leader and have the right to dismiss him as well.

Muddying this murky picture further, the umbrella Islamist movement, that took over power after the 1979 revolution, is rent by ideological differences. The Ayatollah Khomeini's charisma and authority as well as the rallying effects of the eight-year war with Iraq kept in check these ideological differences among the leadership but they became more apparent and pronounced subsequently. Here, it must be qualified that simplistic labels such as conservatives/

hardliners/radicals *versus* reformers/moderates/modernists do not adequately capture the various ideological strains present. Buchta[3] provides useful insights into the various ideological strands, which I have adapted here, also taking into account the views of others like Rajaee.[4] Buchta uses the typology commonly used by Iranians themselves, *viz.*, the Islamic left, the traditionalist right and the modernist right. These are broad strands and are distinguished primarily by the key values that the factions emphasise though this may not mean they are necessarily united on other values; institutional preoccupations and questions of power, patronage and personal rivalries often divide each faction into subgroups and contribute to some blurring on many of the positions that their leaders hold. Indeed, some of the factions have split and evolved over the years and, on occasions, have formed alliances with one another for reasons of expediency.

The traditionalist right gives primacy to the Khomeinist notion of *velayat-e-faqih* or guardianship of the judiciary. As noted by Buchta, the survival and dominance of Islam and of the clergy takes precedence over the idea of the people's will for this faction.[5] This is as much an institutional interest as an ideological one, the traditionalist right comprising mainly clerics and the *bazaar* traders, who have historically been its supporters. The traditionalist right, therefore, opposes any political reform and is conservative in sociocultural matters. Economically, it supports private property and private enterprise but accepts the need for subsidies and handouts in order to avoid the redistributive policies of the Islamic left that could threaten *bazaari* interests while also sharing some of the leftists' economic nationalism. But Rajaee[6] notes that the traditionalist right has been less dogmatic and revolutionary in its foreign policy than the Islamic left. Supreme Leader Khamene'i is the main representative of the traditionalist right.

The modernist right comprises mainly technocrats and places emphasis on economic development and industrialisation and on transforming Iran into a modern state by pursuing open door policies. Former President Rafsanjani is its key flag bearer although he is also its link to the traditionalist right. Indeed, Rafsanjani was initially identified with the traditionalist right but seems to have made a pragmatic shift towards the modernist right. As president of Iran until 1997, he adopted the China model of tight political controls and *laissez faire* economic policies, recognising that as the public became disenchanted with the Islamic revolution and the economic toll of the war, the regime needed to seek legitimacy through economic performance. But being a *bazaari* (and *mullah*) himself, Rafsanjani's economic policies are said to have benefited mainly the *bonyads* and bazaaris with connections. Rafsanjani also sought to normalise relations with Iran's neighbours and especially Europe in a pragmatic pursuit of a more secure international environment and with a view to securing trade and investments. In the sociocultural sector, the modernist right is somewhat more liberal than the traditionalist right.

The Islamic left is what might rightly be called radical. It is to a large extent similar to the traditionalist right in that it advocates strict adherence to the concept of *velayat e-faqih,* strict sociocultural controls and export of the revolution. It is in its economic approach that the left differs most markedly from the traditionalist right. It strongly upholds the notion of social justice and supports statist and redistributive policies. However, by 1992, the left became politically marginalised as the post-Khomeini years ushered in a period of pragmatism in economic and international policy.

A sizeable number of the original Islamic leftists underwent a phenomenal transformation from radical Islamists to modernist Islamists by the early 1990s. This group of modernists gives primacy to the rights and will of the people as well as the rule of law and civil society. Its flag bearer is former President Khatami, who captured power in 1997. During his presidency, Khatami put into effect a dormant constitutional provision for elected municipal government, which significantly increased the number of representative offices throughout the country. However, it is a mistaken notion to assume that the modernists wish to see the complete unravelling of the Islamic fabric of the state and society. As Rajaee puts it, the modernists are a moderate strand within Iranian Islamism that is "trying to negotiate modernity with tradition."[7] Unlike many of their student backers and other liberals who want a secular system, the modernists seek to reform the system from within and merely curtail the unbridled power of the unelected traditionalist clergy; they do not wish to subvert the system altogether.

While many of the Islamic left abandoned their revolutionary past, a younger generation among them remained unreconstructed radicals. This group includes senior members of the Islamic Revolutionary Guard Corps, the force that was formed out of the various revolutionary paramilitaries to counter the influence of the regular forces, and members from the Basij volunteer force. Taheri notes that, over the years, the Guards have built a grassroots network throughout Iran and created two political organisations comprising younger generation of military officers, civil servants and intellectuals.[8] The leftists made a significant political comeback when this network captured the Teheran city council in 2003, wielding populist slogans that resonated better with those Iranians affected by the widening income gaps in the country and unable to relate to the political reform agenda of the modernists. And, it was the same network that propelled Ahmadinejad to power.

Notwithstanding the recent resurgence of the leftists, it is clear that President Ahmadinejad has limited room for manoeuvre. He has already found the Majlis obstructive, contrary to the sweeping view that the 'conservatives' had consolidated control over the Majlis in the February 2004 elections. Iranian *émigrés* Mehdi Khalaji and Mohsen Sazegara note that only 120 of the current parliament's 290 members back the president. These are largely from the Abadgaran Iran-e-Islami (Developers of an Islamic Iran) bloc, comprising a number of former members of the Revolutionary Guard. Of the remainder, 90 members are what Khalaji and Sazegara consider traditionalists (rightists in my typology), 15 are allied to Khatami and the remaining 65 independent.[9] The Majlis' preferred candidate for the presidency was, in fact, Ali Larijani rather than Ahmadinejad. Possibly under Majlis pressure, Larijani has since been appointed secretary of the powerful Supreme National Security Council and is responsible for the nuclear negotiations. The Majlis has also already begun trying to circumscribe Ahmadinejad by rejecting several of his proposed cabinet appointments. It even rejected his nominations for oil minister three times. Much of this obstructive behaviour may be due to concerns over Ahmadinejad's populist platform of bringing about 'social justice' and especially his drive to crackdown on those who abused state power, moves that threaten powerful rightists. In rejecting some of Ahmadinejad's nominations, Majlis members complained that he placed revolutionary credentials and political loyalty above competence. Former president Rafsanjani, who lost to Ahmadinejad in the run-off to the presidential elections largely because he is perceived as tainted by power, has, for his part, openly criticised the new administration for purging what he claimed were competent officials who had long served the country. Earlier, both Rafsanjani (as well as

Khatami) obliquely criticised Ahmadinejad's anti-Israeli outbursts in part to contain any international fallout but probably also, as Taheri suggests, to discredit Ahmadinejad ahead of any attempt on his part to call them to account for mismanagement.[10]

Thus, a serious rift appears to be brewing within the ranks of the so-called conservatives themselves, between the leftists, on the one hand, and the traditionalist rightists and modernist rightists, on the other. On the surface, Ahmadinejad seems to have the support of Supreme Leader Khamene'i, without whose endorsement he could not have attained the presidency. Khamene'i has come out recently to publicly defend Ahmadinejad against his detractors. Ahmadinejad, in turn, has pledged his loyalty to the Supreme Leader. However, there is at least one indication that Khamene'i feels the need to rein in Ahmadinejad. In what was widely regarded as an effort to curb the excesses of Ahmadinejad and his fellow extremists, in October 2005, Khamene'i ordered that the role of the Expediency Council be enhanced to oversee the "correct implementation" of the policies of the executive, legislative and judicial branches of government. The extent of these enhanced powers is not clear yet but the Expediency Council is ironically headed by none other than Rafsanjani.

How the power struggle plays out may be critical to how Iranian policy evolves, especially policy in the international arena. In this regard, the Revolutionary Guard, the main powerbase of the leftists, may tip the scales. This force has served since the time of the revolution as a praetorian guard, the ultimate guarantors of the regime's survival. It would seem that the Guards are now flexing their muscles and may want a greater share of political power. In a dramatic move in May 2004, the Guards stormed and shut down the newly-opened Imam Khomeini International Airport in May 2004, citing security grounds. The airport was reopened only one year later.

Several observers (Taheri, Khaliji and Rajaee[11]) note that it was Khamene'i himself who had facilitated the political rise of the Guards and the leftist faction. For instance, Khamene'i appointed ex-Guards officers to politically-sensitive positions such as director of state radio and television. Over the years, Khamene'i, who lacks the theological weight and charisma of Ayatollah Khomeini, has recognised that the key to preserving the current system and maintaining his own hold on power is to play a careful balancing role between the various factions. Khamene'i may have, in fact, empowered the Guards and the leftists to check what many traditionalists perceived as the far-reaching political and social reforms attempted by Khatami and the modernists in recent years. Notwithstanding his conservative instincts, Khamene'i has had on several occasions to show sensitivity to public pressures generated by the changes Khatami and the modernists had engendered. For instance, owing to protests from the Majlis and civil society, both in the elections to the Majlis in 2004 and in the presidential elections, Khamene'i had to step in to reverse some of the candidates' disqualifications undertaken by the Guardians Council. Khamene'i may have thus, calculated that he needed a force to counter the modernists and also bring about some policy equilibrium.

Whether the Guards will now turn on Khamene'i and challenge his preeminence, however, remains to be seen. The Guards do not appear to be as cohesive a force as they were once assumed to be. The lower ranks, at the least, are believed to be divided in their political sympathies, and may not form a solid leftist bloc. Many of the rank and file were said to have voted for Khatami earlier.

The next event to watch in the struggle for power is the election later this year for the Assembly of Experts, the elite religious body that elects the Supreme Leader. The incumbent assembly is believed to be solidly behind Khamene'i. However, some sources (Bill Samii and EIU[12]) suggest that a tussle has already begun between Khamene'i's supporters and those of Ahmadinejad in preparation for the assembly election. The source of emulation or spiritual guide of Ahmadinejad is known to be Ayatollah Mesbah Yazdi in Qom, who is quoted as having recently claimed that "Ahmadinejad's administration is the first Islamic administration in the history of the Islamic Republic." EIU[13] suggests that Yazdi may be able to influence the election process for the assembly through his recently acquired ability to control the certification of clerics. However, there are also various calls for changing the composition of the Assembly, including proposals for bringing in non-clerics and for lowering the qualifications for clerics. Ultimately, however, the Guardians Council would have the final say in vetting the qualifications of the candidates, and as long as at least those members of this body picked by Khamene'i remain loyal to him, his position appears secure.

Implications for International Security

Given Khamene'i's political instincts and balancing role, the power struggle underway in Teheran may not necessarily lead to one faction prevailing over the other. Nonetheless, the political fragmentation in Teheran may not have a significant impact on foreign policy issues. The Iranian leadership shares common threat perceptions that cut across ideological persuasions and, to a large extent, has thus far sought fundamentally similar goals. There is consensus on preserving the country's territorial integrity and national sovereignty and on maintaining stability in its international environment. Indeed, although Ahmadinejad has adopted a hostile public stance on the nuclear issue since assuming power, this does not signal any fundamental shift in Iran's policy; tension on the nuclear issue had already been building up well before and is not a reflection of the election of Ahmadinejad or the rise of the Revolutionary Guards. Any differences that arise in foreign policy are likely to be in approach rather than on the fundamentals. At the same time, however, Iranian leaders have a penchant for using foreign policy issues, notably the question of relations with the US, as a tool for conducting their domestic battles against their political rivals. This tendency to use foreign policy issues as "political football" makes it often hard to distinguish between manoeuvres for one-upmanship at home and genuine foreign policy stances.[14]

The Iranian leadership is bound together by suspicions of the US and Israel. Although there is often a religious overlay to many of the anti-US and anti-Israeli statements emanating from the Iranian leadership there are fundamental security concerns that underpin these. Iran's suspicions of the US agenda are fed by memories of the CIA-backed coup of 1953 that toppled the Mossadeq government and reinstated the monarchy. However, the question of whether the best way of dealing with the perceived US threat is by maintaining its tough stance or by seeking normalisation with Washington is one question that divides the factions. In this regard, the clerical and leftist elements fear that contacts with the US could eventually erode their hold on power. Iran's suspicions of Israel are rooted largely in the reported close relationship between the Mossad and the Shah's notorious Savak intelligence service. Even as Iran covertly bought weapons from Israel during the war with Iraq, there has been strong concern regarding Israel's undeclared nuclear arsenal.

Ahmadinejad's recent anti-Israeli outburst may, in fact, reflect a conviction on his part that Israel may attack Iran's nuclear facilities and his comments may have been intended to rally the Iranian public for such a dire eventuality. In any event, his comments are not significantly different from the sort of pro forma diatribes that senior Iranian leaders occasionally engage in against the "Little Satan" (Israel) and the "Great Satan" (the US) in order to shore up the regime's legitimacy. Ahmadinejad certainly appears keen to distance himself from his predecessors Khatami and Rafsanjani, whose administrations he has criticised for being 'soft' and leading to a decline in 'revolutionary spirit'. As some observers suggest, Ahmadinejad's comments, coming just as Iranian and US officials acknowledged resumption of their backchannel contacts over Iraq, may have also been intended to sour the atmosphere for an improvement of ties with the Washington. Iran's position on Israel, in fact, appears to have shifted over the years. Even as Iran's leaders maintained their public hostility towards Israel, Iran's foreign minister indicated publicly that if the Palestinian people could find a way of coexisting with Israel, Iran would abide their decision. Khamene'i has even welcomed as 'positive' Israel's withdrawal from Gaza even though Ahmadinejad dismissed it as a joke.

The fact that aspects of Iran's nuclear programme were developed undeclared for nearly 20 years, Iran's occasional breach of its safeguards obligations under the NPT, and its less than cooperative behaviour with regard to the IAEA's investigation efforts give strong cause for suspicion that Iran may have intentions beyond pursuing a purely civilian nuclear programme. It is possible that Iran would like to develop a nuclear capability but has left the decision on whether to actually build a nuclear weapon for the future. Whatever the case, Iran has strong security concerns for seeking a nuclear deterrent. That Iraq unleashed a war against Iran within a year of the revolution contributed to a siege mentality in Teheran, especially as the Western countries were covertly assisting Iraq in spite of their official neutrality. Even the Gulf countries ganged up against their non-Arab Shi'a neighbour. (Admittedly, the Islamic Republic was in part responsible for its own isolation as its triumphalist, self-righteous rhetoric in the heady initial days of the revolution and the terror unleashed against the US embassy in Teheran unsettled its neighbours and contributed to a deep-seated hostility on the part of the US) Iran's conventional military capabilities were substantially degraded in the course of the war with Iraq and Iran was also subjected to chemical weapons attack. It may well have therefore, calculated that resuming the nuclear programme begun in the Shah era would serve as a more effective deterrent than any buildup of conventional forces. Iran also felt threatened by the Soviet invasion of neighbouring Afghanistan although the fact that Soviet forces were subsequently, bogged down in that country for a decade would have eased some of Teheran's anxieties. For more than a century, the Russian empire had posed a threat to Iran's territorial integrity and just after World War II, the Soviets occupied large swathes of northern Iran. Iran has also felt hemmed in by Turkey, a US ally whose close ties with Israel, Teheran saw as directed at itself.

Since the events of 9/11, Iran clearly feels further threatened as a result of the global realignment of forces around the US war on terror. It is now encircled by US forces in almost all of its neighbouring countries: Afghanistan, Pakistan, Oman, Bahrain, Qatar, the UAE, Saudi Arabia, Turkey, Azerbaijan, and most recently Iraq. Iran has also felt vulnerable to the spillover effect of the crises that its neighbours such as Afghanistan, Pakistan, Tajikistan, Azerbaijan, and Iraq are engulfed in. The Bush administration's stance towards Iran has not helped in assuaging Iran's concerns. Since the end of the Khomeini era, Iran has clearly pursued a more pragmatic foreign policy by engaging its neighbours and the European countries and curtailing its subversive behaviour and involvement in terrorist activities abroad. Even though then President Khatami's effort to reach out to the US, notably by expressing regret for the hostage crisis, was stymied by elements in Iran that were not prepared for such a shift, the Iranian leadership nonetheless showed a realistic approach and even provided behind-the-scenes humanitarian assistance to the US during the invasion of Afghanistan. With respect to Iraq, although Iran has legitimate concerns about a spillover of the conflict in its neighbouring, country and also has a stake in the political dynamics of Shi'a-dominated Iraq, it has nonetheless refrained from unsavoury behaviour thus far. There has been no evidence of any Iranian orchestration of the Iraqi resistance even though Iran may have an intelligence presence in the country and may continue to provide political and material support to Iraqi groups such as the Supreme Council for Islamic Revolution in Iraq. The US has even held backdoor talks with Iran, recognising that the Islamic Republic has some levers in southern Iraq that it could exercise to bring about peace.

Nonetheless, neoconservative elements in the US ratcheted up their rhetorical pressure about Iranian 'meddling' in Iraq as well as Afghanistan. Bush's 'axis of evil' characterisation, based on the infamous Karine-A affair,[15] the Bush doctrine of preemptive strikes against countries deemed to represent a security threat, and the neo-conservatives' blatant agenda of pursuing regime change in Iran have had a pronounced impact on Iran and led to a strong nationalist backlash from all factions alike. For example, Khatami's reaction to the axis of evil label was that it was "bellicose and insulting." He declared that as long as the US was "threatening, insulting, and humiliating us, neither myself nor the nation is ready to accept any relations."[16]

Moreover, US administration officials' public assertions that Khatami and his supporters were "too weak, ineffective, and not serious" and that the administration would henceforth "seek to support the Iranian people directly" as well as the US government's launching of a radio service targeted at young Iranians certainly could not have endeared it even to the modernist administration of Khatami.

Whatever Iran's past intentions, the Bush administration's stance will certainly have heightened Iranian suspicions of the US and convinced Iran that, given the asymmetry of power, it can obtain leverage *vis-à-vis* the US only with a nuclear arsenal. Rajaee cites recent analyses that suggest that the desire to deter the US now appears to be "the main driver of Iran's nuclear ambitions," regardless of whether or not Iran can actually overcome the vast conventional and unconventional advantage of the US.[17] Iran may in fact be following the North Korean mode of negotiations in its recent dealings with the IAEA by maintaining the mystique of a nuclear weapons programme in order to extract strategic concessions from the US.[18]

Prospects

The current standoff between Iran and the IAEA comes at a delicate juncture in Iran's political evolution, with factional conflict intensifying. None of the factions will find it easy to offer major concessions lest it be criticised at home for selling out. Even the Iranian public is touchy about Iran's sovereign right to pursue a nuclear research programme. Imposing economic sanctions—assuming the US and the EU calculate that the world economy can absorb the shock of a further rise in oil prices that this would entail—would only heighten nationalistic feelings and serve to entrench in power Ahmadinejad and militants of his ilk. Sanctions will also hurt the Iranian economy and consequently lead to a further widening of income gaps, the very factor that Ahmadinejad capitalised upon in the elections, with his populist platform, at the expense of Rafsanjani, who was hobbled by his reputation as a rich, corrupt *mullah-bazaari*.

Isolating Iran also risks a resumption of its past confrontational behaviour abroad. Although pragmatic considerations have led to a rollback in Iranian involvement in terrorism since at least the 1996 Khobar bombings in Saudi Arabia, Iran has continued to maintain support for the Lebanese Hezbollah and a variety of Palestinian militant groups confronting Israel. For Iran, such groups are not just freedom fighters engaged in legitimate resistance but also strategic allies as they serve to check Israel. Even if these groups have less incentive today to do Iran's bidding owing to internal political developments in Lebanon and the Palestinian territories, Iran may be able to find support among Shi'a militant elements in troubled Iraq. Even if Iran's key leaders exercise restraint, the shadowy politico-religious organisations and security-intelligence apparatus tied to radical elements could engage in mischief. Equally, if the Israel or the US decide on a military showdown by attacking Iran's nuclear facilities, Iran could use such proxy forces to strike at US and Israeli interests.

The West should understand the political complexities of Iran and should fashion policies that do not isolate it and revive the xenophobia of the past. Treating Iran like a rogue country run by mad mullahs can only be self-fulfilling. Civil society is still very much a force in Iran, and the Iranian political system, however flawed, is characterised by a far greater degree of pluralism than in much of the Middle East. As Rajaee notes, Iranian society is undergoing fundamental social changes in Iran rooted in the continuing demographic boom. A younger public is demanding greater democratisation and less radical Islamism in government; this inexorable trend, as well as the challenge of delivering economic performance, has even contributed to an ideological evolution among the leftists and traditionalists, as discussed in the first section of this paper.[19] As the International Crisis Group suggests, the international community should find a way of accommodating Iran's right to operate a small nuclear research facility under strict international surveillance or joint Iranian/international management of nuclear sites while undertaking confidence-building measures that address Iran's legitimate security concerns.[20]

Notes and References

1. Ahmadinejad is said to be allied to a cult whose mission is to pave the way for the return of the Shi'a Muslim *messiah*. There is evidently an attempt at character assassination by Ahmadinejad's detractors, who portray him as a fanatic, notably by circulating a tape in which the president is said to relate an other-worldly experience during his UNGA address in September 2005. Nonetheless, the English version text of that address found on the UN website does hint of an idealistic millenarianism, specifically, the lines "0 mighty Lord, I pray to you to hasten the emergence of your last repository, the promised one, that perfect and pure human being, the one that will fill this world with justice and peace. 0 Lord, include us among his companions, followers and those who serve his cause."
2. Buchta, Wilfried (2000). *Who Rules Iran? The Structure of Power in the Islamic Republic,* Washington Institute for Near East Policy, Washington, DC, p.47.
3. Ibid.
4. Rajaee, Bahram (2004). "Deciphering Iran: The Political Evolution of the Islamic Republic and U.S. Foreign Policy After September 11", *Comparative Studies of South Asia, Africa and the Middle East,* Vol 24, No 1.
5. Buchta *op. cit.*
6. Rajaee *op. cit.*
7. Ibid.
8. Taheri, Amir (2005). "Ayatollah Ali Khamenei's New Generation Consolidates Power in Iran", *Democracy Digest*, June 28, Volume 2(7).
9. Khalaji, Mehdi and Mohsen Sazegara (2005)., "Challenges Facing Iran's New Government", Web article, Washington Institute for Near East Studies, August.
10. Taheri, Amir (2005). "Signals from Tehran", Arab News, July 5, reproduced in *http:/ /www.frontpagemag.com/Articles/ ReadArticle.asp?ID=18645*
11. Khalaji and Sazegara *op. cit;* Rajaee *op. cit.,* Taheri *op. cit.*
12. Bill Samii, Iran Press Service, and Economist Intelligence Unit (EIU)
13. EIU
14. Ehteshami, Anoushiravan (2004). "Iran's International Posture After the Fall of Baghdad", *Middle East Journal,* Volume 58(2), Spring.
15. This involved the Israeli capture of the ship Karine-A, which was believed to have on board a consignment of Iranian weapons intended for Palestinian militant groups
16. Cited in Rajaee *op. cit.*
17. Rajaee *op. cit.*
18. Mostashari, Ali (2005). "Iran: Rogue State?", *Iran Analysis Quarterly*, Volume 2(4), July-September.
19. Rajaee *op. cit.*
20. International Crisis Group (2005). "Iran: What Does Ahmadi-jad's Victory Mean?", *Middle East Briefing*, No 18, August.

13

Threats to Stability in Asia

Special Focus on West Asia

MARIAN ABISHEVA

The Near East or Middle East or West Asia, whichever way one chooses to call the region stretching from the eastern coasts of the Mediterranean till Indus in the East, has come to influence the world policy in many ways. The region has attracted the attention of the world powers ever since the discovery of the large reserves of hydrocarbon in early 20th century. The world wants the region to stay stable and peaceful for uninterrupted flow of the energy resources. However, the nagging problem of Palestine as well as the rise of militant Islam in recent years, with their impact on the security calculus of the region, has compelled the world to stay engaged and address the security concerns effectively.

It is the key issue of access to hydrocarbons or oil and gas that primarily determines the strategy of the main global actor, United States towards this region. The US continues to influence the behaviour of the oil-producing countries in the Arab world, through inducements, military action or threat of it, and persuasions of political reforms. The oil factor, as well as situation in the region in general, concerns policies of EU, Russia, Iran, China and all the consumers of the oil flowing from the region. The issue of Islamism (radical Islam) has also, of late, concerned international observers and in the post-9/11 days, it is being considered an immediate threat to international security. The region has become part of the US global strategy on reformation of global order. This US military operation against Iraq in March 2003, conveyed a message to the countries in the region that the dominant order in the region would have to be refashioned in way that would suit the needs of the US. The people in the region strongly believe that the main element of US strategy in the region consists of its unquestioning support of Israel and ensuring flow of energy resources.

Besides this the region is haunted by a number of economic, demographic and cultural-civilisation problems. The main problem facing the people of this region is their inability to deal with challenges of globalisation. Globalisation brings with it liberalising influences and makes it compelling on the part of states and societies to take a departure from tradition and emphasise on values like individual rights, liberty, gender equality and democracy, which are considered universal. The societies in the region are steeped in traditional values and on their part such changes are often perceived as corrosive and destabilising. The ruling regimes, seeking preservation of the *status quo* to perpetuate their hold on power, find this argument quite suitable to their survival and are seen to be insisting on stage-by-stage incremental development when there is external pressure on them for political reform.

The region is also being perceived as the epicentre of instability in Asia and the world. The core of the conflict, many analysts suggest, is the unresolved problem of Palestine and the Palestinian-Israeli (or Arab-Israeli) confrontation. The issue at hand is how to complete the exercise of removing the Israeli settlements from the Gaza Strip and partially from the Western bank of the Jordan River, which began in 2005 and then how to build up the Palestinian state as part of the two-state solution in the area.

By 2004-05, the recent developments in the region since 9/11, seemed to have brought about a change in the perception of the external powers. The military operations in Afghanistan and Iraq, the issue of nuclear proliferation in Iran, the rising tide of violent resistance in Iraq as well as the developments in Palestine have, perhaps, convinced the West that one important cause of popular antipathy towards the US and the West was the festering problem of Palestine where a solution was absolutely necessary to stem the tide of Islamist radicalism. It remains to be seen how the west will handle Israel and the Palestinian extremists in the coming days to resolve the most nagging problem in the region. Iraq is posing yet another challenge for the west and the developments there will also greatly impact the security environment in the region. It will be useful to enumerate all these and other issues while assessing the evolving security situation in Western Asian region.

Threats to Stability

Iran

In 2006, the most probable cause of destabilisation in the region may well become the search for nuclear weapons by Iran. The US views Iran as a hostile regional power and a 'sponsor' of international terrorism. It also considers Iran one of the main sources of proliferation of weapons of mass destruction (WMD). Therefore, it holds that will be even more difficult to handle if the suspected Iranian attempt at nuclear enrichment is not reversed. The result is the policy of containment of Iran, as well as its economic and diplomatic isolation.

Discussions on the possibility of the Israel attacking the Iranian nuclear objectives supported by the US administration have sprung up in the meanwhile. Such operations, the analysts in the US believe, could meet the US interests in the region. However, such aggressive posture could also give rise to further radicalisation of the Islamist terrorism in the region and around the world.

The American apprehensions are further strengthened by the aggressive and irresponsible flaunts by the Iranian leadership. One may say that in the summer and autumn of 2005, the region was on the brink of another attack. There is a popular apprehension in the region that the Israelis are itching to attack the Iranian nuclear facilities and would just require a nod from the US to embark on it. This is supported by the earlier Israeli attack on Iraqi facilities at Osirak in June 1981.

The Eastern Mediterranean

The key country for the US in the region is Israel, which is passing through acute internal-policy crisis at the moment. There is clearly no consensus over the question of Palestine and the hawks among the political elite have managed to regularly swing public opinion in favour of a position not conducive to any lasting political solution. Political elite in Israel is either represented by aged politicians, veterans of party intrigues, more anxious about their place in history and international recognition than about the future of own country, or by political light weights, seeking to whip off popular passion to build their constituency and popular support-base.

Besides this, the politics of Israel is also influenced by the politics in the US. Actually, Israeli military-industrial establishment and political establishment are subordinated to the American interests. Contrary to the popular thinking and the prevailing rumour of a powerful Israeli lobby controlling American policy towards West Asia, the American politicians are entirely devoted to the American interests and as a democracy, they are more controlled by the public opinion on strategic issues. The recent case of Israeli withdrawal of steps to seek military or economic rapprochement with China and Russia is a case in point.

One can foresee the following developments within Israel in the coming years: internal instability, growth of authoritarianism, a possible shift to presidential system of government and worsening of relations between Israel and EU, UN and other international organisations.

As about other states, the US pressure on Syria is mounting and developments in neighbouring Lebanon may engage the attention of the world in the coming days. A conflict between Israel and Syria is not far off if Syrian influence on Lebanon persists. This conflict may be provoked by confrontation in the Israeli society on the problem of removal of settlements from Gaza, as a diversionary tactic because the leadership can very well project a victorious war in the north as a visible success and it is certainly more preferred option for the Israeli generals than participation in internal conflict in the South.

The Persian Gulf

It is very likely that the activities of radical Islamists may increase in Saudi Arabia that may threaten stability of the ruling Royal family. The country may very well shape up as a 'Sunnite variant' of Iran. Successful 'democratisation' of Iraq may give rise to demands for democratic form of governments in other countries which are ruled by dynastic rulers. This may pose a threat to the ruling regimes. At the other extreme, the Islamists in West Asia may make use of such democratic movements to come to power and bring their movements together in the direction of a 'new caliphate', by further integrating their agenda with the Islamists of northern Africa and central Asia. This may destabilise the region as well. From the economic, military and geopolitical points of view, any such alliance may be able to unleash major conflict with the West, and with Israel. It is more probable that such an Islamist union with possible nuclearisation with assistance of non-state nuclear black marketeers operating through the network enabled by Pakistan may pose a big threat for the West and the world. If they would come to power in the Arabian Peninsula and undertake successful hostile measures against the US then military confrontation with the US and the West will be a regular feature of international politics.

The Saudi provinces like Hijaz, Asir and Nedjd are, in fact, fast shaping up as nerve centres of Islamic radicalism. An Islamist resistance is also simmering within Jordan. If the Islamists of Saudi Arabia and Jordan come together then they may stake their claims over the Saudi province of Hijaz (within which the two most important places of worship, Mecca and Medina lie). It has to be remembered that this area was in the traditional possession of the Hashemite dynasty of Jordan.

At present, the monarchies of the Persian Gulf depend on the US in their confrontation with Islamists at the internal level. The only stable state in the Arabian peninsula, from geopolitical and military points of view, is the state of Oman. The dynastic ruler, called Sultan, of Oman has strictly and effectively stopped dissemination of the Islamists ideology in his sultanate. In Yemen, the hold of the central authorities on the peripheral areas is weak and ineffective. Its main problem is the irreversible process of depletion of water resources. Probability of a water-related conflict with Saudi Arabia is very great.

Iraq

Another important source of destabilisation of the entire region is Iraq, which has become the territory of civil war of 'all against all', in the aftermath of the ouster of the Saddam regime. In this war, Islamist volunteers from the entire world are descending on Iraq to fight it out against the forces from US and the West, as well as against each other. Actually, Iraq has become the testing ground for, or the laboratory of, global *jehad*. The Iraqi parliamentary elections, ignored and boycotted by the Sunnis—the most militant part of the country's population and until the overthrow of Saddam the ruling community—and subsequent developments do not give any room for optimism about the future of Iraq.

The worsening of the public security situation and absence of normal social order has seriously affected the economic reconstruction of Iraq. Eighty-five per cent of the Iraqi families experience constant power crises, 45 per cent do not have access to safe drinking water and more than half of the population is unemployed. The Iraqi citizens consider that the condition of their life has become worse than what it was at the beginning of the war.

The anti-American groups are destroying water stations and gas pipelines with increasing frequency. Their attack on foreigners is also making recovery works in Iraq difficult. On the contrary, the phenomenon of brain drain has increased in Iraq and many engineers have gone abroad for better and calmer pastures. With the unavailability of indigenous skilled workforce and the reluctance of workers from outside to come in, the task economic reconstruction in Iraq has become immensely difficult.

The possibility of splitting up Iraq along ethno-national and sectarian lines is being discussed worldwide as a more practicable option even if the Shiites and Kurds have expressed their commitments to work towards a unified Iraq. If this option works out then it will lead to the establishment of a Shiite state in the Arabian Peninsula close to the other Shiite state of Iran in the Persian region. There will also be the first sovereign Kurdish state in modern times and this is likely to destabilise the Kurdish areas in Turkey and Iran.

Observers note that the Americans are fast learning how effectively they can play a role in the region. Thanks to American efforts, in 2005, there were parliamentary and presidential elections in Egypt, the presidential elections in Afghanistan and Iraq, the presidential and local government elections in the Palestine. The US responded most actively on the parliamentary elections in Iraq on December 15, 2005. On the eve of the Iraqi elections till the results were announced, the US president George Bush made five speeches, devoted to events in this country. During these elections in various countries of the region, the Americans, indeed, played a significant role.

The urgent task awaiting the newly elected government in Iraq regime is restoration of real sovereignty of the country. It has to ensure internal security also. But here there is a paradox. Restoration of sovereignty is impossible without the withdrawal of the foreign forces from its territory. And at the same time maintenance of internal security is impossible at the moment without the assistance of foreign forces. There is an influential section in the US which wants the US to stay on in Iraq till it stabilises. This section believes that otherwise it will set a bad precedent and certainly affect the international image of the US on the one hand and strengthen the resolve of the forces opposed to the US at the international level, on the other. The Iraqi leadership is also extremely diffident about its ability to bring order to the country and even if it were to make ritual statements seeking withdrawal of foreign troops it is extremely aware of its dependence of these very forces. Till the end of the year 2005, even if the US had allocated about US $80 billion for Iraqi reconstruction, most of the allocation was spent on the Iraqi military.

Undoubtedly, prospects of stabilisation in Iraq will depend on those measures, which the Americans and new Iraqi leadership will take to gain support and confidence of all groups in Iraq. The violent assertion of the opposition in Iraq is directed not only against the American occupation, to a certain extent it is directed also against provisional government. If moderate forces in Iraq would be able to assume responsibility for policy of appeasement in the country and achieve it, this would significantly increase prospects for general stabilisation. However, at another level, it is very important on the part of the political forces and representatives of ethnic-religious minorities to learn to carry on a dialogue with one another, without resorting to the method of settling their differences through armed conflict.

In Palestine, the Hamas broke the monopoly of the PLO and the Al Fatah. In Egypt, the elections that brought H. Mubarak back to power was criticised the world over for its lack of openness and transparency and the inclination of the radical Islamist group, the Muslim Brothers, to participate in the democratic process was brought to the notice of the world.

Afghanistan

Along with Iraq, Afghanistan still remains important as a military-strategic component of the American strategy in Asia. Afghanistan, which is controlled partially by the NATO troops and partially by former Talibs, has become the largest drug supplier of the world markets in 2005-06. Military-political situation in the country was characterised by sporadic outbreaks of armed actions, conflicts between different factions, and attacks of groups on international forces.

In 2004-05, situation in Afghanistan remained very complex. The greatest political event of this period was Presidential elections in

October 2004, where the Head of provisional administration, Hamid Karzai, was elected. However, internal military situation in the country has remained quite problematic. The US succeeded in convincing the leaders of ethnic minorities that there was a need to have a President from the Pushtun majority. The subsequent election of Hamid Karzai as President came as an assurance for the US but the parliamentary elections in September 2005, showed that the traditional structures of power continued to persist and it would take a long time to bring normalcy to Afghanistan. During the parliamentary elections, an additional factor of tension was drive on disarmament, demobilisation, reintegration of local armed units, by the ISAF, which was directed against the leaders of the Tajik and Uzbek communities.

The subsequent resurgence of Taliban from December 2005, poses yet another problem for Karzai and the NATO-led International Security Assistance Force (ISAF). In the central and Southern-Eastern provinces subversive activity of regrouped Talibs and other extremists' groups began to increase by the end of 2005 and in areas bordering Pakistan, Taliban forces managed to capture and hold large territories periodically.

The economic situation in Afghanistan remains critical. It is aggravated by drought, devastation, inability of the government to carry out large industrial projects and by lack of external support. The country fully depends on external aid. The donor funds for reconstruction are utilised ineffectively. Great part of them is spent on salaries of officials and staff of various organisations. Some revival is observed only in activities of small private enterprises in the field of trade but that is not sufficient to revive the economy.

As before, drug production and peddling remains the only means of income for a large group of people. The Afghan regional strongmen, or the warlords, are also critically dependent on the drug economy and until there is an alternative to this, it is difficult to counter this menace. The drug problem has not only internal but also international implications. After the proscription of the Taliban, Afghanistan has broken all records in the production of poppy and heroin, almost up to three quarters of all world production. Accordingly, profits from the sell of heroin, estimated at billions of dollars, greatly exceeded funds invested by the US in the development of this country. Not less than two quarters of all peasant population of Afghanistan is involved in drug production. In other words, more than one and half millions of the Afghans grow poppy.

The main problem of economics of Afghanistan is that agriculture is still oriented at drug production; turnover of opium poppy growth, drug and heroin production amounted to monstrous scale.

At present, the armed units of the former *mujahideens* form part of the so-called Afghan National Army (ANA). They are nominally subordinated to the Ministry of Defense, but in fact they are formed by the regional principle and are controlled only by their commanders. Of serious concern is the existence of several units sympathetic towards the Talibs. Terrorist acts are quite regular in the country. For the last year, the casualty figure in Afghanistan was nearly one thousand including sympathisers of the Karzai government, international aid workers and military forces and civilians. In general, the strategy of the Taliban has been fragmentation of the coalition forces, by engaging them through small combatant groups in different areas. With minimum expense they keep anti-Taliban forces under continuous stress.

According to many sources, cells of Al-Qaeda may have been still operating independently within Afghanistan. It is also believed by analysts in the region that the Taliban is still receiving enough funds regularly to continue its military actions against the US. And these funds come from the countries considered as allies of America, in particular, from Kuwait, Saudi Arabia and United Arab Emirates. This is also being pointed out in the mass media. Apart from this the role played by Pakistan in containing the Taliban in the areas bordering Afghanistan is being discussed widely. In fact, Karzai himself has, on so many occasions, alleged that Pakistan is insincere about its commitments to contain Taliban. The Pakistan army's role in arresting the growth and resurgence of the Taliban in the tribal areas and northern Balochistan has not been quite convincing.

In September 2005, Pakistan declared that it intends to build defensive fence almost along its 2.5 thousand kilometre-long border with Afghanistan. This was aimed at preventing border-crossing on both sides by the Islamic extremists and smugglers. The Afghan Government rejected the Pakistani proposal to build such a fence as the Durand line that divides the two countries has not been recognised by Afghanistan as legitimate. The Afghan foreign ministry noted that Kabul will never agree to such a proposal until the border demarcation was made according to the international law.

Washington has not been able to carry out large-scale military actions and keep a large contingent in Afghanistan primarily because of the Iraqi factor. By the end of 2006, the US contingent delegated the control over the whole territory of Afghanistan to the NATO troops. However, the US still remains the most powerful military and political force in Afghanistan even if the ISAF has undertaken the task of internal security. In practice, the US follows a contradictory policy in Afghanistan. On the one hand, they support the central government of H. Karzai. On the other, the American and international military forces work with the local strongmen or field commanders/warlords who have taken control over local departments of the central government, including not only military-police sphere, but public health service, transport, etc.

On the other hand, Washington is still very interested in its military-strategic presence in Afghanistan (and correspondingly, in Central Asia). The Iranian factor remains important too.

Pakistan

In the immediate aftermath of the attack on twin towers, the US turned its policy towards Pakistan on its head and realised that it had to ensure Pakistan's support in its war on terror. As such, the US reversed its previous policy of applying sanctions on account of the nuclear tests and accorded the status of "Major Non-NATO Ally of USA" to placate the military leadership in Pakistan. It has ever since turned a blind eye to the undemocratic policies of the Musharraf-led military administration in Pakistan. From 2002, Musharraf has put in place a sham system of democracy by allowing only yes-men to participate in the elections. The US has allowed Musharraf to continue the democratic façade and perpetuate his hold on power in Pakistan. It is largely because the US thinks that the only alternative to Musharraf is chaos in Pakistan. There is an overblown apprehension in the US that if Musharraf goes then the Islamist opposition will take over and, as a result, a nuclear Pakistan in the hands of the *mullahs* will be terribly unsafe for the world.

Many American researchers are of the opinion that non-state actors are playing a predominant role in Pakistan and the authority of the state is being eroded day by day. They would ague that assertion of multiple ethnic groups, radical Islamists, the unearthing of the Khan nuclear network and the prolonged engagement of Pakistani troops in the tribal areas suggest the declining capacities of the state power and there is a need to boost the authority of the state by strengthening the Musharraf administration in Islamabad. Musharraf administration in its turn has been quite obliging in certain areas. In response to American allegations of the nuclear black-marketing by its leading nuclear scientist, A.Q. Khan, Musharraf arrested the scientist, forced a confession and apology out of him and later in May, 2005, Pakistan agreed to cooperate with the IAEA and USA in their endeavour to track down and dismantle this illegitimate network.

There is an increasing pressure on Musharraf to allow real democracy in Pakistan. The opposition has demanded that Musharraf has to shed his uniform if he wants to continue as president of Pakistan. They have also demanded that the leaders of the two principal political parties, Nawaz Sharif and Benzir Bhutto, should be allowed to come back to Pakistan and contest the elections. But Musharraf is dilly-dallying and the future of democracy in Pakistan is uncertain.

Central Asia

From 2005, a new geopolitical project is being discussed in strategic circles in US–the so-called project for "the Bigger Central Asia." The core of this plan is to bring together Afghanistan and the Central Asian republics in a single military-strategic and geopolitical whole, and possibly include some other neighbouring regions.

The aim of this project is isolating this wide region and bringing it out of the (monopolistic) influence of other great powers–Russia and China. The project may have been dictated also by the necessity of bringing Afghanistan out of the destabilising influence of such neighbours as Pakistan and Iran and 'binding' this country to more stable and western-oriented region of Central Asia.

In general, "the Bigger Central Asia" (BCA) project is in many ways a continuation of another geopolitical project—"the Bigger Near East" (BNE), and it has in principle the same strategic aims–diversification of strategic interests and stabilisation of the territory and under dominant influence of USA.

At the same time, according to this plan, Washington seeks to maintain an appearance of 'pluralism' by allowing Moscow and Beijing some role in the region by granting them the status (along with the West) of guarantors and donors in the modernisation process attempted by the US and giving them some illusion of control in the region. However, for the American strategists, the role of China and Russia as 'benevolent observers' is quite acceptable. However, anything beyond that will be countered by the US. It is also being predicted that the US may facilitate the entry of India and Turkey into the region to balance out China and Russia.

The developments in Andizhan in 2005 and the drastic change in foreign policy of Tashkent have jeopardised the prospect of implementation of the BCA project. The main aim of the BCA was to transform Uzbekistan into the centre of the integration processes by persuading it to sign special agreements with Pakistan, cooperate with Japan to build railway to Afghanistan, create transport corridor to the Indian Ocean and free trade zone around the Fergana Valley with participation of other countries of central Asia.

In order to realise this project, the US will have to take following measures: (i) convince Russia and China that they (together with USA) have shared interests in the region; (ii) persuade states of the region to implement economic and social programmes which will liberalise the economy of the region and ensure flow of funds; (iii) prevail upon the leadership of the Central Asian countries to take measures, especially in the field of security, which will meet the interests of all the powers; (iv) convince Moscow and Beijing that the BCA project is not directed against their positions in the region, but contributes to the strengthening of regional stability.

This project proposes implementation of a number of organisational-technical and diplomatic measures to realise the American strategy. They concern strengthening of the role of Pentagon and State Department for more effective US presence in the region. It is also considered necessary to strengthen the role of NATO as one of the main instruments of the American strategy forming part of special BCA Council, through which USA could coordinate and influence regional policies on continuing basis. Further it is necessary for the US Secretary of State to undetake regular visits to the region (annual).

In practice, the BCA project has a concrete geopolitical aim, i.e., to separate Central Asia from the Eurasian area and thus, isolate the states from Russia and CIS in general and form a buffer between Eurasia and China and change the European vector in the development of Central Asia and Kazakhstan. It also hopes to solve problems of reconstruction of Afghanistan through its economic integration with Central Asia and at the same time it seeks to realise a number of communication projects from the Caucasus to Pamirs, having anti-Russian and anti-Chinese directions. The US also wants to integrate the BNE and BCA projects from the tactical point of view.

Thus, the US policy in Central Asia enters a new phase. In short-term this plan of "the Bigger Central Asia" looks both complex and adventurous and its implementation in present conditions is quite difficult.

Prospects

At the moment, there will be many interconnected factors influencing the politics of the region: demographic, economic, technological, social, political and military. High rates of population growth will bring to focus the problem of poverty and unemployment. If there will be no opportunity for gainfully employing the unemployed youth then they will be available for use by the radical forces. Growth of population may also sharpen the water problem in the region, which, in its turn, will increase possibility of conflicts between the states for possession of water sources. The Palestinian-Israeli and Turkish-Syrian-Iraqi competition for water may be the most probable.

Another Important factor affecting the security of the region will be the role of religion (first of all, of Islam) in politics. In most of the states, one may find the ruling regimes reluctant to share power with popular representatives and only bring about rudimentary changes in prevailing political systems, for example introducing representative mechanism at the local level and ensuring perpetuation of dynastic rule at the central level. At the local level too, the influence of tribal and clannish leaders will also affect the nature of local politics. The prospect of a liberal civil society is very remote. This is not to deny that there could be deep structural changes in certain states. For example, in Iran and Saudi Arabia the regimes may be compelled to be more liberal and representative.

Russia is likely to follow a policy aimed at minimisation of consequences of regional conflicts and optimisation of its presence in the Near East and improving its relationships with the states of the region. Russia is likely to avoid involvement in internal problems of the region except in cases, where they will directly infringe on its national interests. Moscow is not quite comfortable at the moment with the worsening of situation around Syria and Lebanon.

In view of the presence of large hydrocarbon reserves in the reserves and the dependence of the US and the west on these resources, the US will continue to play a dominant role in the region and will have the significant military presence too as per the plans of the present administration in Washington. The US will continue to play a major role in the Arab-Israeli relations and security of a number of states will be heavily dependent on the US. On the other hand, all this may lead to the strengthening of anti-Americanism in the entire region spanning from Morocco to the northern India and Western China, including Central Asia. It is beyond doubt that this region will be the epicentre of global geopolitics in the coming days.

Section III

Energy and Security: The Political Economy of Energy

14

Gas Pipelines

Drawing Fresh lines on the Map

TALMIZ AHMAD

Energy is the fuel that drives the economy and provides nations with the annual growth rates essential for their economic development. Most of the world's energy comes from hydrocarbons (oil and gas) which account for 65 per cent of the world's energy requirements. While oil accounts for 42 per cent in the global energy-mix, the other sources of global energy are coal (24 per cent), natural gas (22 per cent), nuclear energy (6 per cent) and renewable and non-conventional sources (7 per cent). World energy demand increased by 95 per cent over the last 30 years and is expected to rise by 60 per cent over the next 20 years. During this period, the demand for oil will increase by 42 per cent while the demand for gas will increase by 97 per cent.

Natural gas, being a 'clean' fuel, is increasingly seen as the fuel of the 21st century. Between 1980 and 2003, the share of gas in the world energy-mix rose from 18 per cent to 22 per cent. The demand for gas is expected to increase at 2.4 per cent per annum till 2020, when it will constitute 25 per cent of the world energy-mix and consolidate its position as the number two fuel in the world's energy-mix. Since 1980, proven world gas reserves have increased at 3.6 per cent per annum, with volume tripling from 77 trillion cm to 179 trillion cm in 2004.

The location of gas is more diversified than oil: OPEC has 50 per cent of world reserves (as against 75 per cent of oil); the FSU states have 32 per cent of world reserves (against 10 per cent of oil), while the OECD has 10 per cent of gas against 7 per cent of oil. Three-fourths of world reserves are in Russia, Central Asia and the Gulf. The principal gas producers today are Russia, Iran and Qatar. Qatar had begun investments in LNG facilities over 10 years back, and is presently reaping the benefits as the world's major supplier of LNG.

However, the principal supplier to Western Europe is Russia, which provides gas through transnational pipelines. Russia is already the largest supplier of gas to Europe; by 2030, it will meet 50 per cent of Europe's needs.

On the supply side, the prognosis relating to gas is quite comfortable: present resources can meet current demand for 60 years. With new discoveries, reserves could meet demand for 150 years at present rate of consumption. According to industry forecasts, international trade in natural gas is expected to increase significantly in coming years, accounting for one-third of world output, reaching 1.35 trillion cm per year by 2020.

This increased trade will cover both LNG and piped gas. International trade in LNG is expected to grow by 7 per cent per annum till it becomes 38 per cent of gas trade by 2020. The US is expected to emerge as a major buyer of LNG, with terminal capacity reaching 140 billion cubic metres after the completion of 10 new terminals. Piped gas carried by transnational pipelines will grow to become 50 per cent of international trade by 2020.

The problem pertaining to gas is not with the reserves; it has to do with the need for fresh investments over the medium-term to develop new fields, albeit in the traditional areas where production in the old fields is declining. It is estimated that investments of about $ 1–1.2 trillion would be required over the next 10 years to develop the potential to meet global requirements.

Transnational Pipelines

Transnational oil and gas pipelines are not only able to transport large quantities of hydrocarbons across hundreds, and even thousands, of kilometres, but given their reach and range and the terrain they traverse, they also have significant geopolitical implications and even the ability to influence bilateral relationships and regional cooperation scenarios. The Baku-Tbilisi-Ceyhan (BTC) oil pipeline seems to be a manifestation of US attempts to exclude Russia and Iran from the Central Asian hydrocarbon calculus. However, this policy is unlikely to succeed over the medium-term as Central Asia will seek to reach the lucrative markets of Asia through pipelines across Russia and Iran.

Kazakhstan on its part had already indicated its interest in utilising its hydrocarbon potential to firm up engagements with all its neighbours: if it has appeased the Americans by participating with its oil in the BTC pipeline, it has also participated with the Russians in the Aktau-Novorossisk pipeline, and with the Chinese in the 2000 km long Kazakhstan-China pipeline, under construction at present. The outlet for Kazakh oil to the Persian Gulf, through Iran, was the first project considered by the Kazakhs, in the late 1990s, and had to be put on the back burner under US influence. There are indications that this project could be revived as a gas pipeline project to take advantage of the huge market for Kazakh gas in South Asia.

Pipelines have had certain curious implications for India's quest to diversify its sources. Thus, the BTC pipeline, aimed at carrying Caspian oil to Europe, has also raised the possibility of Caspian oil being shipped to India from Ceyhan, either through the Suez Canal (in case of small shipments) or through the newly revived Ashkelon-Eilat pipeline in Israel which, through an arrangement of double shipment from Ceyhan to Ashkelon and then from Eilat by crude carrier to India, has for the first time made Caspian oil available to India.

In case the plan for a sub-sea pipeline from Ceyhan to Israel *via* Turkish Cyprus would be realised, then India and other Asian countries would be able to obtain Caspian oil at Eilat on the Red Sea. Similarly, if, the proposed pipeline from Alexandria to the Red Sea were to be realised, India would be able to obtain access to oil from North Africa, particularly Algeria and Libya, which today can reach India only by going round most of the African continent.

In order to meet its gas requirements, India is vigorously pursuing gas pipeline projects both on its eastern and western land frontiers. The Iran-Pakistan-India pipeline is expected to bring to India nearly 90 mmcmd of gas which will be utilised to fuel power and fertilizer projects in North and North Western India. In the East, the Myanmar-Bangladesh-India pipeline will not only bring Myanmar gas, but would also possibly carry Tripura gas to India, and Bangladesh gas from its eastern part where it is produced to the western part where it is required.

India has also agreed to participate in the Turkmenistan-Afghanistan-Pakistan pipeline, initially as an observer. Taking into account concerns relating to Turkmenistan reserves at Daulatabad, India proposes to suggest to participating countries that Daulatabad be seen not just as the sole source of gas in the project but as the junction to which gas from neighbouring countries, such as Uzbekistan, Kazakhstan, Azerbaijan and, possibly, Russia could be injected.

All these pipeline proposals, being transnational in character and involving neighbouring countries with complex histories in terms of bilateral relations, are fraught with political and security related problems that would need to be addressed. If these projects are to be realised, we must first accept that they are extremely important, indeed critical, for India's energy security interests. Once this is understood, international best practice can readily yield arrangements that would be put in place in regard to all aspects of the projects – technical, financial, commercial and legal, that would serve to insulate the projects from the vagaries of day-to-day politics and provide the desired level of comfort to our policy makers.

The Asian Gas Grid: A Feasible Proposition

While the world's gas map depicts numerous gas pipelines moving across thousands of kilometers from Russia, Central Asia and the North Sea to Western Europe, hardly any pipelines move Eastwards and Southwards.

This is now set to change due to two important factors:

(i) the increasing Asian demand for gas; and

(ii) the ability of Asia to transport gas economically from producers to consuming centres.

Over the next 25 years, the energy requirements of Asia are expected to increase 2½ times, an increase of an additional 2-2.5 billion toe. Gas will have a significant place in this scenario. At present, Asia has much less share in gas demand than the world average, (6 per cent *versus* 12 per cent).

Hence, to meet Asia's rapidly increasing energy equirements, consumption of gas will have to increase; the expectation is that it will do so from 210 Mtoe in 1997, through 600 Mtoe in 2020, to 800-900 Mtoe in 2030. In short, natural gas demand growth in Asia will be between 5-10 per cent per annum, depending on the economy, about double the growth in oil demand.

The principal sources of global gas also lie in Asia. Russia has 27 per cent of the world's proven reserves, followed by Iran (15 per cent) and Qatar (14 per cent). In fact, North and Central Asia and the Gulf between them have over 70 per cent of world reserves. As against this, the principal consumers of Asia—China, Japan, Republic of Korea and India—together have less than 2 per cent of global reserves, with Japan and Korea having no reserves at all. At the same time, in 2004, the latter two countries imported just over 100 billion cubic metres of gas as LNG as against a total global LNG trade of 178 billion cubic metres.

The Economics of the Asian Gas Grid

Gas is transported globally as LNG and through transnational pipelines. By 2020, one-third of the world output, i.e., 1.35 trillion cubic metres, will be internationally traded, of which 50 per cent will be transported as piped gas and 38 per cent as LNG. Given the rapidly increasing Asian demand and the availability of huge gas reserves in Asia, piped gas and LNG will both be utilized, and, indeed, will complement each other in the energy-mix of the principal consumers.

The Asian Gas Grid envisages the setting up of a series of pipelines that will carry natural gas to the various consumption centres in East and South Asia. According to current estimates, the additional pipelines required to be set up would be about 22,500 km., costing about 22 billion dollars. The Asian continent, particularly Russia, the principal Asian consuming countries, and the major producing countries of the Gulf are readily able to provide the financial and technological resources for the project.

Besides contributing significantly to the growth and prosperity of the continent as a whole, the project would have other benefits such as developing the electricity, petrochemical and fertiliser industries; promoting transcontinental cooperation in trade and industry; upgrading local skills and expanding employment opportunities; and, above all, generating the resources that would enable Asian governments to fund poverty alleviation and other welfare programmes.

In simple financial terms, there would be very obvious advantages for all the countries involved in the grid, the producers, the consumers and the transit nations. According to an energy consultant, total financial benefits in the first year of the project (2006) would be $ 1.5 billion. Over the next few years, the benefits would increase significantly every year, amounting to $ 55 billion in 2025.

Transcontinental pipelines are particularly difficult and complex ventures since:

i) they involve different countries with widely different interests;

ii) being transnational in character, and involving neighbouring countries, they invariably bring on board a substantial and complex political baggage of disharmony and discord; and

iii) the projects are beset with serious technical and financial difficulties, requiring the mobilisation of huge resources from domestic and international services in a environment of mutual trust and confidence.

It is true that these problems are particularly daunting in an Asian environment which has been the theatre of considerable intra-continental discord and conflict, and has very few success stories in regard to regional and continental cooperation. It is also true that some of the issues that divide Asian countries, particularly neighbours are fairly complex and are unlikely to be resolved in the near future.

Set off against this bleak scenario is the overwhelming common interest of oil and gas producers to have stable markets for their products and for Asian consumers to have assured supplies to maintain their economic development programmes. These complementary interests in energy constitute the strongest element in enabling policy makers to replace contemporary political discord in favour of energy-motivated cooperation.

Regional Cooperation

Meeting the global demand for oil and obtaining the financial resources to ensure supplies require the rejection of political contentions based on narrow national considerations and, instead, an integrated global effort to pool together the world's human, financial and technological resources to explore and develop these difficult and sensitive areas in a spirit of cooperation based on considerations of mutual benefit.

India took the first significant step in promoting this cooperation at regional level by convening a Round Table in New Delhi in January 2005, of the four principal Asian oil-consuming countries—China, Japan, Republic of Korea and India—getting into dialogue with the principal oil-producing countries of West Asia and South East Asia. The 11 assembled ministers agreed on the importance of this first dialogue between Asian consumers and producers and in a consensual statement, identified substantial commonality of interest as also areas of cooperation. They also recognised that, for the interests of the Asian consumers and producers to be pursued effectively, the knowledge-base of Asian countries would have to be expanded even as the Asian producers and consumers develop policies and programmes linked with promoting crisscross investments in each others' hydrocarbon sectors as also in the areas of conservation and efficiency and environment protection.

This regional dialogue has thrown up a number of specific areas for cooperation. These include: reform of the Asian oil markets, promotion of crisscross investments in hydrocarbons between producers and consumers, development of strategic reserves, development of the Asian gas pipeline grid, development and transfer of R&D and technology, and development of capabilities to promote energy conservation and efficiency and environment-friendly fuels. The ministers have agreed to meet annually to pursue their consensual plan of action.

Separately, in order to complement the earlier Round Table, India took the initiative to bring together at another Round Table the four principal Asian oil-consuming countries in dialogue with oil-producers of North and Central Asia, including Russia, Kazakhstan, Uzbekistan, Azerbaijan, Turkmenistan and Turkey, in November 2005.

There is now a slow but steady acceptance that national energy security interests are best served by pursuing policies of cooperation so that energy resources can be harnessed efficiently for regional and global development. The oil market is already integrating in significant ways: there is a clear trend in favour of oil companies integrating across the hydrocarbon value-chain, from exploration to production to transportation, to refining and to petrochemicals. E&P proposals in producer countries are increasingly being linked to refinery proposals and, on occasion, to other infrastructure development proposals such as roads, railways, mining, and port development projects. Hurricane Katrina, by damaging US facilities across the entire supply system in the region, has redefined energy security to mean, as Yergin has noted, "the security and integrity of the whole supply chain and infrastructure from production to consumer." Above all, the surge in global demand for hydrocarbons represents, in Yergin's words, "the success of globalization—the best global economic performance in a generation."

The challenge before India is to understand the impact of these myriad developments and effect the required adjustments in our politics and policies and, above all, in our mindset, so that we can effectively respond to and participate in these global trends.

15

India's Energy Security

Concepts and Measures

A.F. ALHAJJI

India's main energy problem is that supplies are limited while demand continues to grow. Its domestic fossil fuel resources are far from the market, its coal is of low quality, and the renewable energy industry is still growing out of its infancy. India faces tremendous political, economic, natural, legal, and technical challenges to meet the growing domestic demand for energy. These serious challenges have given birth to the concept of 'energy security' in India.

While almost all concerned officials and organisations talk about 'energy security', very few understand what this concept means and how different it needs to be for India than for other nations. In the absence of a clear definition of energy security, India has not been able to develop procedures to identify the data it needs; identify the links between energy security, economic security, social security, environmental security, and national security; measure energy security; assess the energy situation at any given time; and develop policies that can respond to various circumstances. The lack of such identification, measurement and assessment prevents India from developing a sound energy policy.

This paper develops concepts, measures, and assessments to define India's energy security that will enable policy makers to take concrete steps to improve India's energy security. It applies some of these measures to assess the current situation and concludes that India should develop bilateral and regional cooperation with the oil producing countries in the short- and the medium-term. In the long-run, the only way that India can secure its future energy supplies is to develop clean coal technology and build more nuclear reactors.

Introduction

India's economy has been one of the fastest growing in the world in recent years. The economy has grown at an average of 7 per cent annually while India's energy consumption has grown by more than 3 per cent. Oil consumption more than doubled between 1989 and 2004. India's oil imports increased fourfold in the last 15 years on increasing demand, flat domestic oil production since 1995, and the inability of other energy resources such as coal and hydropower to keep up with increasing energy consumption. Policy makers and experts are worried that such growth is not sustainable unless adequate energy supplies are available, especially since India needs high economic growth rates to eliminate poverty and create jobs for its growing and young population.

Therefore, India's main energy problem is that supplies are limited while demand continues to grow. Domestic oil and gas resources are not abundant enough to meet growing demand. These resources are mostly located offshore, which make their development vulnerable to changes, not only in weather but also in the dynamics of world oil markets, especially given that recovery rates are low by world standards. Indian coal is of a poor quality, which limits its future growth unless new, cheap clean coal technology is developed. In addition, productivity in the coal sector is notoriously low. Most of the domestic fossil fuel resources are located in remote areas, far from the major markets. Nuclear power is costly and difficult to finance. Wind and hydro power are limited and intermittent. Renewable energy is still in its infancy, experimental, and on a small-scale. Generally speaking, the power sector suffers from inefficiency and large transmission losses. India faces tremendous political, economic, natural, legal, and technical challenges to meet the growing domestic demand for energy. These serious problems and challenges have increased India's dependence on foreign oil imports and have lent strength to the concept of "energy security".

Since the only way to cover the growing gap between energy production and consumption is to import more oil, natural gas, and LNG, oil and gas dependence have dominated the 'energy security' debate in India. As a result, actions taken to enhance energy security have been mainly directed toward 'filling the gap' by securing upstream oil contracts around the world, encouraging domestic exploration for oil and gas, and exploring the possibility of transporting natural gas *via* pipelines from Iran, the Caspian, and Myanmar. Recently, in a strange twist, several government officials stated that the government's objective is to eliminate energy dependency. If that is the case, why was the visit of the Saudi monarch to India termed "historic"? Why does a country that is worried about dependency on oil imports make such dependency legal and binding by signing exploration and production contracts with some oil producing countries? Why does a country, that is worried about the security of oil supplies, sign contracts with the least secure countries in the world? Does investment in the upstream sector of oil producing countries enhance the energy security, of India? How? Why does a country that is worried about energy security not have the means to measure energy security at any given point of time? Why does a country that is worried about energy security not have the data necessary to assess the energy situation and devise the necessary policy to avoid an energy crisis? In fact, if energy dependence is a threat to energy security, a threat to economic growth, and a threat to prosperity, Indian officials who want to eliminate energy dependence have to answer the following questions:

1. How can they explain the economic miracles of Japan and Germany, which have been 100 per cent dependent on foreign oil since World War I?
2. How can they explain the high rates of economic growth in India in the last four years despite record-high oil prices?
3. How does the security of oil supplies differ from the security of the investments of Indian oil companies in other countries, some of which are not at all secure?

This paper explores the concept of India's energy security, provides an overview of energy security measures, identifies some of the data needed, and provides a general framework for assessing energy security at any given time. Section II explores the concept of energy security. Section III lists various security measures. Section IV summarises the conclusions.

The Concept of Energy Security

What is energy security? The concept of energy security is elusive. There are at least 17 different definitions of energy security in the literature, which range from uninterrupted oil supplies to the physical security of energy facilities. The literature is divided between those who believe that energy security is "an empty concept used to perpetuate bad, self-serving public policy"[1] and those who believe that energy security is a real issue that governments have to devise polices to deal with.[2] If energy security is a real issue, how does one deal with it? Should governments intervene in energy markets to ensure energy security? Or should they leave energy security to market forces, since markets are more efficient in allocating resources than governments? The literature is divided on this matter with some experts advocating government intervention while others advocate the market approach.[3] Experts, who believe in government intervention, do not agree on how and why. Experts, who believe in market forces, do not agree whether governments should interfere during energy crises or not. They also do not agree on the extent to which government should become involved in the promotion of competitiveness in energy markets.

Historically, researchers limited the concept of energy security to consuming countries. Three major issues have convinced experts and policy makers that energy security is the responsibility of both consumers and producers. These issues include political unrest and violence in several oil producing countries, the increased cooperation between oil producing and oil consuming countries, and the possibility of an oil production peak in some oil producing countries.

These differences in opinion make the concept of energy security so diverse and elusive. However, a quick review of the energy security literature indicates that the division among experts cuts along political and philosophical lines. Therefore, any definition of energy security is self-serving, even for those who believe that it is an 'empty concept.'

The Saudis once defined energy security as "maintaining and enhancing access to where the oil exists in such obvious abundance."[4] Some members of the US oil industry have their own special views that serve their interests: "US energy security is best maintained by ensuring that the United States is, and is perceived to be, supportive of free trade and of the use of market forces on a global scale."[5] However, the Bush administration view is summed up in the title of its 2001 energy report, "Reliable, Affordable, and Environmentally Sound."[6] Almost all OECD countries have designed their energy policies around three dimensions: security of supplies, the environment, and economics.[7] However, the statements and actions of various countries indicate that there is more to the concept of energy security than these three dimensions.

To reach an acceptable concept of energy security, researchers and policy makers have to think about it within the framework of energy security *vs.* energy insecurity, independence *vs.* interdependence, energy security *vs.* reciprocal energy security, energy security *vs.* national security, energy security *vs.* economic security, energy security *vs.* environmental security, foreign investment *vs.* formalisation of dependence, secure areas *vs.* insecure areas for imports and investments, market prices *vs.* administered prices, and high prices *vs.* low prices. Researchers have also to investigate the link between energy policy and other policies and decide whether energy policy, and, therefore, energy security, is a subset of economic, environmental, foreign, and social policies or a policy of its own.[8] Regardless, it is very clear that every nation needs some form of integration among these policies to achieve energy security, especially given that policy contradictions are among the major threats to energy security.

Energy Crisis, Threats to Energy Security, and Prices

To avoid the political fault lines that separate the various concepts of energy security and to gain a deeper understanding of energy security, it might help to understand the meaning of energy crisis and the various threats to energy security. It is worth noting that energy insecurity does not mean energy crisis, and it does not necessarily lead to an energy crisis. The existence of threats to energy security increases the risk of an energy crisis, but does not necessarily lead to one.

An energy crisis is "a situation in which the nation suffers from disruption of energy supplies accompanied by rapidly increasing prices that threaten economic and national security."[9] Therefore, an energy crisis has two elements: a decline in GDP growth, usually a recession, and a threat to national security.[10] While the first element is very clear and can be measured, it is difficult to determine the second or to measure it. Does a threat to national security means a reduction of energy supplies needed to fight a war or defend a country against foreign aggression? Or does it mean restrictions on foreign policy options? The US considers any decrease in oil exports that limits US foreign policy options to be a threat to national security. For example, a decrease in Nigeria's oil exports might limit US foreign policy options in dealing with the current government of Iran. If the US has four ways to deal with Iran's nuclear programme, the fact that Iran is a major oil producing country might limit the US options to two. This reduction in options is viewed as a threat to national security even if the US does not act on it. The question is, can India use this concept? If not, what is the connection between energy security and national security for India?

Even if we try to define energy security without linking it to national security, the concept remains unclear. Some experts define energy security as a "reliable and adequate supply of energy at reasonable prices."[11] Both concepts focus on the economic dimension of energy security. The term 'adequate supplies' is very clear: it means 'uninterrupted supplies.' But what are 'reasonable prices'? Are they from the consumers' point of view or producers' point of view? In fact, the only way that reasonable prices can enhance energy security is when they are high enough to guarantee adequate return on investment for oil producers and low enough to stimulate economic growth in the oil consuming countries. Low oil prices are as dangerous to energy security as high prices.

Experts who believe that market forces are the best guarantee for achieving energy security say that considerations of reasonable prices should not be part of the equation. In fact, it is unclear how 'reasonable prices' can enhance energy security when most OECD countries impose hefty taxes on petroleum products.

Thus, the only clear dimension of energy security is the economic dimension. But what about the other dimensions? Can the principles of energy security, that various OECD countries employ, help us understand the concept of energy security? Almost all industrial countries, especially the US, base their energy policies on four principles:

- Diversity of energy supplies.
- Diversity of oil imports.
- Reduced dependence on Middle Eastern Oil.
- Low price volatility.

Since oil is the most tradable energy source, it is no surprise that most energy security discussions are limited to oil. However, the increased trade in natural gas and LNG have already brought those two commodities into recent discussions of energy security. What is unique about the above mentioned principles of energy policy is that they focus on the availability and vulnerability of supplies, but do not consider 'reasonable prices.' Rather, they consider price volatility. Low price volatility, while it might looks a straightforward concept, requires 'reciprocal energy security.'[12] In fact, the only way to lower price volatility is cooperation between energy-exporting and energy-importing countries to ensure the security of supply and demand simultaneously. Reciprocal energy security implies paying a fair price for the resource so that producers can continue to invest and production continues to grow to meet growing demand. In other words, 'low oil prices' or 'reasonable prices' from the consumer's point of view will not ensure steady supplies and will increase price volatility. All of these principles apply to India, with certain differences. For example, India may wish to maximise its energy security through diversification of its oil imports, but its location may force it to do so within the Middle East. However, these four principles ignore several important issues that have become the focus of energy policies around the world, such as the environment.

Threats to energy security can help us clarify further the concept of energy security. These threats include the use of oil as a weapon by one or a group of oil producing countries. In this case, even if India is not the target, India will experience higher oil prices. An embargo that targets the US, for example, will have a limited impact on India. However, if that embargo is associated with a production cut, India will suffer as much as the US, or may be more if it cannot afford to pay the highest prices in the market. In a global oil market, supplies will go to the highest bidder. In fact, given the opportunity cost, even Indian suppliers may be tempted to ship Indian oil overseas if the domestic price is regulated by the government.

Threats also include a sudden peak in world oil production, government intervention, and conflicts among various government polices. Price controls distort the relationship between supply and demand and might lead to shortages. Contradictions among government policies might sometimes favour, for example, a foreign policy that leads inadvertently to energy shortages.

Examining India's *Hydrocarbon Vision-2025* and recent statements and concerns by politicians and experts around the world, including some in India, allow us to identify six competing dimensions of energy security.[13] Balancing these dimensions within an energy policy is a very difficult task. While these dimensions are general and universal, the weight of each dimension and the level of interaction among these dimensions differ from country to country and from time to time. One of the main weaknesses of the energy security literature is the lack of discussion regarding the impact of contradictory policies on energy security. Other weaknesses include the lack of discussion regarding the interactions among various dimensions of energy security and the weight of each dimension of energy security in energy policy.[14]

The Economic Dimension

Countries strive to achieve steady economic growth and low rates of inflation, unemployment, and price volatility. Economic growth usually generates higher energy consumption, and, naturally, higher oil imports in the consuming nations. Focusing on the economic dimension of energy security ensures that scarcity of energy resources does not stall economic growth, increase inflation, raise unemployment, weaken the balance of payments, and reduce the value of the county's currency. Ironically, the solution in this case does not lie in the energy sector of the economy or in the hands of policy makers in Ministries of Energy. One of the major solutions requires the use of fiscal and monetary policies. Therefore, fiscal and monetary polices should be integrated into any energy policy. In fact, among the main reasons why high oil prices have not affected India's economic growth in the last three years are its expansionary fiscal and monetary policies and various economic reforms. The positive effect of a combination of lower interest rates, increased government expenditure, increased exports, and a large influx of foreign capital has outweighed the negative effect of increasing oil prices.[15]

However, fiscal and monetary policies are not enough to enhance India's energy security. India ranks among the top in the world in the strength of the relationship between energy and economic growth. Figure 15.1 illustrates this relationship between GDP and energy consumption. The ultimate goal will be to break the link between energy consumption and economic growth.

Figure 15.1

Relationship between Energy and GDP India GDP (B Rs) versus Primary Energy Consumption (mtoe)

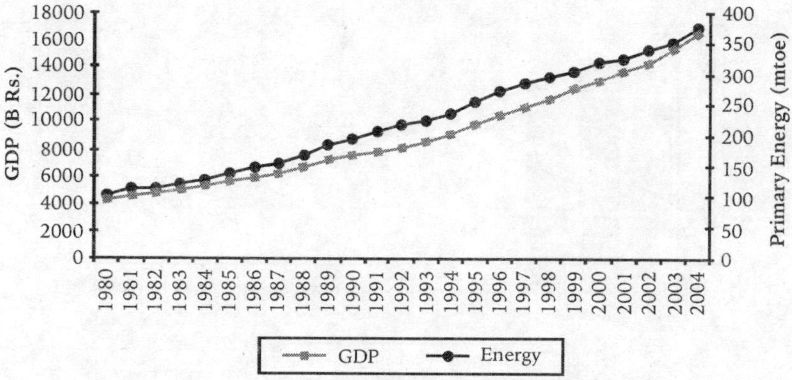

Source: IMF, 2006.

To improve the economic dimension of energy security, the Government of India should limit the relationship between energy consumption and GDP by increasing efficiency on one hand, and increasing the size of the service sector on the other. Figure 15.2 shows the energy intensity in various sectors *versus* the share of each sector. The figure shows an improvement in recent years: while the share of the industrial and service sectors in the GDP increased, their energy intensities decreased. The Figure 15.2 illustrates that to generate one additional unit in the GDP from the industrial sector, India needs 60 times the energy needed to generate the same addition from the service sector, and 10 times the energy needed to generate the same addition from the agricultural sector. It also illustrates that the recent decline in energy intensity is the result of both sectoral shifts and increased efficiency. Therefore, India can improve the economic dimension of its energy security by reducing the link between GDP and energy, reducing energy intensity in the industrial sector, and shifting its economy toward a service oriented economy.

Figure 15.2

GDP Share versus Energy Intensity, India (Selected Years)

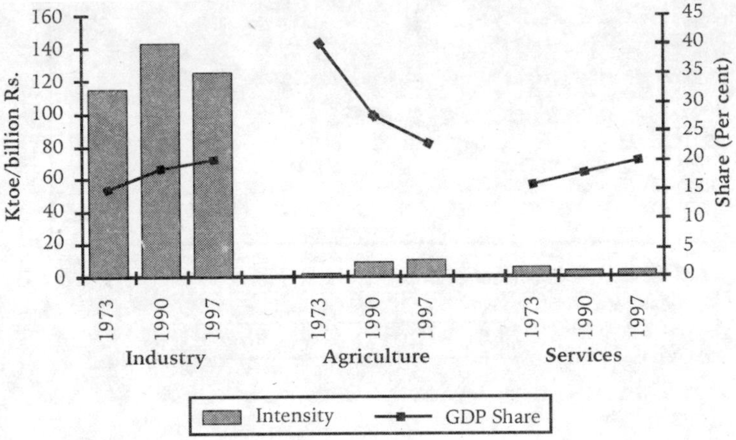

Source: IEA, 2004.

The Environmental Dimension

The international community as a whole is worried about the environmental impact of the extraction, transportation, and consumption of energy resources, especially the burning of fossil fuels. Since the environmental impact of burning fossil fuels has health and economic consequences, several countries have incorporated environmental objectives into their energy polices. Cleaner environment was among the main objectives of *India Hydrocarbon Vision-2025*: "To enhance quality of life by progressively improving product standards to ensure a cleaner and greener India." The Vision also emphasised that exploration and production activities must achieve as near as zero impact, as possible, on environment. The problem that India faces is how to raise income levels while improving living standards, which includes improving air and water quality. It is obvious that a reduction in energy use will decrease pollution, but this will come at the expense of the economic dimension of energy security. Only improvements in the efficiency of cars, equipment, and machines can reduce pollution and ensure economic growth.

Therefore, the environmental dimension of energy security is not related to energy resources *per se*, but to the technologies that use these resources. To ensure energy security, the government should integrate its industrial and technology policies into its energy policy. Such policies will improve efficiency, but they will also affect fuel choice, which may result in contradictions with other dimensions of energy policy.

Low energy prices may benefit the economic dimension of energy security but they worsen the environmental dimension. Some environmentalists contend that higher energy prices will reduce energy consumption and consequently reduce pollution. They advocate a substantial tax increase on fossil fuels to reduce CO_2 emissions. This approach not only contradicts the economic dimension of energy security, but also contradicts the environmental dimension. Low-income families, who constitute the largest segment in India's population, will not be able to buy the heavily-taxed fuel. Instead they will burn wood, hard coal, and animal dung, which will lead inevitably to deforestation and increased pollution. Therefore, taxation and higher energy prices will not enhance energy security for India.

Social Dimension

The energy security debate in all countries, including India, focuses on external factors that are related to security of supplies and political unrest in the oil producing countries. The only internal issue that has gained some attention in recent years is the environmental impact of burning fossil fuels. Energy policies have historically ignored domestic political problems that may arise when the gap between the 'energy rich' and 'energy poor' increases.

Countries strive to reduce social tension through various policies that include the reduction of income and energy inequalities. Any factors that increase income and energy inequalities are viewed as a threat to society. Higher energy prices might at times, increase both. The social dimension of energy security does not directly relate to the availability of energy resources, but rather to the ability of the poor to get these resources, which might explain price controls and subsidies. The larger the energy gap, the more insecure the country becomes. The larger the proportion of the poor who are not able to get at energy resources, the more energy insecurity the country experiences. While the social dimension is not highlighted in most energy policies, it becomes part of the energy debate every time a country experiences an energy crisis. This is true of the most developed country, the United States, as well. When energy prices are high, the gap between the rich and the poor becomes more obvious. The rich can afford to heat their homes and drive their cars, while the poor will suffer from the cold and will be forced to walk. While schools in rich areas are open during frigid seasons, schools in poor areas have to shut down. The result of such an 'energy gap' could well be political unrest, which will reduce economic growth in the process.

Low energy prices for all segments of society may not enhance the social dimension of energy security. Low prices of petroleum products might end up helping the rich who own several cars and several homes, while the poor will not benefit as much. The government can enhance the social dimension by adopting fiscal policies that reduces the energy gap between the poor and the rich.

Foreign Policy Dimension

Oil and politics are intertwined with one another in an unending dance. The evidence of this connection abounds. As the significance of natural gas and LNG in international trade increases, gas and politics will develop a similar relationship. India needs both oil and gas, which puts it at a disadvantage when it negotiates with oil and gas producing countries. Fear of oil shortages may force India to cooperate with countries with which it would not otherwise do so. India's need for oil or gas might force it to make foreign policy decisions that would compromise it on other important issues or principles. Realising this problem, Indian policy makers stated in the *Hydrocarbon Vision-2005* the following objective: "To ensure oil security for the country keeping in view strategic and defence considerations." The word 'strategic' refers to the foreign policy dimension of India's energy security.

The oil producing countries might not use the oil chip when they negotiate political issues, but their politicians understand that the other side sees all the cards, including oil. Explicitly or implicitly, even powerful countries such as the US and UK have oil on their minds when they negotiate political issues with the oil producing countries. The need for oil might affect India's foreign policy and international decisions as well.[16]

The need for oil and gas might force India to limit its foreign policy options. It is up to the politicians on how to view this matter. However, there could be cases where India's need for oil and gas will limit its options.

Technical Dimension

Energy has another eternal partner beside politics, technology. When it comes to energy security, policy makers must be aware of this relationship to balance the various energy dimensions of energy security. A government's push to improve energy security by supporting technologies that facilitate the production of renewable energy resources may actually lead to energy insecurity if the country becomes a hostage to spare parts that come from other countries. For example, fuel cell technology requires the use of palladium, which is exported by only two countries in the world, Russia and South Africa. In some cases, supporting certain technologies might inadvertently switch demand from domestic energy resources to foreign resources. Supporting new energy technology may require an increase in imports of the energy sources that the new technology is intended to replace if this imported source is necessary to operate this new technology. For example, the development of clean coal technology would require India to import additional quantities of oil and gas. Supporting a new technology may also increase energy prices, which, in turn, will affect the economic and social dimension of energy security. A country that adopts a new technology that improves energy security may end up with energy insecurity if it has to import skilled labour force to operate this technology.

On the other side, adopting new technologies might improve energy security. The use of biomass energy in electricity generation in India has proven to be very successful, especially in this high oil prices environment. Since low oil prices could choke these power plants, low oil prices would not be beneficial to India. In this case, the quest for low energy prices would kill certain technologies.

Security Dimension

While the foreign policy dimension focuses on diplomatic and trade relations, the security dimension deals with two issues: the physical security of energy installations and the energy needs of a nation's military and police to protect the country or quell domestic uprisings, terrorist attacks, or any other violent threat to energy production. Threats to physical security of energy installations include terrorist attacks, human errors, natural disasters, and technical malfunctions.

It is outside the scope of this paper to discuss ways to protect energy installations. However, measures of the security dimension include the degree of concentration of domestic energy resources and installations. They also include the geographic location of energy resources and installations relative to: (1) the location of the market, (2) the locations of various religious, ethnic, and political groups in the nation, (3) the location relative to natural disaster prone areas. In addition, these measures include frequency of attacks, and the timing of the attacks relative to the daily rhythms of energy usage and seasonality in energy consumption. Energy security measures are also related to energy consumption by the military and the police force, the security of supply lines to the military and police, and the speed of delivery of energy sources to them.

Dependence, Interdependence, and Energy Security

Top Indian officials stated recently that their objective is to eliminate India's dependence on oil imports. This far-fetched fantasy not only distorts the facts, but compromises energy policy and energy security. Those who want to eliminate dependence have to explain why they are also pushing for increased Indian involvement in upstream operations in the oil producing countries. They have to explain why they are negotiating with other countries to build gas and oil pipelines to India. They argue that self-sufficiency is a long-term objective. In the short- and medium-term, India has to invest overseas and has to get oil and gas from neighboring countries *via* pipelines. This argument has several built-in problems. If investors are convinced that India is serious about self-sufficiency and can make it happen, they will not invest in these projects and the pipelines, which cost billions of dollars. If the oil producing countries are convinced that the investment of the Indian oil companies is only for a limited period, they will not allow these companies to invest.

Self-sufficiency in energy does not shield the country from world supply disruptions and the resulting price increases.[17] Even if India is self-sufficient, a disruption of Venezuelan oil supplies will increase oil prices worldwide, even in India. In this case, it does not matter whether India imports 1b/d or 3 mb/d, prices will go up by the same percentage. The effect on India's economy will be the same whether it is dependent on foreign oil supplies or not. Figure 15.3 demonstrates that eliminating dependency may not improve energy security. The Figure plots oil prices *versus* the ratio of world excess capacity to OECD stocks between 1999 and 2005. The Figure clearly illustrates a strikingly negative correlation between prices and the ratio of excess capacity to stocks. Prices increased when excess capacity and stocks were low, regardless of the level of dependence on oil imports. For prices to stabilise, excess capacity and stocks have to increase. Excess capacity exists in the oil producing countries. Stocks exist in the oil consuming countries. Therefore, only interdependence and cooperation can enhance energy security. India's focus on eliminating dependence on oil imports is misplaced. India's energy security is nested in interdependence and cooperation, not isolation and self-sufficiency. In fact, if other countries thought in the same way about their dependence on India's IT industry, India's economy would plunge into an abyss within a few months.

While low oil prices are not good for the oil producing countries, they are not good for India either. Low oil prices increase consumption, increase dependence, chock alternative energy resources, increase wastage and increase pollution. In other words, while low oil prices contribute positively to the economic dimension of energy security, they destroy the rest of the dimensions. Since both producing and consuming countries suffer, only interdependence and cooperation can eliminate this suffering. The same argument can be made when oil prices are high.

Measures of Energy Security

Policy makers can prevent an energy crisis or reduce its impact on the nation if their experts can execute the four steps that are required to deal with energy security issues: data collection, measurement, assessment, and policy recommendations. However, to know which data that the government or others should collect requires knowledge of the objectives and the various measures used to assess energy security.

Measuring energy security at any given time enables policy makers to make informed decisions and act quickly to avoid energy crises. We need to use several measures to assess each of the six dimensions of energy security. Developing these measures requires an extensive list

Figure 15.3

Oil Prices versus the Ratio of World Execss Capacity to OECD Stocks

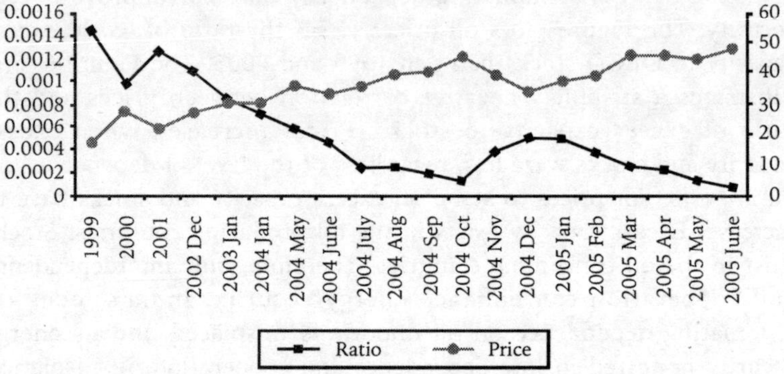

Source: *EIA*, 2006.

of timely data. Lack of timely data prevents India from measuring energy security in a meaningful manner. However, the objective of this section is to list these measures and identify the needed data. Assessment involves answering questions such as: Is there an energy problem? If it is there, in which sector? What are the causes of the problem? What actions should be taken? Is there a contradiction among various government policies?

The first measures that India needs to develop are energy supply measures, especially for oil and gas. These measures are related to domestic reserves, domestic production, domestic commercial and strategic stocks, import dependency, import concentration and import vulnerability, and investment in the energy sector. For example, given current available data, India can develop measures such as petroleum imports as a percentage of total petroleum consumption, the number of days total petroleum stocks cover petroleum imports, the number of days total stocks cover consumption and the percentage of petroleum in total energy consumption.

The second set consists of energy/economy measures. These are related to variables such as energy/GDP ratio, income/energy price ratio, energy efficiency, per capita energy consumption, sectoral energy consumption, sectoral production/energy ratio, fiscal policy variables/ energy price ratios, monetary variables/energy prices ratios, and the exchange rate relative to the US dollar.

The third set consists of energy/environmental measures. These are related to CO_2 and other greenhouse gases emissions by each energy source, region, and time. They also include various measures of oil spells and energy-related deforestation.

The fourth set consists of social and domestic political measures. They are linked to the availability of energy sources, the energy gap between rich and poor, the types of energy used by the rich and the poor, the degree of subsidies, the extent of price control of energy products, the difference between domestic energy prices and energy prices in neighbouring countries, and the geographic distribution and location of domestic energy resources.

The fifth set of measures pertains to foreign policy. They are related to major, identifiable foreign policy objectives on major international issues in relation to energy production and consumption, level of dependence on oil and gas imports, the availability of substitutes, the level of foreign investment in India and India's foreign investment in the countries in question, the political and diplomatic relations with oil and natural gas producers, the political and diplomatic relations with countries that might provide transit services to transport energy products, the number of Indian expatriates in the oil producing countries, the amount of expatriate remittances, the regional impact of expatriates within India, and membership of those countries in certain world organisations that might affect India such as the UN Security Council. These measures involve questions related to India's stand on Kyoto Protocol, nuclear proliferation, human rights, terrorism, and other issues.

The sixth set of measures consisting of measures related to the technical dimension of energy security include the relative cost of each energy technology to oil and gas prices, the level of energy-technology imports from foreign countries, the level of concentration of these imports, the level of interdependence among various energy technologies, the number of experts and technicians in each technology, the ratio of domestic to foreign workers, and the availability of educational and training institutions that will produce the skilled workers to develop and maintain these technologies.

The only timely data available in India are those related to the macroeconomy, oil and gas production, and oil imports. While data for most of the above measures are available, they are out of date.

Conclusion

No energy policy is complete without focusing on energy security. Any discussion on energy security is futile without understanding its meaning or measuring and assessing it. Despite the recent formulation of an energy policy in India, the contradictions between the concept of energy security and the actions of both the Indian government and the Indian oil companies, on one hand, and the call by the top political leadership to eliminate dependence on foreign oil imports, on the other hand, indicate that most politicians do not know what energy security is. Even if they are aware of its meaning, India lacks the measures needed to assess energy security. It even lacks the data required to build such measures.

This paper provided a general framework in which it identified six dimensions of energy security and introduced a number of measures to assess these dimensions. The critical problem is the lack of timely data to measure and assess the energy security situation in India. Without measurement and assessment, decision makers cannot make the correct policy recommendations to avoid an energy crisis.

Given the data that is available, India faces serious energy insecurity problem in all dimensions. Energy insecurity is apparent in the level of dependence on one or two sources of energy, import dependence, import concentration, dependence on unstable countries, energy investment in countries that suffer from rogue regimes, low levels of stocks, high levels of energy related emissions, the distance of domestic energy resources from main markets, concentration of domestic energy resources, the attacks by Assam rebels on oil and gas installations, old nuclear reactors, and a strong link between energy consumption and GDP, to name only a few. The reasons why India's economy has not suffered from high oil prices in recent years include the expansionary monetary and fiscal polices accompanied by economic reforms and freer energy markets.

While energy independence might improve some aspects of energy security, it does not shield the country from energy shocks. The petroleum market is global. Any shortage in any part of the world will increase petroleum prices worldwide. Therefore, India's policy makers can enhance India's energy security by fostering interdependence rather than agonising about dependence.

To improve its energy security, India needs to collect relevant, up-to-date data, measure the various dimensions, and assess energy security. Only then can policy makers take informed decisions and protect India from future energy crises. Keep in mind that even a correct policy will not be effective if it contradicts another policy response, policy responses must be integrated in a way that maximises energy security. The experience of the last three years indicates that policy makers can enhance the economic dimension of energy security through the implementation of a combination of fiscal and monetary policies. While a free energy market is among the best tools to enhance energy security, the government can pursue policies that can improve market competitiveness and make corrections for market failures. The social dimension of energy security might require the government to intervene to reduce the energy gap between the rich and the poor. Fear of oil shortages might also force the government to build strategic petroleum reserves. Recent press reports indicate that India has planned to establish strategic stocks in cooperation with Saudi Aramco.[18] Building strategic oil and gas reserves, to be used during energy shortages, is an effective tool only if strategic reserves do not replace commercial stocks.

Notes

1. From a presentation by Vito Stagliano, Deputy Assistant Secretary of Energy for Policy Analysis during the first Bush Administration, at the *Energy and National Security in the Twenty-First Century* conference in 1995. The Institute for National Strategic Studies, National Defense University.
2. There are numerous books and articles that support this view, see APERC 2002, for example.
3. For more information, see Constantine, 2005.
4. A statement by Hisham Nazer, the former Saudi Arabian Minister of Petroleum and Minerals Resources, at the 60th Annual Meeting of the Independent Petroleum Association of America in San Antonio on 31 October, 1989.
5. A statement by long Time Amoco vice President John Lyman at the Energy and National Security in the Twenty-First Century conference in 1995. The Institute for National Strategic Studies, National Defense University.
6. The report is posted on the White House website at: *http://www.whitehouse.gov/energy/index.html*
7. The infamous Japanese three Es are: Energy security, Environmental protection, and Economic efficiency.
8. For more information see Kohl, 2004.
9. Williams and Alhajji, 2003.
10. It is worth noting that national security is not among the three dimensions of energy security mentioned above.
11. See, for example, "RIS Background Paper on India's Energy Security" which was conducted by the center for Research and Information System for Developing Countries, an Indian government backed organization.
12. From a speech by Hisham Nazer, the former Saudi Oil Minister, at the 60th Annual Meeting of the Independent Petroleum Association of America in San Antonio on 31 October, 1989. "reciprocal energy security was echoed in recent years by the Saudis and the articipants in the International Energy forum.
13. The Government of India released *India Hydrocarbon Vision-2025* in 2000.
14. For discussion on the impact of contradictions of US foreign, environmental ,economic, and energy policies on the world oil market see Alhajji (1999).
15. For more information see "Why Haven't High oil Prices Affected India's Economic Growth" a presentation by A.F. Alhajji at *www.oilcorner.com*
16. For information on the relations between India, the Gulf region, and other oil and gas producers and the influence of oil and gas on India's foreign policy, see Ahmad (2005); Khan (2004), and Pant (2002).
17. For more information on the link between petroleum imports and national security see Lichtblau, 1994.
18. "India Plans to partner with Saudi Aramco to increase storage capacity", *India Daily*, January 19, 2005.

References

Ahmad, Talmiz (2005). "Geopolitics of Oil", Symposium on Oil, November, *http://www.india-seminar.com/2005/555.htm*

Alhajji, A.F. (2004). "The Failure of the Oil Weapon: Consumer Nationalism *vs.* Producer Symbolism", *Bridges*, Spring/Summer.

———. (1999). "US Energy Policy Contributed to Lower Oil Prices", *World Oil*, Vol. 220(11), November.

Asia Pacific Energy Research Centre (2002). "Energy Security Initiative: Emergency Oil Stocks as the Option to Respond to Oil Supply Disruptions", *APERC Background Report*, Tokyo, Japan.

Constantin, Christian (2005). "China's Conception of Energy Security: Sources and International Impacts", *Working Paper* No. 43, March, Centre of International Relations, University of British Colombia, Canada.

EIA (2006). *Monthly Energy Review*.

Government of India (2000). *India Hydrocarbon Vision-2005*, February.

International Monetary Fund (2005). *IMF World Economic Outlook, 2005*, Washington DC.

International Energy Agency (2001). "India-A Growing International Oil and Gas Player", March, *www.iea.org*.

———. (2001). "Curbing Energy Demand in India? Where to Start?" Findings from the IEA Feasibility Study, *www.iea.org*.

Khan, Javed Ahmad (2004). "India's Search for Natural Gas in the Persian Gulf." A Monograph. Academy of Third World Studies. Jamia Millia Islamia, New Delhi.

Kohl, Wilfrid (2004). "National Security and Energy", *Encyclopedia of Energy*, Elsevier, CA, Vol. 4, March.

Lichtblau, John H. (1994). "Oil Imports and National Security: Is There Still A Link?", *The Energy Journal*, Special Issues: The Changing World Petroleum Market.

Pant, Girijesh, (2002). *India's Energy Security: The Gulf Factor*, Gulf Studies Programme, Center for West Asian and African Studies, Jawaharlal Nehru University, New Delhi.

Ramachandra, T.V., Yves Loerincik, and B.V. Shruthi (2005). "Intra- and Inter-Country Energy Intensity Trends", *Journal of Energy and Development*, Vol. 31(1), Autumn.

Williams, James, and A.F. Alhajji (2003). "The Coming Energy Crisis?", *Oil and Gas Journal*, Vol. 101(5), February.

16

Asia's Energy Quest

Redefining the Security Dynamics of West Asia

GIRIJESH PANT

Energy has been critical factor in defining the security dynamics of West Asia. The large reserves of oil and gas at very low cost have created enormous stakes for energy consumers in the region. The growing dependence on the regional resources brought security into the ambit of energy trade, pushing it into the strategic domain. Thus, the leading consumers, in consultation with the regional regimes, worked out security architecture to ensure energy supplies. The script of West Asian security regime is today under revision. It could possibly be rewritten too. The reason being that after 9/11, the principal author has recognised the limitations of the *status quo* that it has been sponsoring in the region and has felt the necessity of change. As articulated by the Secretary of State, USA, Dr. Condoleezza Rice, the fundamental character of regimes matters more today' and West Asia is the region where, "the problems, emerging from the character of regimes, are more urgent than in any other."[1] That possibly explains the invasion of Iraq. It is a message to all friends and foes, not only in the region, but across the globe having stakes in the region.

Till recently, the energy stakes in the region were limited to USA, Europe and Japan. The US administration factored access to the Gulf energy to its national security and defined the doctrine of energy security in terms of access to energy at a reasonable price. Any attempt to intervene is seen as a threat to the American national interest, hence, provides justification for sending troops to capture the oil wells if required. In the Cold War context, America could push its doctrine with least resistance. Europe and Japan did not protest the American control over the West Asian energy even if that meant dependence. They even endorsed American conceptualisation of energy security doctrine and joined by institutionalising it in the form of IEA. Today, the global energy map and regional energy context has changed. The energy market is globalising and there many more consumers and suppliers. Besides, there is a strong demand in OPEC to shift oil transactions to the Euro. It is even argued that the reason for US invasion on Iraq was for, "preventing further OPEC momentum towards the Euro as an oil transaction currency standard."[2] Further, Europe and Japan are under no compulsions to accept the American position and they are demonstrating differences at least in nuances. The point made here is that today there are many diverse stakeholders whose energy frontiers have expanded to West Asia. They too depend critically on Gulf energy and would like to defend their interest. These stakeholders are also the ones emerging as principal players at the global level. Among the new stakeholders, Asia is the most prominent.

Today, Asia consumes much larger volume of Gulf energy than the West. In the global context, the Asians are contesting the unipolarity of the world polity. They are eloquently arguing for a multipolar world. This could be seen as a hindrance, if not a challenge, by the hegemonic superpower. As observed by Dr. Kissinger, "The rise of China—and of Asia—will, over the next decades, bring about a substantial re-ordering of the international system. The centre of gravity of world affairs is shifting from the Atlantic, where it was lodged for the past three centuries, to the Pacific. The most rapidly developing countries are in Asia, with a growing means to vindicate their perception of the national interest." It may be mentioned in passing, that the United States, consumes 25 per cent while China only 6 per cent of the world's production. The paper asserts that the American policy in West Asia to induce, if not impose, the 'democratic change' is unfolding new uncertainties in the region having direct security implications of consequence for Asian energy stakes. This provides a new dimension to America-Asia engagement. Ironically, while the former perceives the rise of Asian demand as a threat to its energy security, the latter would like energy supplies to be emancipated from American influence. Significantly, the suppliers themselves are not comfortable with the prevailing mode of energy security relationship, and might as well look for its restructuring with larger players in the field. This unfolds new kind of encounters between Asia and USA, America and the region and the region and Asia. Apparently, it will have a bearing on security dynamics in West Asia beyond Pax Americana.

West Asian Energy-scape and the Multiple Stakeholders

West Asian energy is moving from regional boundaries to global energy-scape. The imperatives of geoeconomics are impinging on the energy market with much larger number of sources of supplies and demand of different sizes and volumes enhancing mutual interdependence and sensitivities. On the supply side, the new players range from Russia, Central Asia to Angola, eroding the supremacy of OPEC. Similarly, the corporate clout of the seven sisters is contested by the presence of many new identities like Gazprom, CNNOC, ONGC-OVL, and Petronas, to mention a few. The fact that along with supply, the demand security has also become the central concern for the energy stakeholders demonstrates that the days of binary relations between producers and consumers are over. The national and the international hydrocarbon companies are engaged in collaborations of different dimensions blurring the past prejudices of resource sovereignty. The transnationalisation of security threat to energy supplies is necessitating a transnational response, making it imperative to redefine the energy security parameters beyond the adversarial relationship. Yet it cannot be ignored that the geopolitical concerns are impinging on the geoeconomic transition of the emerging global energy-scape. The present analysis is located in this phase of transition and confined to its regional dimension.

The West Asian energy-scape could be divided into four clusters: Saudi Arabia, UAE, Kuwait politically moderate but major oil players, Qatar politically ambitious building a niche for itself on the strength of being the leading gas exporter, Iran and Iraq richly endowed with oil and gas but politically constrained as at present, are more of potential spoilers, Egypt, Libya, Algeria, the incremental energy players having crucial political role in defining the security matrix of the region. Clearly in terms of resource endowment and the export capability the first five are the key players. The paper, thus, will be focusing mainly on these countries.

Asia consumes nearly 23 million barrels oil per day, which accounts for over 30 per cent of the global oil produced. It is projected to go up by 50 per cent by 2010 and more than double by 2020. GCC exports two-thirds of its oil to Asia. The rise in demand by Asian countries, China, Japan, India and South Korea for secure energy sources is going to recast the prevailing network of relationships, transforming regional relations in the next 10 years. While the GCC region is the major global supplier of oil and gas, its export pattern has undergone a significant shift. According to one study, the "Gulf oil exports to Western Europe peaked at 10.3mn b/d in 1973, representing 41 per cent of the area's total primary energy consumption, but by the year 2002, Gulf exports had fallen to 3.2mn b/d, representing only 10 per cent of Western Europe's energy consumption—a dramatic decline of 76 per cent in its dependence on Gulf oil. Over the same period Japan, a natural market for Middle East oil, reduced its energy dependence on Gulf oil from 63 per cent to 38 per cent—a drop of 57 per cent. The declining energy dependence of these two areas on Gulf and Middle East oil will inevitably continue."[3]

Table 16.1

Breakdown of 2002 Middle East Producers' Crude and Condensate Exports to Asia-Pacific (in per cent)

	Saudi Arabia	*Iran*	*UAE*	*Kuwait*	*Oman*	*Iraq*	*Qatar*
Australia	2	-	1.1	-	-	-	-
China	7	16	-	11	20	6	1
India	11	22.1	7.3	15.7	6	45	-
Indonesia	3	-	-	-	1	15	-
Japan	33	37.9	53.7	36.1	26	9	59
Pakistan	3	-	3.3		-	-	-
Philippines	4	2.5	3.9	-	1	-	3
Singapore	8	1.9	4.7	9.4	6	-	16
South Korea	21	10.3	15.8	15	14	-	15
Taiwan	5	9.7	1.6	11.6	7	19	-
Thailand	3	-	8.6	1	19	6	6
Total ('000 b/d)	**3,120**	**1,360**	**1,810**	**1,070**	**760**	**140**	**670**

Note: Table compiled by MEES from a number of pie charts in the report. Percentages may not add due to rounding.

Source: "Middle East Crude Oil Production and Exports", *MEES* Vol. XLVI(40), October, 2003.

With the Chinese market prospects rated fairly high, GCC countries are recognising their stakes in that market. *World Energy Outlook, 2002* has projected that by 2010, Chinese imports will be 5mn b/d in 2010 and over 10mn b/d in 2030.[4] What is going to be crucial is that the old players see the emergence up of the new players with apprehensions. The rise in demand by Asian countries such as

China, Japan, India and South Korea for secure energy sources is likely to redefine the energy relations between West Asia and countries from west and east in the coming decades. The energy relations between Saudi Arabia and the West are undergoing changes in a subtle way. After 9/11, American views and assessment of their energy-security relations with Saudi Arabia did come under stress. Though the two have made efforts to recover ground since then, it no more remains the same. Saudi Arabian energy trade is moving towards Asia. China, Japan, India and South Korea are the major customers, accounting for 60 per cent of its crude oil and even refined products. As pointed out by the Saudi Oil Minister, "Asia has become our number one customer. Today, we ship more than (4.5) million barrels of petroleum per day to Asia, or about 60 per cent of our exports. This is about 20 per cent of Asia's current petroleum consumption. We are fully committed to all our customers, and, in particular, Asia."[5] But the Kingdom also ranks among the first four-oil suppliers to the USA. Both the countries are keen to maintain and develop the strategic partnership.[6] In case of Kuwait too, 60 per cent oil goes to Asia, namely, Japan, India, Singapore, South Korea, Taiwan, and Thailand.[7] Japan is the largest consumer of oil from UAE accounting for 62 per cent of its exports and meets one eighth of need for gas.[8] Qatar exports almost all of its oil production to Asia, with Japan by far its largest customer. Though Japan and South Korea were the lead customers, India has emerged the principal taker of LNG from Qatar.[9] Iran exports around 2.7 million bbl/d, with major customers including Japan, China, South Korea, Taiwan, and Europe. Iran contains an estimated 940 trillion cubic feet (Tcf) in proven natural gas reserves—the world's second largest, next to Russia.[10] The stakes in Iranian market could be better appreciated by looking at the investment made by the Asians like China, India and Japan. Asian energy relations with West Asia are transcending the commercial mode. The restructuring of oil industry is said to have opened new investment possibilities like Saudi investment going to Korea, the Philippines, China and Japan.

"Such partnerships provide these countries with reliable, timely delivery of products that meet the requirements of their markets as well as other benefits. For instance, when Saudi Arabia invested in S-Oil Refining Company of South Korea 14 years ago, the company's capacity was about 180 thousand barrels per day. Since partnering with Saudi Aramco, its capacity stands today at more than 550 thousand barrels per day. In the Philippines, the joint venture, Petron, is performing better because of new programs made possible through Saudi investment. Petron is now running at about (180) thousand barrels per day compared to about 120 thousand barrels per day a few years ago. Last year Saudi Aramco signed an agreement with Sumitomo Chemical of Japan to up-grade and build a large refining and petrochemical complex on the west coast of Saudi Arabia. The value of this project is about 4.3 billion dollars. When finished, this refining and petrochemical complex is expected to be one of the most competitive in the world. Additionally, Saudi Aramco became a partner in Showa Shell Sekiyu Co. by buying about 15 per cent of Shell's stocks in this company. This makes Saudi Aramco the second largest shareholder in the company. Like all Saudi Aramco joint ventures this will ensure crude supply to Showa Shell at all times. We are certain that this deal will lead to strengthening Showa Shell's competitive position in the local market and increase its production capacity. Moreover, Saudi Aramco is a partner in a new multi billion dollar refining, marketing and petrochemical project in Fujian Province, in Southeast China, holding a 25 per cent stake with Sinopec and Exxon Mobil. We are also in negotiations with several a local companies in Asia for joint investment in refining and marketing sectors with the aim of increasing partnership and capacity abroad for demand security to ensure that our partners have reliable petroleum supplies. In indicating our stance to strengthen our ties with Asia, we will continue looking for partners across the continent for refining and marketing investments. We have the capacity to supply crude to our joint ventures throughout Asia, and let me assure you that supplies are and will remain secure. We believe that the relationship that binds Saudi Arabia and the countries of Asia is not one based on buyer and seller but rather mutually beneficial partners. Geography, good relationships, and the absence of political misunderstanding between our countries is helping in this matter, and we expect to see more Asian joint ventures over the next few years."[11]

Asia's trade with the Gulf region has jumped up phenomenally. What is more important is the new recognition and assertion of Asian identity of interest between the consumers and exporters from Asia as was demonstrated in the two meetings of Asian consumers and exporters organised in India last year.

The Asian quest for energy is perceived with apprehension by America. As observed by Dick Cheney in *Energy Task Force*, "The Gulf will be the primary focus of US international energy policy." Though the approach of America's West Asia policy has changed, its essence remains the same. America would like to have full control over the regional hydrocarbon resources though it could be a big question whether the changed regime would ensure it or not. Iraq is going to be the test case. Looking at the American assessment of terrorism being sourced in the Saudi system, America would be interested in moulding it drastically if not inducing any regime change. This could be possible only if it has control over Iraqi oil to shift its supplies from Saudi Arabia in case it is required. Moreover, it also argued if 2010 is going to be the peak year for oil demand, then America would like to have Iraqi oil before that.[12] Iraq's oil is needed by America to reduce its stakes in Saudi Arabia, to redefine its relationship for which it still remains not very sure as the recent hearing of the Senate Committee on the Judiciary on "Saudi Arabia: Friend or Foe in the War on Terror?" and The September 2005 report by the US Government Accountability Office, brings out the that the spread of Islamic extremism, rather than Al Qaeda, is the "pre-eminent threat facing the United States." American interest in the regional energy is also defined by its global policy where emerging China is counted as a principal factor.

China's increasing presence in the region though essentially emanates from its energy needs but has a bearing on the regional security architecture. It has the potential to contribute in reducing the dependence of the region on the West and provide it not the alternative but certainly a substantial leverage. It is observed that oil could be the principal source of tension between USA and China. "The U.S. and China are now the world's number one and two consumers of oil respectively, with China's need growing as rapidly as its economy is. This could lead to Sino-American confrontations over oil that could in the years ahead threaten our national security and global security unless each of our nations—two great nations—develop and employ new technologies that will reduce our dependence on oil" says senator Joseph Liberman.[13] Whether the two will push their interest to the point of collision could be matter of debate but the fact remains that the China factor in the region could undermine American plans in the region at least that is how it is seen by a section of American strategic community. The region has been responding to Chinese overtures positively. The scale and volume of China's engagement with the region is certainly big enough to reconfigurate the security parameters, not only in relation to the west, but even within Asia also. With increasing energy stakes in the region, China is building and consolidating strategic ties with the countries along the energy route including building infrastructure and capabilities like the blue water navy, having security consequences. Gwadar Port, southwest coast of Pakistan, near the mouth of the Persian Gulf and some 250 miles from the Strait of Hormuz, developed by China, is one such illustration which enables China "to ensure the security of its energy-related shipments along existing routes and could also monitor U.S. naval activity in the Persian Gulf, Indian activity in the Arabian Sea, and future U.S.-Indian maritime cooperation in the Indian Ocean."[14] The key point here is that as the Chinese energy frontiers expand in the region, America is perceiving it as an encroachment to its energy influence/space. The American senate's reaction to the Chinese attempt to purchase of Unocal is a pointer to the psyche guiding its foreign energy policy in the changed context.

China and the Gulf states are planning to expand their bilateral trade over the next five years to 100 billion dollars from the present level of nearly 40 billion dollars. It is observed that, "the Chinese partnerships with major producers such as Saudi Arabia and Iran have increased exponentially. Whereas in 1994, Iran accounted for just one per cent of China's total imports, less than a decade later, Beijing purchased $ 2 billion (US) of oil from Tehran, representing more than 15 per cent of its total 2002 oil imports. Beijing and Tehran has signed $ 100 billion, 'the deal of century' enabling annual export of 10 million tons of Iranian liquefied natural gas (LNG) for a 25-year period, as well as the participation, by China's State Oil Company, in exploration and drilling, petrochemical and gas industries, pipelines, services and the like. With Saudi Arabia, the Chinese are building strategic partnership. The 1999 visit of then–Chinese president Jiang Zemin was the defining event of their relationship. The agreement on 'strategic oil partnership' provided opening of the domestic oil and gas market to China. In return, China agreed to open its downstream sector (refining products from crude oil for marketing to end users) to Saudi Arabia. This basic bargain was positive for Beijing in both of its dimensions. China quickly expanded its imports of Saudi oil. Even though Chinese refining capability was not well suited to heavier Saudi crudes, the Saudis shifted some of their lighter crudes to the Chinese market from other customers with more developed refining infrastructures. By 2002, the kingdom had become China's leading foreign supplier of crude oil. At the same time, China has attracted much needed Saudi investments in joint ventures to expand and upgrade Chinese refining capacity."[15]

China's increasing presence in West Asia has implications for other leading Asian consumers, namely Japan and India. Given the historical baggage, the rise of China is seen cautiously by Japan. On the energy theatre the two are engaged in competitive relationship as demonstrated in the case of their pipeline project with Russia. In West Asia too, the Chinese reach is seen with concern by Japan. The failure of Japan's Arabian Oil Co. (AOC) to renew its contract with Saudi Arabia in the divided zone was a big setback, the impact of which was to some extent moderated by a contract with Kuwait. According to a study by the Centre for Safety and Security Research (CSSR) of Japan, there could be two possible scenarios of the Chinese's energy policy having a bearing for Japan. "The first scenario is a battle over energy sources. It assumes that if China reinforces its procurement of energy without taking cost efficiency into consideration, the world will be plunged into a situation in which each country competes for oil by ignoring international market mechanisms. As a result, political tension between the two countries over resources in the East China Sea will mount. The second scenario assumes the isolation of Japan. If China succeeds in concluding free-trade agreements (FTAs) with Southeast Asian countries, their reliance on China will increase, leading to the isolation of Japan."[16] However, while concern of Chinese aspirations, Japan would not like to take a head-on policy. It would like to protect and promote its energy interest by deploying diplomatic resources. Japan has been successful in getting the pipeline deal clinched against the Chinese proposal with Russia. Apparently, while Japan would like to develop partnership with other consumers to moderate the Chinese energy campaign, its new energy policy has laid emphasis on reducing "oil-dependency rate to 40% or less by 2030 from the current 50%, promotion of nuclear energy, and securing of energy resources abroad through the fostering of more powerful energy companies."[17] It may be underlined here that, in the changing context, Japan too is pursuing its energy relations with the region distancing with America. "In early 2003, a consortium of three Japanese companies acquired a 20% stake in the development of the Soroush-Nowruz offshore field in the Persian Gulf, a reservoir thought to hold 1 billion barrels of oil. One year later, the Iranian Offshore Oil

Company awarded a $ 1.26 billion contract to Japan's JGC Corporation for the recovery of natural gas and natural gas liquids from Soroush-Nowruz and other offshore fields."[18]

India's Expanding Energy Frontiers

India is emerging as a leading and dynamic player in the energy market. It is likely to be amongst the first five economies of the world in the next three decades.[19] The acceleration in the pace of the economy is bound to change its energy profile. The new energy demand is likely to be so huge that it has to be met by both domestic and external sources. In either case, India has to engage itself at much wider scale with the world energy market. It is precisely for the growing needs that, today, it conducts energy transactions with as many as 47 countries.[20] These countries are far and near, from all the continents of Asia, Africa, Europe, South and North America.[21] Its energy trade is increasingly becoming dynamic. Though a net importer of hydrocarbons, India exports petro products. It is planning to gear up its oil refining capacity to earn from exports of petroleum products to meet its import bill. It has the refining capacity to process a range of crudes. India's has doubled its refining capacity in the last six years and is expected to rise to 140 million tonnes a year, or about 2.8 million bpd, by 2007 and to 3.6 million bpd by 2012. Consequently, its exports of petroleum products were 17.6 million tonnes in 2004-05. In terms of value, it went up from Rs. 16,397 crores in 2003-04 to Rs. 30,518 crores in 2004-05, registering a rise of 86 per cent.[22]

It is developing the energy infrastructure of pipelines to link Central Asia to South East Asia and possibly even to China. India is developing LNG terminals on its western and eastern coasts over the next five years as it expects that LNG imports may touch 25 mm tpy. With its huge market and geographical location, it has all the attributes to become the energy hub between West and East Asia. Clearly this makes the country susceptible to external factors and it needs to hedge against the probable and possible risks. While large volume makes India a vital market, but to play a role in the global or regional energy arena it has to conceive a comprehensive doctrine and strategy of foreign energy policy. It is obvious that security is vital but energy relations have to grow beyond that. Indian stakes in West Asian energy, particularly in the Gulf region, could be appreciated by the fact it provides more then 60 per cent of its imports. Saudi Arabia alone accounts for nearly 25 per cent of imports. Given the fact that

India will be meeting more than 80 per cent of its oil demand from imports, the region becomes strategically important for it. For Gulf suppliers too, with India constituting the leading component of the Asian demand, it has become the important destination. India is emerging as important market. No wonder, it has become a key market for Qatar. The point is that India—West Asia energy relations are also transforming as India is emerging along with China as the key architect of the Asian renaissance. This brings out the significance of Sino-Indian relations both at the global and in the context of regional energy-scape.

The prospects of Asian energy market critically hinge upon the dynamics of the evolving energy relations between China and India, the two largest energy markets of Asia. It is acknowledged that these two are going to be among six key global energy players to influence the world energy scene. Saudi Arabia, Russia and Iraq are identified on the supply side as 'Givers' to the market, and the U.S, China and India as major consumers or 'Takers.'[23] China and India have been locked in a competition despite the fact that there is asymmetry in their profile. China has the comparative advantage in its favour. In Angola, Indonesia, Kazakhstan and Myanmar, India lost due to Chinese leverage.[24] China is the second-largest energy consumer in the world after the United States and by 2020, its energy consumption will equal that of all the Organization for Economic Cooperation and Development (OECD) countries combined. India currently depends for more than 70 per cent on imports China needs only 30 per cent. China is better endowed with 18 billion barrels of reserves; Indian reserves are 5 billion barrels. China National Petroleum Corporation has invested more than $ 40 billions; Indian company ONGC has invested about $ 3.5 billion. Besides geography too is more favourable to China than India in accessing the regional energy. Thus, gradually the two have been sensing the advantage of converging their energy stakes.

A concept paper from the Indian government says, "The world's two largest oil importers, which compete and cooperate in different countries, would benefit more if they could pool their resources to maximise returns from the efforts being made by them separately at home and abroad—the Sino-Indian competition to secure equity oil merely increases the asking price—to the detriment of both sides. Consultations between our companies would ensure that we understand each other's requirements and attempt to develop opportunities, which would be beneficial to both of us."[25] The visit of the Indian petroleum minister to China in January 2006, was primarily aimed at to explore such possibilities. Reportedly the two have agreed to promote common understanding on their foreign energy policies.

It cannot be ignored that the pressure for energy is triggering the old differences into tensions and some times conflict as well. The Japanese move to initiate natural gas production in a disputed area of the East China Sea, led massive anti-Japanese protests in China on April 16, said to be "the worst outpouring of such animosities in over 30 years. Although leaders of both countries sought to diffuse the crisis by promising fresh efforts at reconciliation, neither side has backed off its claims to the offshore territories."[26] Indonesia and Malaysia have been contesting for claim over Ambalat, richly endowed with hydrocarbon. Both nations claim sovereignty over that area in Sulawest Sea near the border. In February 2005, Petronas of Malaysia awarded oil exploration rights to Shell but Indonesia awarded the rights to Unocal in 2004. Similar dispute is between Malaysia and Brunei. "The scattered islands of Southeast Asia have become hotbed of territorial busts up, especially in search for oil has moved offshore involving heavy weights China, Australia and most countries in between."[27]

However, it is important to notice that as buyers while China and India are crucially dependent on import of energy from the Asian sources, the same is not true for USA. In fact the view is gaining currency in America that high cost presence in the West Asian and central Asian region is not worth the returns. "So when the United States sends its military into the Gulf to protect all that oil, it's clearly America's blood—but is it really their oil?"[28] Yet it will be naive to assume that the America would be vacating the area for narrow cost calculations. In fact, in the emerging equation, America would like to use its energy leverage to define its relations with China possibly with India too. Energy has emerged as critical component of their bilateral relationship. It is well recognised that, in emerging Sino-US global diplomacy, India is seen as 'positive' factor by America. China too is aware that America could use its energy dependence to checkmate its growth. So their strategy is to underplay the conflict dimension and promote a policy of win-win at least with those countries, which could bleed it to meet the American challenge. China would therefore take a positively a framework which promotes cooperation with countries like India in safeguarding its energy interest. Yet it would be contesting vigorously for each and every oil well. This could place India and China on adversarial positions in some cases which could go against India. Chinese are also aware that India could be a proxy to America in restricting its interest in Asia. So their attitude could be more to play safe than build high stake ties. India has little choice but to engage China in a wider energy frame pan Asia to minimise the intensity of conflict, competition and the consequences flowing from divergence of their energy interests. At the same time, the Indian energy interest would demand creating a space in the region where Chinese overwhelming profile is perceived with apprehension. While Indian anxiety of Chinese aggressive oil diplomacy is understandable, it does not make much sense to follow China in each and every energy jaunt/hunt. Besides, the gains are likely to be more pronounced for either if the two could synergise their moves and postures. This becomes more urgent in view of growing Chinese necessarity to safe guard its oil trade, more than 50 per cent of Chinese oil imports come from the Persian Gulf via the Indian Ocean and nearly 80 per cent of all imported oil passes close to India's southern coast through the Strait of Malacca. This has made China to strengthen and enhance its naval presence, which could, if not appreciated by either party, lead to tension. The point made here is that due to expanding boundaries, hydrocarbon trade brings the two countries to face many new questions—with regional and global ramifications. Thus, for China, a better strategy would be to have energy engagement than competition. India too is aware as pointed by the Prime minister, "China is ahead of us in planning for its energy security—India can no longer be complacent." Chinese would not like that India, empowered by American support, toughens the competition.

The Emerging Dynamics of the Volatile Gulf

The oil rich Gulf region has been described as the region in turmoil or turbulence. Yet, the fact remains that despite episodes of war and conflicts, the region has been able to meet its obligations. Besides, it would not be fair to ascribe the burden of perceived uncertainty to the region alone. Extra regional factors too have their share to contribute. However, since 9/11, the regional dynamics is undergoing qualitative changes. The Gulf countries, "are finding their domestic and foreign policy interests increasingly at odds with U.S. strategic considerations, especially in the light of rising Arab-Israeli tensions and U.S.military intervention against Al Qaeda and other terrorist groups. A continued trend toward anti-Americanism could affect regional leaders' ability to cooperate with the U.S. in the energy area."[29] Consequently, the policy of preemptive regime change, beginning with Iraq, has escalated tension in the region affecting the prospects of security of energy flow from the region. Control over Iraqi resources will change the US-Gulf energy relations.

While Asia's increasing dependence on the Persian Gulf, make it more susceptible to regional dynamics, the American diversification of energy supplies, reducing reliance on the Gulf, allows it to take adventurist policies including an attack on Iran. It can be argued that, in the short run, the region will continue to pass through the phase of uncertainty and remain volatile due to external and internal divergence of understanding on the re-mapping of the region. The Persian Gulf countries are looking towards the East but it would be naïve to argue that they could cultivate their ties beyond the American influence. The initiatives from Saudi Arabia could be the pace setter. The visit of King Abdullah to Asia is a pointer to the shifting orientation of the region's approach to global affairs. Iran too will be critically defining the space for Asian consumers in the region. Similarly, Central Asia and Russian resources too will cast their spell in determining the dynamics of Asian competition. Further, with gas acquiring new proportion in energy trade, the imperatives of logistics, whether pipeline or LNG, would have moderating influence on competition because none of the consuming Asian countries could take the burden on its own. It may not be ruled out if the pressure of logistics pushes the competition to cooperative mode. This is also supported by the new concerns and sensitivities of the Gulf suppliers following the Western attitude towards the region since 9/11.

"The Persian Gulf states, particularly Iran, are looking to diversify their diplomatic and economic bases of support and see Asia, particularly China, as a potential counterweight to U.S. power and hegemony in the region. China is strengthening its ties in the region, particularly with Iran, but is also concerned about not antagonising the U. S., with whom it needs good relations and open markets for its exports. The Saudis, although traditionally dependent on the U.S. for security, are seeking to diversify their support and build stronger ties with China and Asia through diplomatic efforts as well as stronger energy link-ups, in the form of cross investments in refineries and natural gas. Japan is not taken seriously by Saudi Arabia or Iran because of its lack of 'hard power', i.e., military strength. But Iran sees Japan as an important source of low-cost investment. There is a significant potential for growing competition between China and India for Gulf oil and gas supplies, along with diplomatic influence."[30]

Mutual Asian Stakes and Interdependence: Towards Asian Energy Security

While forging energy alliance for bidding could be problematic, initiatives towards cooperative framework could always be explored which contributes to strengthening energy transport and transit security infrastructure. The strategic choice to go for cooperative approach can be appreciated by the fact that the energy relations, despite moving to markets, are still governed by geopolitical concerns. The global energy space continues to be dominated by America. Besides a leading market, America is the only country to have the capabilities to mobilise wherewithal to hedge against risks associated with energy transactions, be it safety of the trade route or checkmating terrorist attacks. Thus, the clout that USA enjoys could impinge upon both the Chinese and Indian interests. While the two Asians have been the beneficiaries of American security network, but in the changed context, where the pressure is building on the American government to reassess the loss-gain matrix, the two Asians have to take care of their interests. It can also be argued that, by depending on US security, the two Asian are incurring the risk of mortgaging autonomy of their energy policy, which might not necessarily be compatible to their national interests. US hostile reaction on India-Iran gas pipeline project is indicative of the possible divergence and potential friction between the two. American efforts to retain control over Asian energy oil supplies might not be seen as a threat but certainly does make India uncomfortable. American administration is looking for energy engagement with Asian consumers. "Bush Administration is 'very seriously worried' about the shifting energy balance and wants to 'fix it' before it goes out of hand. Washington is engaging New Delhi in an energy dialogue at levels far higher than ever before because it is concerned about growing consumption by India and China and what it could do to overall energy balance."[31]

"Rising absolute levels of Asia oil imports oil will triple the flow of oil eastward from the Gulf, across the Indian Ocean, through the Straits of Malacca and across the South China Sea to East Asia's major oil consumers. Nearly 15 per cent of the world's supertanker capacity already routinely transits Southeast Asian waters en route from the Gulf, with 1,100 fully-laden supertankers passing eastbound through the Straits of Malacca every year. More than 15 per cent of world cross-border trade, worth more than $600 bn annually, passes along that route, and the degree of importance attached by China and Japan and Korea to maintaining open sea-lanes between Northeast Asia and the Gulf is a vital geo-strategic consideration."[32]

The shift of hydrocarbon market from Atlantic to Asia means not only the enlargement of volume in the direction of trade flows but enhancement of Asian vulnerability and therefore need for a mechanism to address them. It would demand infrastructure to facilitate the incremental flows and safeguard mechanism against any interruption. With larger hydrocarbons coming from West Asia, the consumer countries of Asia have much large stakes in the stability of the supplier countries, in the West Asia. In other words, countries like India and China are going to be more vulnerable to any war, embargo or political change in the West Asia. It means the enhanced energy engagement also requires preparation to meet contingencies either individually or collectively. It may involve building reserves, security of sea-lanes and engagement with the region as troubleshooters. Certainly, like America, none of the Asian countries could talk of occupying the oil fields nor can Asian forces be deployed for rapid action. The western mode of thinking and mechanism on the subject, are neither feasible nor ideologically acceptable. A more serious dilemma for the Asian consumers could be posed by the prospects of the preemptive intervention for regime change in oil producing countries of West Asia leading to disruption and uncertainties. Iraq, a leading exporter, is an example and if Iran is put in a similar situation, the context could be dangerous for the Asian consumers. The American pressure for so-called democratic changes has escalated uncertainty, which in the short run could put the region's hydrocarbons on high risk. Equally alarming is the way that Chinese and Indian energy growth is being projected as the possible cause of tension with West—to be more precise, America. Asian endeavour to build energy ties in Central Asia, West Asia, Africa and Latin America is portrayed as new rivalry to be dealt with. The Asian countries have neither individual capacities and capabilities nor a system of collective response. In the absence of any attempt for collective endeavour, the Asian buyers including China and India are stepping into the high-risk phase.

Thus, the shift of global energy trade to Asia has its geostrategic implications. The global energy security architecture, geared to the needs of the western market, has to undergo change. So far USA was the custodian of energy security regime. It arrogated to itself the power to police the oil rich regions to ensure security and safety. It has been deploying its military forces to protect the strategic assets. The context has changed with Asia emerging as the major component of global energy market. The energy security architecture will have to be redesigned to the new context. In the context of maritime security, some Asian countries are exploring the feasibility of developing Regional Maritime Security Initiative which aims at, "to develop a partnership of willing regional nations with varying capabilities and capacities to identify, monitor, and intercept transnational maritime threats under existing international and domestic laws. This collective effort will empower each participating nation with the timely information and capabilities it needs to act against maritime threats in its own territorial seas. As always, each nation will have to decide for itself what response, if any, it will take in its own waters."[33] Its present focus is Strait of Malacca and the Singapore Strait. Clearly this needs strong naval forces but countries facing maritime terror does not have adequate navy. Indonesia, Malaysia, and Singapore's navies have agreed for Operation Malindo, which provide year-round anti-piracy and anti-terrorism protection in the Strait of Malacca from on 20th July 2004. Each participating navy is to provide 5 to 7 warships. For rapid communications provision was made to have a hotline particularly when a warship of one nation is in hot pursuit into waters of another partner."[34] In the emerging context it is becoming necessary that regional initiatives be attempted because externally dependent approach could have destabilising bearings. However, a search for regional capabilities should not necessarily be seen as hostility towards western and America. The point made here is that since China and India are the leading energy consumer they have to take initiative to develop the energy engagement where their maritime forces strengthened the maritime security by mutual cooperation.

The world is not running out of oil and gas. Yet energy is not secured. The uncertainties are unfolding the dimensions making the global energy regime fragile. It could be made robust provided it is rescued from the geopolitics of dominance. This in policy term means locating energy in multipolar world where price of energy is not escalated by the risk premium imposed by the politics of one. The energy trade by 2015 is expected to increase to two-thirds. That would mean more than 65 million barrels a day of oil traded across an international border before it reached the consumer. Security of such enormous task requires a common understanding and vision. Rise of Asian energy market could be a modest but defining beginning in that direction.

Notes and References

1. Rice, Condoleezza (2005). "The Promise of Democratic Peace: Why Promoting Freedom is the only Realistic Path to Security", *Washington Post*, Sunday, December 11, B07.

2. Clark, William *(2003). Revisited-The Real Reasons for the Upcoming War With Iraq: A Macroeconomic and Geostretegic Analysis of the Unspoken Truth*, January, *http://www.ratical.org/ratville/CAH/RRiraqWar.html.*

3. Chalabi, Fadhil (2004) "The World's Waning Energy Dependence On Gulf Oil", *MEES*, Vol. XLVII(18).

4. Ibid.

5. "Fundamentals of the Saudi Petroleum Relationship with Asia", H.E. Ali I. Al-Naimi, Minister of Petroleum and Mineral Resources Kingdom of Saudi Arabia, Roundtable Conference, New Delhi, January 6, 2005 with Asia.

6. *http://www.eia.doe.gov/emeu/cabs/saudi.html.*

7. *http://www.eia.doe.gov/emeu/cabs/kuwait.html.*

8. *http://www.government.ae/gov/en/biz/industry/oil.jsp.*

9. *http://www.eia.doe.gov/emeu/cabs/qatar.html.*

10. In February 2004, a Japanese consortium led by Inpex signed a final agreement on the $2 billion Azadegan oilfield development project. Azadegan was discovered in 1999, representing Iran's largest oil discovery in 30 years, and is located onshore in the southwestern province of Khuzestan, a few miles east of the border with Iraq. Reportedly, Azadegan contains proven crude oil reserves of 26 billion barrels, but the field is also considered to be geologically complex, making the oil more challenging and more expensive to extract. *http://www.eia.doe.gov/emeu/cabs/Iran/Oil.html.*

11. Fundamentals of the Saudi Petroleum relationship with Asia. Ali I. Al-Naimi Minister of Petroleum and Mineral Resources Kingdom of Saudi Arabia. Roundtable Conference, New Delhi, January 6, 2005.

12. Importance of Iraq could be gauged from the fact that out of the 80 known oilfields in the country; just 17 are currently in production. A further 63 underdeveloped fields have an estimated 75 billion barrels of oil, while some industry experts believe that between 100 billion and 200 billion barrels lie in unexplored fields.

13. "China-U.S. Energy Policies: A Choice of Cooperation or Collision", Remarks by Senator Joseph I. Lieberman (Rush Transcript; Federal News Service, Inc.).

14. Goodenough, Patrick (2005). "China Gains Strategic Foothold Near Persian Gulf", *CNSNews.com*, International Editor, February 23.

15. Leverett, Flynt and Jeffrey Bader (2005). "Managing China-U.S. Energy Competition in the Middle East", *The Washington Quarterly*, Vol. 29:1 pp.187–201, Winter 2005-06.

16. Asia Times Online: Japan News and Japanese Business and Economy, *http://www.atimes.com/.*

17. Masaki,Hisane (2006). "Japan's New Energy Strategy", *Asia Times Online:* Japan News and Japanese Business and Economy, January 13., *http://www.atimes.com/atimes/Japan/HA13Dh01.html.*

18. Klare, Michael T. *Iran, War and World Energy*, Japan Focus, *http://japanfocus.org/products/topdf/1785.*

19. According to CIA by 2015, combined GDPs of China, India and Japan would surpass that of the United States and the European Union at US$19.8 trillion, US$14 trillion and US$11.6 trillion respectively in 1998 dollars. By 2050, Goldman Sachs has projected, the largest economies in the world will be China, United States and India respectively, with Japan a distant fourth.

20. ONGC-Videsh Limited (OVL) and other National Oil Companies are already involved in 14 countries and the Indian presence is in the process of being established in a further 33 countries world-wide. Committed investment so far is in excess of $ 5 billion. During the

last one year, OVL acquired exploration blocks in Australia, Egypt, and Qatar; OVL & Oil India obtained interests in a block in Ivory Coast while the consortium of OIL–IOC acquired one exploration block in Libya. *http://pib.nic.in/release/release.asp?relid=9285.*

21. India is currently examining specific expressions of interest from Ecuador, Columbia, Trinidad and Tobago and Surinam in South America, Chad, Niger, Sao Tome-Principie, Gabon, Congo-Kinshasa and Tanzania in Africa, Saudi Arabia, Qatar, Uzbekistan, Turkey and Kazakhstan in Asia. Attempts are also being made to add Turkmenistan and Azerbaijan. Proposals being pursued include 15 E&P projects in 14 countries, eight refinery and pipeline projects in six countries, five marketing projects in five countries besides four gas projects in three countries including transit through Pakistan and Bangladesh. *http://www.financialexpress.com/fe_full_story.php?content_id=88937.*
22. Directorate General of Foreign Trade, *http://dgft.delhi.nic.in/.*
23. Verrastro, Frank A. (2005). "Emerging Global Energy Trends and Their Implications for U.S. Energy Needs, Security and Policy Choices" Center For Strategic And International Studies, February 3, *http://www.csis.org.*
24. Mukul, Jyoti (2005). "India, China Set for Big-rig Oil JV", New Delhi, *http://www.rediff.com/money/2005/feb/23oil.htm.*
25. Aiyar favours Hindi Chini bhai bhai, *http://in.news.yahoo.com/050513/48/5yjex.html.*
26. Michael T. Klare, "Asia Energizing New Wars", available on *http://atimes.com/atimes/Asian_Economy/GE18Dk01.html.*
27. *Petroleum Economists*, April 2005.
28. Barnett, Thomas P.M. (2004). "Whose Blood for Whose Oil?", *The Globalist*, Monday, June 14, *http://ww.theglobalist.com/DBWeb/StoryId.aspx?StoryId=3964.*
29. Post-September 11 Update Report, *Political, Economic, Social, Cultural and Religious Trends in the Middle East and Their Impact on Energy Supply, Security and Pricing*, The James A. Baker III Institute For Public Policy of Rice University.
30. "Asian Energy Security and Implications For The United States" co-sponsored by the National Bureau of Asian Research and the Pacific Northwest Center for Global Security, September 28-29, 2004.
31. *http://www.hindustantimes.com/2005/Apr/28/181_1339997,00050001.htm.*
32. Mehmet Ögütçü, "China's Energy Security: Geopolitical Implications for Asia and Beyond", *http://www.gasandoil.com/ogel/samples/freearticles/article_15.htm.*
33. Regional Maritime Security Initiative. *http://www.globalsecurity.org/military/ops/rmsi.htm.*
34. Ibid.

17

Geopolitics of Energy Resources in the Arabian Gulf

MOHAMED NOMAN GALAL

The Strategic Setting

The Arabian Gulf is of critical importance to the USA,[1] Europe and Asia. It is an important waterway for the world trade in oil, gas and commodities. So the free flow of commerce is a yardstick for the security of the Gulf. The security of the Gulf is part of regional security; to its west is Arab security, and to its east the security of the Indian subcontinent, consequently the security of this region is intertwined with the security of the world outside.

The countries in the Arabian Gulf contains two-thirds of the world's known oil reserves and one-third of the world's known natural gas reserves and they are mostly known around the world for their hydrocarbon resources. However, there are other important issues which are paid little attention, yet they have the potential to augment the role of the region in world politics in near future. The world hardly knows that it is the same area that has a huge capital to be invested. The revenues of the GCC countries from oil alone reached US $280 billion in 2005, while it is estimated that the wealth of individuals in the Gulf countries has reached US $1.5 trillion. This represents three per cent of the total wealth worldwide which is estimated at US $48 trillion.[2]

It is an area of low population density yet with a high rate of economic activity, particularly in the fields of construction and industrial projects such as petrochemicals and aluminum smelting plants. A new business-oriented class is emerging in the region and this is a hybrid class comprising of different nationals. The immigrant workforce in the gulf region has diverse social habits, customs, life styles and work ethics. They come from various nations in search of career opportunities and bring with them the habits and customs of their homelands. At the same time, they mix, voluntarily or involuntary, with others and thus, a relatively new type of business, industrial and service community is emerging gradually.

It must be noted that around 90 per cent of the labour force in UAE is from outside the UAE. Similarly in Bahrain two-thirds of the workforce is from abroad. The percentage of workforce from outside is 70 per cent in Saudi Arabia, 90 per cent in Qatar, 72 per cent in Kuwait, and 79 per cent in Oman. The members of this community (of foreign workforce) are an important source of foreign capital for the countries of their origin. The combined strength of such capital was estimated to be around US $27.2 billion in 2004. (See Tables 17.1 and 17.2)

Table 17.1

Immigrant Workers in the GCC Countries, 2002

	Percentage of the Labour Force	*Percentage of Total Population*
Bahrain	64.9	36
Kuwait	72.2	63
Oman	79.9	26
Qatar	90	72
Saudi Arabia	70	30
UAE	91	80

Source: *Gulf Affairs*, Quarterly Journal by GCSS, (in Arabic) Autumn 2005. p.55.

Table 17.2

Percentage of Immigrant Labour Remittances to GDP in GCC

Year	*GDP (Billion US$)*	*Remittances (in Per cent)*
1975	74	2
1985	155.3	3
1995	222.1	11
2004	407.9	7

Source: *Gulf Affairs*, Quarterly Journal by GCSS, (in Arabic) Autumn 2005. p.55.

It is, in a way, a capital drain from the Gulf countries, but, in another way, the immigrants, who total around 13 million people and represent 25 per cent of the population, play a very important role in the economic development and prosperity of the region. Needless to say, without the presence of this immigrant workforce, development in the GCC countries could not have been achieved. These immigrant workers also represent a huge consuming force, which not only has its impact on consumption but on the whole economic cycle as well.

The dilemma of the GCC countries, *vis-à-vis* these workers, is multifaceted. They cannot do without them as they occupy many jobs at various levels, from the maidservant at home, to the guard at the gate of the house, to talented experts in business, management, education, finance and technology. The other side of the coin is that the GCC countries are faced with a growing eligible national workforce, many of whom are unemployed for various reasons. The rate of unemployment, for example, in Saudi Arabia, according to official sources, is 9.7 per cent.[3] There is another dilemma that most of the expatriate workers, who are in unskilled or semi-skilled jobs, are underpaid and live in inhuman conditions, which contravene the human rights conventions adopted by the UN system. And now with the increasing presence of the human rights watchdog organisations, the GCC countries are under close scrutiny.

The fourth aspect of the dilemma is that an expatriate living in a European country or in America will become eligible for naturalisation after a certain period of time, however, in the GCC countries, this is nearly impossible. The Bahraini Minister of Labour stated that if this should happen it would be a catastrophe for the GCC countries. [4]

The second important feature of the political dynamics in the Gulf region is the huge presence of foreign military forces from powerful nations as well as regional powers. The Fifth Fleet of the US and British and French navies represent the overwhelming forces. The growing Indian naval presence in the Indian and Arabian seas also has its weight and influence felt *albeit* in an indirect way. Of course, the former Iraqi military power and the Iranian military power have also had their weight felt in the region and sometimes heavily. To highlight the implications of military concentration in the Gulf region, one only has to recall the three great Gulf Wars, i.e. the Iraq-Iran War of 1981-1987, the Invasion of Kuwait and its liberation in 1990-91 and the invasion of Iraq in March 2003. These wars caused huge damage, to the extent that UN reports described their impact on the infrastructure, on the environment and on human lives as being similar to a tsunami.

The third feature is related to energy. In this connection, one could highlight the following points:

The increasing dependency on energy imports by the world's consuming countries represents a vital insecurity issue. The USA consumes 25 per cent of total primary world energy demand. It is estimated that the USA's GDP in 2030 will be more than US $23 trillion and in Europe it will be US $20 trillion. Other countries are far behind in their GDP. And their critical dependence on energy supply from the Gulf is likely to grow in direct proportion to their economic growth.

The gap between the USA's oil supply and its consumption is expected to grow by 70 per cent by 2025. The gap between the USA's natural gas production and its consumption is expected to rise by 28 per cent by 2025. In 2003, the USA was the world's largest natural gas net importer, while Europe will be the world's largest natural gas net importer by 2020. Chinese oil demand was around 8 per cent of the world demand at the end of 2005 and 70 per cent of its oil is imported from outside.

India's vast growth rate in recent years has added stress to the world demand for energy. India will soon become the world's third largest importer of oil after the USA and China. Indian oil and gas diplomacy has affected its relations with the countries in the neighbourhood, i.e., Iran and Pakistan. To diversify its sources of energy, India has also concluded strategic cooperation in the field of energy with the USA.

The inter-regional gas trade in the world will grow three-fold by 2030, with most of the new exports coming from the Middle East, Africa and Russia. The risk of oil supply disruption will grow as trade through key maritime chokepoints and pipelines expand. This situation will lead large consumer countries to follow import diversification policies. Two cases which come to mind are the USA and China, and other countries could do the same. With regard to China see Table 17.3.

Table 17.3

China's Crude Oil Imports Top Three from Each Region
Percentage of Total Supply

	2003	*2004*
Middle East	50.9	45.4
Saudi Arabia	16.7	14.0
Oman	10.2	13.3
Iran	13.6	10.8
Africa	24.3	28.7
Angola	11.1	13.2
Sudan	6.9	4.7
Congo	3.7	3.9
Europe and Western Hemisphere	9.6	14.3
Russia	5.8	8.8
Norway	1.0	1.6
Brazil	0.1	1.3
Asia–Pacific	15.2	11.5
Vietnam	3.8	4.4
Indonesia	3.7	2.8
Malaysia	2.2	1.4

Source: Foreign Affairs, September/October 2005, p.28.

India's peaceful rise on the international scene, due to its fast economic growth and its huge population size, necessitates the existence of a peaceful and stable environment in the Middle East.

The Indian and Chinese approach is one of quiet diplomacy and they are keeping away from involvement in the Middle East's political problems. They are focusing on promoting cooperation with the GCC countries to cater for their economic development needs. India has managed to strengthen relations with Iran and improve its relations with Pakistan, so as to allow a gas pipeline from Iran to India *via* Pakistan.

Conflicts and Crises

The second part of the paper seeks to answer an important question, i.e., Why has the Middle East became a hotbed for major armed conflict, particularly in the last few years?[5] For any convincing answer to this question, one needs to dig deeper into the history, culture and civilisation, as well as the geopolitics of the region. An unconscious slip of the tongue, or perhaps not, by President George W. Bush after 9/11 betrayed the hidden dimension behind the war on terrorism, which is the concept of a clash of civilisations. He did this when he characterised the war against the so-called Islamic terrorism as a "crusade."[6]

Nevertheless, one is inclined to narrow the focus to three inter-related issues:

Oil ⟶ Capital-flow ⟶ Capital recycle

Arms sales ⟶ Economic development ⟶ Prosperity

If one looks at the economic situation in the Middle East, one is struck by a persisting trend, i.e., the import of capital from the consumer countries to producing ones through oil transfers is balanced out by counter flows or capital recycle to many developed consuming nations through arms sales. This capital plays a role in economic development and prosperity in the developed world. There is perception in the region that this is at the core of American foreign policy. In order to have an alibi to sell arms, the local conflicts in the region are allowed to fester and get worse. Many analysts believe that this is realism not conspiracy. It has been the policy of great powers throughout history, and is especially true at the current time.

In the modern world, the concept of security (in the region as else where) revolves around two main issues: national survival and national security. A country like Kuwait, faced with Iraqi invasion in August 1990, had to spend whatever resources it had to regain its sovereignty. When other GCC countries are faced with serious threats to their national security, they also have to do the same. Casting a look at defence expenditure in Arab countries could bear witness to this conclusion.

Table 17.4

Defence Expenditure of Countries Directly Involved in the Arab-Israeli Conflict (US $ billion)

Country	*Egypt*	*Syria*	*Jordan*	*Lebanon*	*Israel*
Defense Spending	2.5	1.8	0.49	0.56	7
Year	1999	2000	1999	1999	2000

Source: Data complied from various official and media sources.

Table 17.5

Defence Expenditure of the GCC Countries and their Neighbours (US $ billion)

	GCC								
Country	*Saudi Arabia*	*Kuwait*	*UAE*	*Qatar*	*Oman*	*Bahrain*	*Iran*	*Yemen*	*Iraq*
US $ in billions	18.2	2.6	2.1	0.12	1.75	0.21	7.5	0.27	1.4
Year	2000	2000	2000	2000	2000	1999	2000	1999	1999

Source: Same as Table 17.4.

One needs to highlight certain aspects of the defence expenditure based on the data in the above tables. Firstly, the Arab countries directly involved in the Arab-Israeli conflict together spend about US $5.3 billion on defence. Secondly, the GCC countries spend about US $30 billion while Saudi Arabia spends US $18.7 billion alone. Thirdly, the total defence expenditure for all the Arab countries has reached US $44 billion, with the GCC countries' spending representing 67 per cent of the total and Saudi alone representing 40 per cent of it. Fourthly, in spite of all this expenditure, the end result is that the region remains insecure and every country in the region remains a soft target for the high and the mighty. Their combined defence posture also cannot be taken seriously if one considers the situation with regard to Iraq or the threat of attack against Iran.

Why such insecurity is rooted in the security dynamics of the region? The answer is twofold. There is no real Arab or regional national security concept; consequently each country develops its own national strategy for defence, and there is no Arab or regional agreement on the concept of the enemy. The letter from Mr. Amre Moussa, Arab League Secretary General to the GCC Secretary General, which was read at the GCC Summit in Abu Dhabi in December 2005, hinted at the lack of unity and integration among the Arab countries as far as their security policies were concerned. Mr. Moussa called upon the GCC countries not to focus only on the Iranian nuclear programme and to ignore the well-advanced Israeli one.

Table 17.6

Military Balance in the Middle East (1999-2000)
(Israeli Calculations)

Military Power	*Israel*	*Syria*	*Egypt*	*Jordan*	*Saudi Arabia*	*Iraq*	*Libya*	*Iran*
Human forces in one thousand	621	512	699	754	185	1082	76	8686
Tanks	3895	2700	2785	872	865	2000	746	1520
Guns	1348	2575	3510	788	404	1950	220	2640
Missile Platforms	12	62	26	-	24	39	128	-
Aircraft	624	365	498	101	346	180	166	145
Choppers	289	295	224	68	160	366	127	242

Source: Same as Table 17.4.

The following observations flow from the Israeli estimates of military balance in the region. Firstly, the Israeli sources overestimate what the Arabs have and underestimate what Israel has. The reason is to pressurise the USA to supply more arms to Israel and to block any arms deals between the US and any Arab country, including even the closest ally of USA, Saudi Arabia. Secondly, although the Arab countries in the region have an advantage over Israel in the field of arms and armaments, a unified Arab military or political decision-making system is nonexistent. Each Arab country takes an independent stand and consequently one cannot be seen to be acting in concert with or as part of a unified Arab military force. Thirdly, the quality of training and preparedness of the armed forces in Arab countries is far too inferior to Israel.

The Interplay of Geopolitics and Economics

The geopolitics and economics of oil and natural gas has led to economic and social transformation in the region and ensured a continuous generation of additional capital for the countries in the region. The oil price hikes, either after 1973 October War or since 2002, have generated a huge amount of revenue in most of the oil producing countries of the region. It is estimated that in the year 2006 oil revenues in the Gulf region will reach US $305 billion, a fivefold rise over 1998.[7] It is interesting to see how they spend this resource. For the oil producing Arab countries, particularly the GCC countries, there are three main ways in which they use their money. Firstly, they spend most of their money on construction, real estate, stock markets and industry. Consequently, gigantic airports, big malls, and petrochemical or aluminum complexes are built or are under construction. There is also a new trend to create new funds from the oil revenue for the coming generations. Kuwait has started this trend and allocated US $75 billion for such a fund during 2004.[8]

Secondly, due to the serious security concerns of the ruling regimes in most of the GCC countries, they stockpile arms and buy the latest military equipment from the West. In addition, they have alliances with major powers such as the USA, the UK and France. Of course, the USA has the lion's share in this connection. Thirdly, they extend a small helping hand financially to their fellow Arab countries or Third World countries. However, the amounts allocated for assistance to other countries are relatively small compared to the available capital of the GCC countries, as well as the needs of the developing countries. Nevertheless, one cannot blame the GCC countries for this situation as the matter depends on many factors, which need not be elaborated in this paper. It must be mentioned, however, that some of the recipient countries have very poor records of capital absorption, red-tape, and low standards of transparency and high rates of corruption. In this situation, no country, including the GCC countries, is ready to donate money or to give loans without adequate guarantees.

As about the large consuming countries such as USA, China, Japan, India and others, one has to highlight the concerns of the producing countries and their dilemma *vis-à-vis* the consumers. The USA and China, in particular, do not disclose information about their oil reserves. This leaves the producing countries in the dark about their future energy policies. This could also create havoc in the price of oil in future, due to uncertainty which leads to economic problems. For example, China's increased demand for oil in 2005 was 3.6 per cent only, thanks to the expansion in the use of coal and the increase in the use of hydroelectric power due to a good rainy season.

At the moment, the USA is also not ready to construct new refineries or to disclose its reserves.[9] The GCC countries spent US $30 billion on investment in the oil and natural gas industry in 2005. It is expected that the investment will be around US $44 billion in 2006. Needless to say, such a huge investment in the field of oil and natural gas necessitates certain assurances from the side of the consumer countries. The visit of King Abdulla of Saudi Arabia to

China and India in January 2006, was an important event that helped to consolidate relations between the major exporter and major importer countries.

At another level, there has been a decrease in oil production and of reserves in a number of producing countries such as the UK, Norway, and Indonesia. The latter is about to become a net importer, while the first two will also become major importers in a few years. This is bound to increase the pressure on the oil producing countries, and may change the energy situation around the world. The instability in Iraq, the tense situation in Venezuela, the stand off between Iran and the West, and the precarious situation in some African oil producing countries also add to the concerns of energy security analysts worldwide.

There could be a crisis in leadership in the region as there is an unclear situation in major GCC countries. The top leaders in Kuwait and Saudi Arabia are ageing and the leadership in UAE is still new, so the future policies of these countries could be described at least as uncertain. Saudi Arabia has cut a deal to buy military planes from the UK worth around US $17 billion and other military equipment that could cost US $70 billion. [10]

Many oil consuming countries in the world are trying to create long-term alliances with the producing countries, either by purchasing contracts or through investment in various aspects of the oil and gas industries. China is very active in this field. India is also trying its best to get closer to the energy-rich countries of the region. Some of the western countries are also seeking to invest in these countries. For example the Netherlands and Algeria concluded an agreement on January 17, 2006 for cooperation in the field of energy and avoidance of double taxation.

Conclusion

One may conclude this paper with the following thoughts: Sustained high oil prices have transformed the relationship between major oil producing and oil consuming countries. This has also sparked an interest in exploration ventures in previously low-profit regions such as West Africa and Central Asia. The main questions today are: what are the main trends in exploration and transport over the coming years? How could investments in various sources of energy particularly oil, gas, nuclear, wind and solar energy transform the geopolitical landscape? The answers to these questions are not easy to find at this juncture. However, it is quite safe to predict that oil and natural gas in the Middle East will remain the main source of energy worldwide at least for the coming 25 years. Thus, the big power game in the Middle East will continue.

Non-traditional threats have started agitating the minds of security analysts in the region today. Such threats include threats from terrorist acts, or from piracy, from civil wars, international intervention, and social disruption such as strikes and political instability. In addition to all these risks, there is also the effect of natural disasters such as hurricanes, earthquakes and climate change, particularly in the oil producing regions.[11] This necessitates four kinds of reactions by governments: first emergency reserves, second diversification of resources, third rationalisation of consumption and fourthly exploring new sources of energy through R&D in renewable energy sources or alternative energy sources, such as sand oil, green oil from plants, and biofuels, etc.

The era of easy and cheap oil has ended. Even conservative estimates suggest that higher oil prices are here to stay. One could recall that three out of four global recessions in post-World War II history were triggered by oil price shocks. The world in general and industry in particular, have to become accustomed to new realities as far as energy is concerned. If one casts a look at the Agenda of the World Economic Forum in Davos which was convened on January 25-29, 2006, one will find that the first theme of the agenda was the emergence of China and India. Consequently many scholars and politicians are now considering that the gravity of economics and politics is shifting from Europe to Asia. This shift in the focus to Asia will be a milestone in economics and politics, and consequently the geopolitics of energy is in the process of being reshaped.

Notes

1. If one goes by the famous statement by Professor Michael Klore (an American professor of international security): "Control Over the Persian (Arab) Gulf Translates into Control Over Europe, Japan and China. It is having our Hand on the Spigot". One could raise the question here: Does this Explain American Policy in the Gulf Region?
2. *Al-Hayat*, Arabic daily newspaper, December 15, 2005.
3. Saudi Minister of Labour quoted in *Akhbar Al Khaleej*, Bahrain, Jan. 7, 2006.
4. *Alsharq-Alawsat*, Newspaper, Jan. 15, 2006.
5. See for example, *The Economist*, July 16, 2005, pp.38-39.
6. On September 16, 2001, just few days after 9/11, Bush said addressing the people of America: "This crusade, this war on terrorism is gonna take awhile. And the American people must be patient. I'm gonna be patient."
7. "The World in 2006", *The Economist*, 20th Edition, p.80.
8. *Alayam*, Bahrain January 7, 2006, Economic Section, p.8.
9. *Akhbar Alkhaleej*, December 13, 2005.
10. *The Guardian*, December 21, 2005.
11. See "The World in 2006", *The Economist*, 20th Edition, pp.127-128.

References

1. Abdel Salam, Mohamed (2002). "Armament issue in Arab Countries", *Shu'un Al-Awsat,* a Quarterly Journal in Arabic, Spring 2002, Center for Strategic Studies, Research and Documentation, Beirut, Lebanon, pp.45-63 and see also the complete file on armament in the Middle East in the same issue from pp.23-144.
2. Ahmed, Gaffar Karar (2005). "Relations between China and Arab Gulf Countries in the Oil Sector, *Journal of Strategic Studies,* Vol. 1(2), December, pp.30-58 (English Section) Bahrain Center for Studies and Research.
3. Cordesman, Anthony H. (2003). *Saudi Arabia Enters the 21st Century: The Military and International Security Dimension,* The Political, Foreign Policy, Economic and Energy Dimensions, Vol. 1, Greenwood Press.
4. Cordesman, Anthony H. (2003). "Weapons of Mass Destruction in the Middle East", Center for Strategic and International Studies", Washington DC.
5. Haskins, Mark S. (1998). "Iran and The Arabian Gulf: Threat Assessment and Response", Air Command and Staff College, Air University, Alabama, USA.
6. Hafez, Saad *et al.*, (1999). *Unemployment in the Arab World,* Center for Developing Countries Studies, Cairo University (in Arabic).
7. Klare, Michael (2001). *Resource Wars: The New landscape of Global Conflict*, Henry Holt and Company.
8. Pamir, Ahmet Necdet (2005). "Turkish Energy Security: Role of the Arabian Gulf and the Caspian and Black Sea Basins", A paper submitted to a seminar on *Turkey and the GCC,* held at the Emirates Center for Strategic Studies and Research, UAE, December.
9. Qing, Ye. "Sino-India Energy Cooperation in the Middle East", at *International Energy Security and Cooperation Conference.*
10. Roshandel, Jalil (1999). "Oil and Strategy, Beyond the US, UK Operation in the Persian Gulf", *The Iranian Journal of International Affairs,* Vol. XI(2), Summer, pp.216-258, Institute for Political and International Studies (IDIS), Tehran, Islamic Republic of Iran.
11. Youyong, Wang (2005). "Cooperation between China and GCC in the field of Energy", *Journal of Strategic Studies,* Vol.1(2), December, pp. 37-44, Bahrain Center for Studies and Research. in the Arabic Section.
12. Zweig, David and Bi Jianbai (2005). "China's Global Hunt for Energy" *Foreign Affairs,* Sept./Oct., pp.25-38.
13. *Agenda World Economic Forum for 2006,* Davos, website.
14. UNDP and Ministry of Planning and Development Cooperation (2005). *Iraq Living Conditions Survey 2004,* Vol. 11: Analytical Report, Baghdad.
15. Bahrain Center for Studies and Research (2005). International Energy Security and Cooperation Conference: *Towards a Sino-Arab Partnership in the Energy Sector,* Papers of a Seminar held in Shanghai, China, on 24-26 June 2004.
16. *Sipri Yearbook 2004: Armament, Disarmament and International Security*, Oxford University Press.
17. *Shoun Kholeejiya "Gulf Affairs"*, Quarterly Journal in Arabic published by Gulf Center for Strategic Studies, Autumn issue 2005, pp. 9-88.
18. The Economist Intelligence Unit, various reports, Monthly and Quarterly on various Gulf Countries. See for the assessment of GCC a number of articles by veteran Bahrain politician Dr. Mohamed Ali Fakhroo published in Akhabar Al-Khaleej during Dec. 2005 under the title *"Majlis Altaawin Al taih (The Bewildered GCC)"*.
19. *The Economist*, Annual Issue for 2006 (The World in 2006, 20th edition).
20. *The Economist*, Weekly Magazine, various numbers.
21. *The Guardian*, British Newspaper, various issues.
22. *The Military Balance 2004-2005*, (Tel Aviv University, Jaffee Center for Strategic Studies) 1999.
23. Various Arabic Newspapers.

Section IV

WMD and the Nuclear Dynamic in West Asia

18

Israel's Nuclear Weapons Policy

CAMERON S. BROWN

Open source estimates suggest that Israel is not only the sole nuclear power in Southwest Asia, but possesses one of the most robust nuclear programmes on the earth.[1] This paper explores Israel's rationale for developing and maintaining its nuclear weapons programme, and considers recent claims that Israel should give up its nuclear arms and become a member of the Middle Eastern nuclear-free zone. The paper begins by laying out Israel's strategic environment, including the conventional and non-conventional threats it has faced in the past and faces now. The paper then describes Israel's programme and the rationale given by Israeli policy makers regarding nuclear deterrence. Looking at other counter-proliferation strategies Israel has developed over the last decade, it considers whether there remains a need for Israel to maintain its nuclear option.

Nuclear Weapons and the Israeli Threat Perception

In order to understand Israel's rationale for developing and maintaining its nuclear weapons programme, it is imperative to consider the threats Israeli security planners have taken into consideration when formulating its nuclear strategy.[2]

Nuclear Weapons in Response to Conventional Military Threats

Firstly, there is the issue of conventional military threat. Since Israel declared its independence in 1948, the country has fought five major conventional wars, three of which involved fighting multiple Arab countries simultaneously. What has historically concerned Israel most is that the Arab side has always enjoyed vast numerical superiority in terms the size of their populations, armed forces, as well as critical equipment, such as the number of tanks and aircraft. In 1948, for instance, Israel's population of 650,000 was surrounded by an estimated 30 million Arabs in Egypt, Syria, Lebanon, Trans-

Jordan, and Palestine. As of May 15, 1948, Israel possessed not a single fighter plane (*vs.* 74 for the Arab side) and only 1 tank (*vs.* the 40 Arab tanks). As of today, Israel has about 7 million people, as opposed to over 100 million in the five surrounding Arab countries.[3] Israel's standing army totals about 186,000 soldiers, while those five surrounding countries have almost 1 million men on active duty. The Table 18.1 gives a sense of the quantitative challenge Israel faces.[4]

In order to overcome this quantitative disadvantage, Israel has relied on its qualitative advantage in order to win wars and deter potential aggressors. Though today it is taken for granted that such a qualitative advantage can be decisive in wartime, this has not always been the case historically. Israeli decision makers in the 1950s and 1960s could not be certain that a qualitatively higher calibre conventional force would always be sufficient to beat numerically superior adversaries. Neither, for that matter, did Israel have any reason to conclude that it would always maintain the qualitative advantage. In 1955, for instance, the erstwhile USSR decided to provide Egypt with first-rate weapons systems while Israel remained under a British, American, and French arms embargo that had been in effect since 1948.[5] Similarly, while Arab countries could generally count on some degree of international support and superpower patronage, Israel often found itself diplomatically orphaned, especially in clutch moments like in 1948 and 1967.

Consequently, nuclear weapons were originally viewed by Israel's decision makers as a way of ensuring the country's survival, no matter how numerically superior its adversaries' forces and no matter how diplomatically isolated Israel may become. Still, the country's decision makers did not take this decision lightly. Even though Israel had struck an agreement with the French government to supply it with a nuclear reactor following the 1956 Sinai War, the decision to pursue nuclear weapons was not made until late 1962 or early 1963.[6]

Nuclear Weapons in Response to Non-conventional Military Threats

Beyond the conventional threat, since the early 1960s, Israel has had to contend with a growing threat from weapons of mass destruction (WMD). The gravity of the situation for Israel lies in a combination of factors that play to Israel's disadvantage. The first factor is the country's small geographic size (just over 20,000 square kilometres) and its small number of highly valuable targets (e.g., three major seaports, one major civilian airport, and four central power

Table 18.1

Military Might: Israel vis-à-vis Neighbouring Countries

Country	*Standing Army*	*Reserves*	*Total*	*Tanks*	*Fighting Vehicles*	*Artillery*	*Combat Aircraft*	*Helicopters*	*Combat Vessels*
Egypt	450,000	254,000	704,000	~3,000	3,680	~3,530	505	~225	62
Syria	289,000	132,500	421,500	3,700	~5,060	2,990	451	225	16
Iran	~520,000	350,000	870,000	1,680	~1,570	~2,700	207	365	28
Jordan	100,700	60,000	160,700	~970	1,815	844	97	83	
Lebanon	61,400		61,400	280	1,235	~335			
PA	~45,000		~45,000						
Israel	186,500	445,000	631,500	3,700	7,710	1,348	518	205	15

Source: *CIA World Fact Book 2005.*

stations). This is then compounded by the country's extremely high population density, with two-thirds of its residents living within a 75 km radius in three major metropolitan areas. These factors would increase the temptation to actually use WMD, knowing that the consequences of their use could have a major strategic effect, possibly even ending the country's very existence. The country's small size might also encourage an adversary to try to destroy any potential Israeli second-strike capacity with a big enough initial surprise attack.

Secondly, in any conflict, Israel is heavily dependent on a system of calling up reserve soldiers to reinforce its small standing army. Should one of Israel's adversaries use a WMD in the first few hours of hostilities—and especially if it hits near a mobilisation centre—it would unquestionably impede any efficient mobilisation of the reserve forces. An attack on its air force bases could be similarly disabling.[7]

Lastly, almost every country that Israel might have to confront in war is believed to possess WMD of some sort; many are attempting to enhance their capabilities, and some have directly threatened Israel with their use. Before the war in Iraq, security planners believed the country to be within range of over 1,000 surface-to-surface missiles (SSMs), mostly Scuds that can be armed with non-conventional (mainly chemical) warheads.[8] Even worse, most of these missiles would be fired from mobile launchers, which (as the 1991 Gulf War demonstrated) are much more difficult to eliminate than fixed launchers.[9] While a few of the following estimates have been amended lately as new information has come to light, a consideration of what Israeli policy makers believed were the WMD threats the country faced is important in understanding why they chose to pursue the policies they did.

Starting closest to Israel's borders, Syria is considered to have an advanced chemical weapons programme (capable of producing sarin and VX nerve gas) and a small biological weapons programme. Syria can deliver these weapons *via* air-dropped bombs or Scud-B, -C, and apparently -D missiles—the last one having a range covering all of Israel's territory, even from Syria's northeast corner.[10] To Israel's south, Egypt was the first country in the region to obtain and use WMD (using mustard gas in Yemen in the 1960s), and it is still believed to possess a significant chemical weapons (CW) programme and is suspected of also possessing a small biological weapons (BW) programme, as well as Scud-B, Project T, and possibly even Scud-C missiles for delivery.[11]

Looking further afield, Libya and Saudi Arabia each have pursued small, but not insignificant WMD programmes. While it possesses anywhere from 40-60 CSS-2 medium-range missiles (with the longest range of any missile force in the region),[12] Saudi Arabia is not thought to possess nuclear, chemical, or biological weapons at present.[13] In the past few years, however, Israeli intelligence has claimed that the Saudis approached Pakistan to provide it with the know-how and technology to make a nuclear weapon—a claim supported by a Pakistani source.[14] Considering the discovery of the extent of Pakistani proliferation activities in January 2004, such reports should be seriously considered.

Before it admitted to the extent of its proliferation activities, most observers considered Libya's WMD programme tiny and rudimentary compared with those of any of the other major regional actors.[15] For this reason, few took seriously the Israeli Prime Minister Ariel Sharon's comments in October 2003 that "one would not be surprised if Libya would be the first Arab country [to] have nuclear weapons."[16] However, when Libya revealed everything in its possession two months later, Libya's progress towards a nuclear weapon was much more substantial than previously suspected, and did indeed fall in line with the prime minister's comment. Similarly, by the end of the decade, the Libyan programme, with cooperation from North Korea, was on its way to producing 50-100 long-range ballistic missiles capable of hitting Israel and most of Europe.[17] Nevertheless, while Libya did possess advanced designs for nuclear weapons and several undeclared sites whose purpose was to create nuclear weapons-grade material, the country was extremely dependent on foreign suppliers, and at the very earliest, was several years away from producing an indigenous nuclear device.[18]

Still, the threats that have concerned Israel most have come from Iraq and Iran. While today's *pundits* claim that Saddam Hussein's WMD programme was all a figment of the George W. Bush Administration's imagination, after the Gulf War in 1991, the situation was exactly the opposite: UNSCOM and International Atomic Energy Agency (IAEA) inspectors were surprised by exactly how far advanced the Iraqi programmes were. The Iraqi nuclear effort was especially noteworthy for its immense size, financial and human resources investment, multidirectional approach, and that despite all of the above, it was chiefly able to remain undetected by Western intelligence throughout the 1980s.[19] With this experience in mind, and with the knowledge of what technical progress the Iraqis had made, almost all Western assessments throughout the 1990s, believed that without intrusive inspections, Iraq could have a nuclear weapon within a few years; if it obtained highly enriched uranium from abroad, Iraq could go nuclear within less than a year.[20] Of course, Israeli planners also argued that Iraq would certainly have come to possess nuclear weapons decades earlier, had Israel not struck Iraq's Osiraq (Tammuz) nuclear reactor in June 1981.

In addition to its suspected nuclear programme, during the 1990s, Iraq admitted to having previously created about a dozen biological agents in vast quantities, most importantly 8,500 litres of anthrax.[21] Reportedly, Iraq even weaponised about 25 surface-to-surface missile warheads and 166 air bombs with BW agents.[22] This amount, as Richard Butler (the former head of UNSCOM) claimed, was enough to "blow away Tel Aviv."[23] In addition to mustard, sarin, soman, and tabun gas, Iraq's CW programme had also developed 3.5 tonnes of VX nerve agent (of which a single droplet can be lethal within minutes), beginning as early as 1985.[24]

In April 1990, not only did Iraq test-flight its Scud missile with a chemical warhead, but Saddam gave a speech where he admitted to possessing binary chemical weapons and threatening that, "By God, we will make fire eat up half of Israel if it tries [anything] against Iraq."[25] Having used chemical weapons without hesitation in his war against Iran and against the Kurds in 1988, these were not considered empty threats. With all this in mind, it is easy to understand why Israelis panicked as Saddam fired 39 Scuds at the country only eight months after his overt threat, and why that crisis caused the country to begin investing enormous sums to counter such a threat in the future.[26]

While Israel can now consider the Iraqi threat to have been eliminated (at least for the foreseeable future), the Iranian threat has only worsened. Recent Iranian admissions to the IAEA have given credence to the claims of Iran's critics that it has been developing a crash nuclear programme despite its NPT obligations; and indeed, Iran appears to be following the Iraqi model of diversifying its nuclear sites as well as its enrichment and reprocessing methods.[27] Recent Israeli intelligence assessments suggest that Iran could cross the so-called "point of no return" (i.e., when Iran would no longer need outside assistance to develop a nuclear weapon) at any moment; after which it is estimated that Iran would be able to deploy nuclear warheads anywhere from 18 months to several years.[28]

Even without Iran obtaining nuclear weapons, it is already considered to have one of the largest chemical weapon (CW) programmes in the Third World, with several thousand tonnes of sulphur mustard, phosgene, and cyanide agents in stock and a capacity to produce an additional 1,000 tonnes of these agents each year. Moreover, Iran has weaponised these agents, putting them into artillery mortars, rockets, aerial bombs, and possibly even Scud warheads.[29] Iran is also believed to possess a small biological weapon (BW) programme that may be capable of producing small quantities of agent.[30]

While Iran reasonably claimed it needed these weapons in order to deter future Iraqi WMD use, what makes Iran's WMD development so worrisome is its intensely resolute efforts to obtain missiles with ranges far beyond any target in Iraq. In addition to its fleet of Scuds, Iran has also been developing several medium range ballistic missiles, including the Shihab-3 (range 1300 km, 700 kg payload), Shihab-4 (estimated range 2000 km, 1000 kg payload), and Shihab-5 (estimated range 5,500 km), all of which have been developed with the aid of North Korean, Chinese, and Russian technology.[31] The Shihab-3 has been tested nearly a dozen times (though some tests were failures), and Israel believes that as of 2004, Iran has at least one to two dozen missiles that are operational.[32] In addition, it is suspected that Iran has recently managed to obtain 12 Ukrainian cruise missiles with a 3,000 km range. The missiles are apparently capable of carrying nuclear warheads.[33]

Israel's Counter-Proliferation Strategy

Given these multiple WMD threats Israel faces, it is important to realise that Israel's nuclear weapons programme is part of a much larger counter-proliferation strategy meant to counter these non-conventional threats.[34] The country's multi-layered strategy rests on four main pillars: non-proliferation, active defensive, passive (or civil) defence, and deterrence.[35]

The first pillar of Israel's counter-proliferation strategy is to deny its adversaries the opportunity to develop WMD in the first place, or, in other words, an active non-proliferation effort. This means intelligence gathering, sharing intelligence with other countries or agencies (e.g., UNSCOM or the IAEA) in order to increase diplomatic pressure on these countries or block certain sales or technology transfers, and when all else fails, pre-emptive attacks (e.g., the strike on Osiraq).

A second pillar of Israel's counter-proliferation strategy is referred to as active defensive. This strategy aims to thwart WMD from successfully hitting its intended target, in particular, by foiling the three delivery methods that are the most difficult to stop: ballistic missiles, unmanned aerial vehicles (UAVs), and terrorist attack. While this can include steps such as improved border control, active defence refers most often to anti-ballistic missile systems, such as the Patriot and the Arrow, which aim to knock down incoming ballistic missiles by shooting a missile at them. Another (though still underdeveloped) method of active defence is called as Boost Phase Intercept (BPI), which aims to destroy an incoming ballistic missile during its initial boost phase. In general, there are three types of BPI: 1) Airborne Interceptor, which fires high-speed interceptors launched from a fighter aircraft; 2) Airborne Laser, which uses a high-power laser carried aboard a converted 747-400F; or 3) a space-based laser.

A third pillar of Israel's counter-proliferation strategy is passive or civil defence, which refers to efforts to minimise the damage of a WMD attack. This strategy has included: distribution of gas masks and atropine to the general population; sealing rooms during times of crisis; changing Israeli building codes to ensure each new apartment included a room specially designed to withstand shrapnel blast and the use of a chemical or biological agent; deploying CBW detectors; stockpiling antibiotics and vaccines; and pre-vaccinating health officials and emergency personnel ('first responders').

While all of these strategies can be helpful, each has its limitations. Given that powerful countries like the United States have had only marginal success with their non-proliferation efforts, it comes as no surprise that Israel is at a severe disadvantage in this area. Besides the general difficulties of preventing other countries from developing WMD, Israel faces two unique challenges that other Western nations do not. First, because of Israel's general diplomatic isolation, it usually is unable to lead major diplomatic offensives and garner support for UN resolutions of its own. Second, it runs the risk of highlighting its own WMD programmes and bringing unwanted attention and criticism whenever its adversaries' arsenals are mentioned.

The biggest weakness of active defence is that most are not operational (especially BPI), and the only ones that are (the Arrow and Pac-3) have yet to prove themselves in the heat of battle. The only war that saw the extensive use of these systems was the 1990-1991 Gulf War, when the US attempted to give its allies some way to deal with the Iraqi missile threat, other than a counterstrike. Although at first the system, was considered a roaring success, with the US army claiming the system scored 45 hits during the 47 engagements over Saudi Arabia and Israel, the real success rate was later found to be somewhere between zero and twenty per cent.[36]

As for passive defence, its weakness is the most obvious: it is used when the country has already absorbed a WMD blow, which even if casualties are minimised, creates the most severe chaos and panic. Moreover, passive defence is only good for handling CBW—there is very little that can realistically be done to lessen the effects of nuclear or radiological weapons.

Israel's own WMD programme today, therefore, is essentially the final pillar in this comprehensive, multilayered counter-proliferation strategy. The point of its WMD programme is to deter hostile countries (like Iran) from launching a non-conventional attack on Israel. Deterrence can be defined as using the threat of severe retaliation to convince one's adversary that it would lose more than it would gain by undertaking a certain course of action. Strategies of conventional deterrence have been the cornerstone of Israeli policy since the state's establishment, and in general, have enjoyed a fair degree of success. Yitzhak Rabin once stated Israel's particular deterrence strategy in the following terms:

> If the worst an Arab leader can perceive as happening as a result of a war he initiates is that he will not achieve his goal, then this is insufficient deterrence on Israel's part. Rather, an Arab leader...must constantly bear in mind that, should he initiate war, his armed forces will be badly clobbered, along with sensitive targets causing disruption to the local population, and in a way that will endanger his regime. Otherwise, our deterrence will be minimal...
>
> What kind of army is required by Israel...to fulfill this formulation of deterrence? To be explicit: armed forces with the greatest offensive potential. For if attacked, we must be able to immediately transfer the war to enemy territory, to destroy as much of the aggressor's armed forces as possible, and to pose a credible strategic threat to that country's very regime...Again, the basic strategy of Israel is defensive, to prevent war, to deter from war. Yet, ironically, the best means for preserving the *status quo* is by possessing this ultimate offensive capability in reserve...[37]

When it came to forming a policy to deter Israel's enemies from using WMD, policy makers have followed a similar logic. Israel is thought to have anywhere from 100-400 nuclear weapons (possibly including thermonuclear weapons), mustard and nerve agents, and possibly even a biological weapons programme.[38]

Israel has also developed an extensive range of delivery systems. Its ballistic missiles include an estimated 50 Jericho-2 missiles (1,500 km range and 1,000 kg payload, on mobile launchers); 50 Jericho-1 missiles (500 km range and 500 kg payload); MGM-52 Lance missiles (130 km range and 450 kg payload); and unconfirmed reports of Jericho-3 programme under development using Shavit space launch vehicle technologies (range up to 4,800 km and 1000 kg payload). Israel also possesses the Delilah, Gabriel-4, and Harpoon cruise missiles (with ranges of 120-400 km), and several fighter jets with ranges exceeding 1,000 km—most notably the new F-15I, whose range is believed to be 3,500 km.[39] Over the past few years, there have also been contradictory reports in the press about whether the new Dolphin submarine has the capability to fire nuclear-tipped missiles.[40] If such reports are accurate, it would mean that Israel now possesses a nearly indefeatable second-strike capability and would mark a major improvement in its strategic security situation.

The other important aspect of Israel's deterrence strategy has been that it has largely left its capabilities and threats opaque. While frequently promising massive retaliation in the event of a WMD strike against it, Israel has never officially said what it has at its disposal to retaliate with, nor has it elaborated specifically how it would retaliate to a WMD strike (e.g., would it use nuclear or chemical weapons in retaliation for a CW strike). One illustration of a typical Israeli warning was given in July 1988 by then Defence Minister Rabin, as a result of the increasing use of CW by Iraq against Iran and its own Kurds:

> One of our fears is that the Arab world and its leaders might be deluded to believe that the lack of international reaction to the use of missiles and gases gives them some kind of legitimisation to use them. They know they should not be deluded to believe that, because it is a whole different ball game when it comes to us. If they are [so deluded], God forbid, they should know we will hit them back 100 times harder.[41]

If there is one serious problem with Israel's deterrent strategy, it is that Israel has failed to build reinforced silos for its missile fleet, and has instead sufficed with storing both its Jericho-2 missiles and nuclear weapons (both warheads and gravity bombs) in limestone caves that cannot be reinforced. Since the missile site covers an area smaller than 24 sq. km, it is possible that just a few nuclear-tipped missiles could neutralise Israel's missile threat and damage the nearby bunkers holding the air force's nuclear gravity bombs.[42] While Israel has most likely found this step unnecessary until today, with Iran on the verge of gaining a nuclear weapon, Israel may now have to reconsider that decision—that is, unless it is true that the Dolphin submarine can fire nuclear-tipped cruise missiles.

Israeli Nuclear Disarmament

Over the years, there have been some who have called upon Israel to give up its nuclear weapons and instead to work for the creation of a WMD-free zone.[43] As Israel's strategic environment has improved, this debate has re-intensified.[44] The most serious and developed of these arguments came from Ze'ev Ma'oz in an article published in *International Security* in 2003.[45]

Ma'oz argues that Israel's nuclear weapons failed to deter the 1973 Arab invasion and Saddam Hussein's Scud attacks in 1991; that they have had little impact on bringing Arab decision makers to make peace with Israel; and that they have fuelled a WMD arms race. Like others, Ma'oz suggests Israel should abandon its nuclear weapons in the framework of a new WMD-free zone.

The Historical Arguments

Without getting into excessive depth on the historical arguments, Maoz arguments regarding both 1973 and 1991 are narrow and unconvincing. The first issue, totally ignored by Ma'oz, is that nuclear weapons have more than just a deterrence function—they can also be used to defeat an enemy in war. Even if we were to accept Ma'oz's analysis that Israel's nuclear deterrence failed (which, as will be clear below, is a most dubious claim), the fact is that had Syria and Egypt continued to successfully advance immediately after the initial Israeli counterattacks had failed, the possibility existed that the only thing Israel could have done to prevent getting annihilated would have been to drop a nuclear weapon. That Israel is said to have armed its nuclear weapons in both 1967 (when Arab forces were not aware of its weapons capability) and 1973 shows that Israel's military planners did consider this scenario as within the realm of possibility. Ma'oz's contention reminds us that while in retrospect it is always clear who will win a war, this was not the case for those making decisions in real time in 1967 and 1973.[46]

In which case, the issue is whether or not Israel's nuclear weapons succeeded in deterring Arab armies in 1973. While Maoz is correct to assert that conventional deterrence did play the primary role in limiting Egyptian war planning, the fact that Egyptians eventually moved from their fortified positions in the western Sinai Peninsula shows that initial successes in breaching Israeli fortifications and repelling Israel's counter attacks did whet Egypt's appetite. Had Israel's follow-up counter attacks similarly failed, it is difficult to know for sure where Egypt's army would have stopped, and that demonstrates the extreme limitations of relying solely on conventional deterrence.

Moreover, while Egypt had intended to capture a small portion of the Sinai in order to restart negotiations, the same was not the case for Syrian planners, a fact that Ma'oz conveniently overlooks. In Patrick Seale's biography of Hafiz al-Asad (which Ma'oz quotes from), Seale makes it clear that Asad "genuinely believed the Arabs could snatch back and hold some if not all their lost land" and that "Syria's aim was the recovery of the Golan."[47] Here it is quite uncontested: Israel's conventional deterrence totally failed.

This said, Syria's war aims were limited and Syria did not consider trying to destroy Israel as it had tried to do only 25 years previously—why? No one could seriously claim it is because Syria had come to recognise the Jewish population's legitimate national rights. Rather, Syria had become rightly convinced that even if they should have overwhelming success on the battlefield (the sort of success they believed possible on the Golan), they could not truly threaten Israel's existence without provoking a nuclear response.[48]

Still, the most difficult period for Israel's deterrence doctrine unquestionably came with the 1990-91 Gulf War. Here Ma'oz argues that because Saddam attacked Israel with Scud missiles, its nuclear deterrent failed, because Israel had promised an overwhelming response to "any kind of attack" on Israel.[49] While most experts claim that Israeli deterrence (at least in combination with American deterrence) effectively prevented Hussein from equipping his Scuds with chemical warheads, as he realised that the price of such a strike might be a nuclear strike on Baghdad, Ma'oz rejects this contention. Ma'oz argues that because Hussein aimed the Scuds at the major cities, he was "obviously hoping for as many fatalities as possible."[50]

This claim is similarly flawed. First, a conventionally-armed Scud is not the way to cause maximum casualties. Quite the opposite: after

39 Scuds were fired at Israel, the total damage amounted to only two direct deaths, one person seriously wounded and nine moderately wounded, and 60 people hospitalised. To demonstrate the extent to which the conventionally loaded Scuds were no real threat, one could note that five times as many Israelis died indirectly from the Scud attacks as did from the actual hit (four from heart attacks and seven from suffocation due to improper use of their gas masks) and many more were wounded (226 people injected themselves with atrophine and 539 suffered from shock and anxiety). Even the total financial losses to property were far from staggering: roughly $40 million.[51]

Likewise, while Ma'oz claims that Saddam did not fire a chemical weapon at Israel because its civilians had gas masks and atropine (and so it was supposedly less likely than conventional warheads to cause significant harm), the Saudis were not nearly as well protected as the Israelis—and yet no chemical weapons were fired at Riyad either. Furthermore, as previously mentioned, the Iraqis had developed 3.5 tons of VX nerve agent, which requires not just gas masks to be protected, but a full body suit—which Israeli citizens were not provided. In other words, the Iraqis would have been fully capable of killing hundreds, if not thousands, of Israelis had they really tried.

A much more convincing explanation for Saddam's behaviour is that Saddam Hussein fired Scuds at Israel precisely because he was hoping that Israel would retaliate as promised, but at the same time, believed that Israel would not dare drop a nuclear warhead on Baghdad if a dozen civilians died from a conventionally loaded Scud. Hussein was following a strategy of limited escalation *vis-à-vis* Israel because he believed that Israeli involvement in the war would likely force the coalition to collapse—or at least force its Arab members to desert.[52] At the same time, while Saddam should have considered that Israel might retaliate for the conventional missile strike, it was highly unlikely that Israel's conventional retaliation would even be felt given the coalition's massive aerial bombardment, then underway.

Ma'oz also tries to point to the Israeli government's decision to pass out gas masks and atropine to its citizens and instruct them to enter sealed rooms as evidence that the policy-makers themselves had realised that deterrence had failed and that the "government believed that a chemical attack was imminent."[53] Such a statement is not only historically inaccurate, but shows how little Ma'oz understands the dilemma of sitting in a chair of responsibility. Israeli officials decided to pass out gas masks mainly due to public pressure and panic, and indeed, they delayed distributing the gas masks for several months precisely because they feared it might send the wrong signal (that a chemical attack was sufferable) and thus reduce Israel's deterrence. More importantly, however, is the fact is that it was more responsible to plan for any contingency rather than only best-case scenarios.

Similarly, that Israel began to seriously develop active and passive defences following the Gulf War does not "reflect a strategic admission that nuclear deterrence cannot be relied on to guarantee Israel's security."[54] Instead, it showed that Israel could not rely solely on its nuclear deterrent. In the same way, Israel's decision to employ passive defences in March 2003, does not prove that decision makers were sure that the Arrow would fail. Rather, a responsible leader must take all possible precautions for all possible scenarios. In more cynical terms, one would say that politicians are keenly aware of the need to cover themselves in case something goes wrong.

This said, the Gulf War did expose two unexpected problems with Israel's defence doctrine. First, Israel encountered a similar dilemma to that faced by America's doctrine of "massive retaliation": doomsday weapons cannot be used in retaliation for non-catastrophic blows. In other words, when Saddam struck with conventionally armed ballistic missiles, he upped the ante in a way that Israel could not legitimately respond with its WMD deterrent, which the international community would have considered overkill. Secondly, while normally Israel might have retaliated with a conventional military strike, the Gulf War also created a situation Israeli planners had not taken into account previously: not all conflicts will be solely between Israel and its adversary, but may include complex international alignments that could be greatly affected by any Israeli military action. In effect, there are times when all offensive operations are not politically-feasible options. This, then, was the primary motivation for Israel's focused drive to acquire an effective active defence option following the 1990-91 Gulf War.

Why Nuclear Rollback Would be a Bad Idea

Beyond the historical arguments, the proposition that Israel abandon its nuclear weapons is even more flawed. First and foremost, because non-proliferation regimes are only useful when all countries involved want to disarm. If nothing else, this is the main lesson one should draw from the recent experiences with Libya, Iran, and North Korea, as well as Iraq in the 1980s. In each case, countries violated their explicit signed promises to forfeit nuclear weapons ambitions, and even succeeded in pulling the wool over the eyes of IAEA inspectors by creating advanced covert programmes. With this in mind, it should be clear that if Israel gave up its nuclear weapons, there is no good guarantee that this would lead to similar permanent disarmament by Arab states, Iran, or Pakistan—even if they declared they would.

To the contrary: although Ma'oz claims that "Israel's nuclear policy paved the way for a massive arms race in the region,"[55] it is important to note that Israel was often not the primary reason why these countries developed or acquired WMD or missile delivery systems. In other words, it is difficult to claim that the alarming WMD proliferation in the region is all the result of a prisoners dilemma whereby Israel's actions have caused its adversaries to seek WMD. For example, when Egypt deployed CW, it did so in Yemen, not in the Sinai (and did so 'before' Israel even decided to pursue a nuclear programme). Iran's main motivation for developing WMD was originally its experience during the eight-year Iran-Iraq War, in which it was hit mercilessly by Iraqi chemical weapons. Today, Iran's primary rationale is its concern about the United States. As the Pentagon's own report concludes, "Iran recognizes that it cannot match U.S. military power and therefore seeks other asymmetric means to challenge the United States... [including] the acquisition and development of NBC [nuclear, biological, and chemical] weapons and missiles... which it views as a means to offset its own vulnerabilities and weaknesses."[56] Iraq and Saudi Arabia have pursued WMD in order to deter the more populous Iran. The problem is that even if the primary motivating spark was not Israel, it is all too convenient to add Israel onto these countries' lists of threats their WMD is meant to counter once they have made the decision to acquire these weapons. In fact, should a comprehensive Arab-Israeli settlement come along, it is highly doubtful that any country will roll back its WMD programmes, because distrust is the common denominator amongst countries in the Middle East. In other words, in a bad neighbourhood, giving up your gun does not lead to disarmament—only building trust and eliminating threat does.

This leads to may be the most important consideration, which is also the most obvious: Israel never wanted to wipe Syria or Iraq or Egypt off the map. Indeed, this is why the majority of Arab countries have not bothered to pursue nuclear programmes: even if there is no strategic balance: Arab citizens do not stay awake worrying that Israel is going to decide to annihilate them with their nuclear weapons. The opposite cannot be said for Israel's neighbours, who have frequently declared as much since 1948. While many governments are increasingly willing to acknowledge Israel, there is still no popular recognition of Israel's legitimate right to exist—even within the 1948 borders. In other words, those who argue for compromise generally do so because they say "there is no choice." The cornerstone of the pragmatists' argument is that the Arabs are not only incapable of destroying Israel now, but it is highly unlikely that with Israel's nuclear option that they ever will have that capability.

In short, the basic truth is that real nuclear disarmament will not come about until a comprehensive Arab-Israeli peace has been reached, and not before enough time has passed that Israelis become convinced the Arab world is truly ready to accept them. Of course, in such a case, few will care if Israel has nuclear weapons. Notice, for example, how disarmament treaties no longer monopolise the attention of Washington and Moscow. When states no longer hate each other, they no longer feel threatened by the other's possession of WMD (so long as it is well safeguarded). Consider, for example, America's total disinterest in Britain's possession of nuclear weapons. This logic is another reason why many Israelis are particularly unsettled by Arab demands for disarmament. If Arab countries really believed in peace, they would be less concerned that Israel is so well-armed.

Conclusion

Finally, it is worth noting that while Ze'ev Ma'oz claims that Israeli public opinion's support for its policy of nuclear ambiguity "has eroded over time," this claim is extremely misleading. While it is accurate that public support for ambiguity has dropped from 80 per cent in 1987 to 61 per cent in 2002,[57] this does not mean that Israelis are increasingly in favour of disarmament. Quite the contrary: in the 1998 poll cited by Ma'oz, 92 per cent of respondents agreed that Israel should develop nuclear weapons—up from 78 per cent in 1987.[58]

So why do so many Israelis agree that their country should continue to hold on to its nuclear weapons capability? It would seem that looking back on the past 40 years, most Israelis believe that possessing nuclear weapons has simply made Israel more secure. If asked, most Israelis would probably respond nearly the same way that then-US Secretary of Defense Donald Rumsfeld did when asked at a press conference about Israel's nuclear programme.

> Israel is a small state with a small population. It is a democracy and it exists in a neighbourhood where many—over a period of time—have opined from time to time that they'd prefer it not be there and they'd like it to be put in the sea. And Israel has opined that it would prefer not to get put in the sea, and as a result...it has arranged itself so it hasn't been put in the sea.[59]

Notes

1. By some counts, its programme may be the world's fourth largest, behind Russia, the United States, China, and France, putting it ahead of the UK, India, Pakistan, and North Korea. All of the analysis below is based on open source material.
2. This paper does not intend to go into great details as a detailed threat assessment would require a paper all its own.
3. *CIA World Factbook*, 2005.
4. Source of chart: Jaffee Center, *Middle East Military Balance*, 2004. *http://www.tau.ac.il/jcss/balance/glance.pdf.*
5. Ma'oz, Ze'ev (2003). "The Mixed Blessing of Israel's Nuclear policy," *International Security*, Vol.28(2), Fall, p.52.
6. Ibid. pp.45-6.
7. Steinberg, Gerald M. (1993). "Israeli Responses to the Threat of Chemical Warfare", *Armed Forces and Society*, Vol. 20(1), Fall, pp.85-101. *http://faculty.biu.ac.il/~steing/arms/chem.htm.*
8. Herzog, Brigadier General Michael (1999). "The Israeli Erception of Missile Defense", in Roger Weissinger-aylon and Anne D. Baylon, *Proceedings from the XVIth International Workshop on Political-Military Decision Making in the Atlantic Alliance*, Hungarian National Parliament, Budapest, Hungary, June 20-23, Center for Strategic Decision Research, Menlo Park, California. *http://www.csdr.org/99Book/1999_chap28.htm.*
9. While Iraq's fixed missile launchers were destroyed within the first few minutes of the 1991 Gulf War, the mobile launchers managed to elude coalition forces for most of the war. Wilkening, Dean A. (2000). "Ballistic-Missile Defence and Strategic Stability", *Adelphi Paper* Vol. 334, Oxford University Press, New York, p.88, Note 14.
10. Carnegie Analysis, "Summary of Syria's Chemical and Biological Weapons Programs", April 15, 2003. *http://www.ceip.org/files/nonprolif/templates/article.asp? NewsID=4671* and Wall, Robert and David Fulghum (2002). "Arrow Fielding Slows as Threat Increases," *Aviation Week and Space Technology*, June 24, p.85.
11. Centre for Nonproliferation Studies resource page on Egypt. *http://www.cns.miis.edu/research/wmdme/egypt.htm.*
12. While the CSS-2 has a long range, its liability is that it is fired from a fixed launch site, greatly increasing its vulnerability to a counterforce strike. Centre for Nonproliferation Studies resource page on Saudi Arabia. *http://www.cns.miis.edu/research/wmdme/saudi.htm.* Wilkening puts the maximum number of missiles at 120, see Wilkening, "Ballistic-Missile Defence and Strategic Stability," p.79.
13. Centre for Nonproliferation Studies resource page on Saudi Arabia. *http://www.cns.miis.edu/research/wmdme/saudi.htm.*
14. See *Jerusalem Post*, October 22, 2003; *Washington Times*, October 22, 2003; *Ha'aretz*, October 23, 2003; *New York Times*, January 4, 6, and February 21, 2004.
15. See, for example, the page on Libya's WMD programme at the Center for Nonproliferation Studies site *http://www.cns.miis.edu/research/wmdme/libya.htm.*
16. *Reuters*, October 14, 2003. The quotation is an indirect one given by an unidentified aid.
17. Wall, Robert and David Fulghum (2002). "Arrow Fielding Slows as Threat Increases", *Aviation Week & Space Technology*, June 24, p.85.
18. The most interesting aspect of the Libyan confession regarding its nuclear weapons programme was not that it was heavily dependent on foreign supply of know-how and material, but the degree to which this was the case. In terms of a weapon design, for instance, while Libya had purchased one from a foreign supplier, Libya apparently had "no national personnel competent to evaluate the (design) and would have asked the supplier for help in the event it had opted to take further steps to develop a nuclear weapon." See the IAEA Director General's report on Libya, *Implementation of the NPT*

Safeguards Agreement of the Socialist People's Libyan Arab Jamahiriya, February 20, 2004, p.6 *http://www.fas.org/nuke/guide/libya/iaea0204.pdf*.

19. By the time Iraq invaded Kuwait during the summer of 1990, it had established an extensive nuclear programme, at a cost of approximately $8-11 billion, mploying approximately 7,000 scientists and 20,000 technicians. Shai Feldman, *Nuclear Weapons and Arms Control in the Middle East* (Cambridge: The MIT Press, 1997), p. 53; Shahram Chubin, *Eliminating Weapons of Mass Destruction: The Persian Gulf Case* (Washington, DC: Henry L. Stimson Center, 1997), p.15.
20. Especially if they already had a warhead design. Author's interview with Pentagon source (Alexandria, VA, January 21, 1998); Chubin, *Eliminating Weapons of Mass Destruction*, p. 16; and Feldman, *Nuclear Weapons and Arms Control*, p.53.
21. It also admitted producing botulin toxin (19,000 L), aflatoxin (2,200 L), gas gangrene, wheat cover smut, ricin, haemorrhagic conjunctive virus, and rotavirus. These numbers were used by various military and media sources throughout the years, and many given again by Hans Blix (former UNMOVIC chairman) in his report to the U.N. Security Council on Iraq, January 27, 2003. Some examples include: George J. Tenet, Testimony before the Senate Armed Services Committee, March 19, 2002. *http://www.cia.gov/cia/public_affairs/speeches/senate_select_hearing_03192002.html*, accessed October 2003; Department of Defense, Proliferation: Threat and Response, pp.38-42; "A Sick Inventory," *Economist*, pril 12, 1997; Arieh O'Sullivan, "Saddam, Again, Pushes the Envelope," *Jerusalem Post*, February 6, 1998.
22. Stockholm International Peace Research Institute (1997). *SIPRI Yearbook 1997: Armaments, Disarmaments and International Security*, Oxford University Press, New York, p.458; Harel, Amos (1998). "Butler: Saddam could destroy Tel Aviv," *Ha'aretz*, January 28.
23. "Report: Iraq Hiding Weapons," Associated Press, January 27, 1998.
24. Chubin, *Eliminating Weapons of Mass Destruction*, p. 17; and O'Sullivan, "Saddam, Again, Pushes the Envelope."
25. Baghdad Domestic Service in Arabic, broadcast 10:30 am, April 2, 1990; found in Ofra Bengio (ed.), *Saddam Speaks on the Gulf Crisis: A Collection of Documents* (Tel Aviv: Tel Aviv University, 1992), pp.55, 60.
26. *BIAF—Israel Aviation and Space Magazine* (Hebrew), No. 72 (Spring 1991), p.31, as quoted by Postol in Robert M. Stein & Theodore Postol, "Correspondence: Patriot Experience in the Gulf War," *International Security*, Vol. 17, No. 1 (Summer 1992), p. 232, note 19. On the number of Scuds that actually hit populated areas, see Rubin, Uzi (2003). "Meeting the 'Depth Threat' from Iraq: The Origins of Israel's Arrow System," *Jerusalem Issue Brief*, Vol. 2(19), March 5. *http://www.jcpa.org/brief/brief2-19.htm*.
27. See Breit, Marshall (2003). "Iran's Programs to Produce Plutonium and Enriched Uranium," Carnegie Non-Proliferation Project, updated December 1. *http://www.ceip.org/files/projects/npp/resources/Factsheets/iransnuclearprogram.htm*; as well as U.S. Undersecretary for Arms Control and International Security John R. Bolton, "The New World After Iraq: The Continuing Threat of Weapons of Mass Destruction," October 30, 2003. *http://www.state.gov/t/us/rm/25752.htm*.
28. *Ha'aretz*, December 1, 2005; August 16, 2005; October 13, 2003. U.S. intelligence services, however, recently said that Iran would not achieve nuclear weapons capability for another decade. See *The Washington Post*, August 2, 2005. An important article by Ronen Bergman showing commercial satellite photographs of Iran's nuclear sites also show that the country has stepped up its efforts to make a breakthrough and has set up dozens of anti-aircraft missile sites to protect these facilities. See Bergman, Ronen (2006). "The Iranians Accelerate", *Yediot Ahronot* (in the 7 Days supplement), Nov. 10.
29. *SIPRI Yearbook 1997*, p. 462; Leonard S. Spector, "Nuclear Proliferation in the Middle East: The Next Chapter Begins," in Efraim Karsh, Martin S. Navias, and Philip Sabin (eds.) *Non-Conventional-Weapons Proliferation in the Middle East* (New York: Oxford University Press, 1993); and Smith, R. Jeffrey (1996). "Chinese Firms Supply Iran With Gas Factories, US Says," *Washington Post*, March 8.

30. Department of Defense, *Proliferation: Threat and Response*, January 2001, p.36. *http://www.defenselink.mil/pubs/ptr20010110.pdf.*
31. Jehl, Douglas (2004). "Iran Reportedly Hides Work on a Longer-Range Missile," *New York Times*, December 2; Oren, Amir (2003). "New Iranian missile threat worries Israel," *Ha'aretz*, July 4; Central Intelligence Agency, "Unclassified Report to Congress on the Acquisition of Technology Relating to Weapons of Mass Destruction and Advanced Conventional Munitions," from January 1-June 30, 2003, released November 10, 2003. *http://www.ceip.org/files/nonprolif/templates/article.asp?NewsID=5621*; and Wilkening, Dean (2000). "Ballistic Missile Defence and Strategic Stability", *Adelphi Paper* 334, International Institute for Strategic Studies, London, May.
32. Wall, Robert and David Fulghum (2002). "Arrow Fielding Slows as Threat Increases," *Aviation Week & Space Technology*, June 24, p.85; Oren (2004). "New Iranian missile threat worries Israel;" and "Iran warns its missiles can hit anywhere in Israel," *Reuters*, August 15.
33. Alon, Gideon (2005). "MI chief: Int'l pressure delayed Iran's nuclear bid by two years," *Ha'aretz*, December 20.
34. In this paper, non-proliferation refers to the policies and efforts aimed at stopping other countries from acquiring WMD, and counter-proliferation refers to the policies enacted to oppose the potential use of WMD.
35. For more, see the author's "Israel and the WMD Threat: Lessons for Europe" *MERIA Journal*, Vol. 8(3), September 2004.
36. 158 Patriots were fired at 47 Scuds, and 39 other Scuds were not engaged as they were outside the Patriot perimeter or the impact was anticipated to occur in an area where damage would be minimal. On Patriot performance, see Postol, Theodore (1991). "Lessons of the Gulf War Experience with Patriot," *International Security*, Vol. 16(3), Winter 1991/2; Davis, Bob (1991). "Patriot Missile, High-Tech Hero in Gulf, Comes Under Attack as Less than Scud's Worst Enemy," *Wall Street Journal*, April 15; United States General Accounting Office, "Operation Desert Storm: Data does not Exist to Conclusively say How Well Patriot Performed," Report to Congressional Requesters, September 22, 1992; *http://www.fas.org/spp/starwars/gao/b250335.htm*; Postol, Theodore (1992). "Letter to Congressman John Conyers Jr., Chairman of the Committee on Government Relations", Dated September 8, 1992, published in *Inside the Army* October 5, p.10. *http://www.fas.org/spp/starwars/docops/pl920908.htm.*
37. Rabin, Yitzhak (1993). "Deterrence in an Israeli Security Context," in Aharon Klieman and Ariel Levite (eds.), *Deterrence in the Middle East: Where Theory and Practice Converge*, Jaffee Center for Strategic Studies, Tel Aviv, pp.9-10.
38. CNS, "Israel: Weapons of Mass Destruction Capabilities and Programs". *http://www.cns.miis.edu/research/wmdme/israel.htm*; SIPRI, *SIPRI Yearbook 2003*, p.627. There is additionally some speculation that Israel may also have non-strategic nuclear weapons as well, including nuclear artillery shells and landmines. See Ibid.
39. CNS, "Israel: Weapons of Mass Destruction apabilities and Programs". *http://www.cns.miis.edu/research/wmdme/israel.htm*; Stockholm International Peace Research Institute, *SIPRI Yearbook 2003: Armaments, Disarmament and International Security*, Oxford University Press, New York, p.627; on the Jericho-2, see Hough, Harold (1997). "Could Israel's nuclear assets survive a pre-emptive strike?" *Jane's Intelligence Review*, September 01. *http://www.janes.com/regional_news/africa_middle_east/news/jir/jir990901_1_n.shtml.*
40. See, for instance *Los Angeles Times*, October 11, 2003; *Ha'aretz*, October 12, 2003; and subsequent denials by former deputy minister of defense Efraim Sneh, *Ha'aretz*, October 13, 2003.
41. Rabin in FBIS-NEA, July 21, 1988, pp.28-9, cited by Eisenstadt, *Sword of the Arabs*, p. 54 and Steinberg, "Israeli Responses to Chemical Warfare."
42. Hough, "Could Israel's nuclear assets survive a pre-emptive strike?"
43. The Egyptian government has long led the Arab world in its demand for Israeli nuclear disarmament, though recently Libya has joined the chorus more loudly since it decided to abandon its WMD programmes. In addition, a group of marginal leftwing Israelis

have also called for their country to disarm, most notably Mordechai Vanunu. In December 2004, one prominent Vanunu supporter, Nobel Peace Prize laureate Mairead Corrigan Maguire, compared Israel's nuclear arsenal to Hitler's gas chambers and urged Israel to give up the weapons, saying "Nuclear weapons are only gas chambers perfected... and for a people who know what gas chambers are, how can you even think of building perfect gas chambers?" See "Nobel laureate compares Israeli nuclear arms to gas chambers," Associated Press, December 19, 2004.

44. For example, Amr Moussa, head of the Arab League, said that Israel should be put under pressure to follow Libya's decision to abandon weapons of mass destruction. "Arab states call on Israel to follow Libya's example," *Ha'aretz*, December 21, 2003.

45. Maoz, Zeev (2003). "The Mixed blessing of Israel's Nuclear Policy," *International Security*, Vol. 28(2), Fall, pp. 44-77. In December 2005, Maoz restated many of these arguments in an article in *Ha'aretz*, "Toward a Different Nuclear Policy", December 21, 2005.

46. Consider, for instance, that American officials were quite surprised that Israel's initial counter-offensive against Egypt failed.

47. Seale, Patrick (1989). *Asad of Syria: The Struggle for the Middle East*, University of California Press, Berkeley, pp.185, 197.

48. Ma'oz attempts to claim that this was not a consideration because Seale—for all his interviews of Asad—never mentions it in his book. Basing his claim on such flimsy evidence is highly problematic: because Asad clearly told Seale what he wanted him to know and no more, and because Seale's narrative in general often reads like a source approved for use in Syrian classrooms, one should not expect that it will include the whole truth and nothing but the truth. In this regard, admitting that Israel's nuclear deterrence is effective and made Syria realise that it must one day come to terms with Israel does not serve Syrian interests.

49. Maoz *op. cit.*, p.62.

50. Ibid.

51. *BIAF—Israel Aviation and Space Magazine* (Hebrew), No. 72, Spring 1991, p.31, as quoted by Postol in Robert M. Stein & Theodore Postol, "Correspondence: Patriot Experience in the Gulf War," *International Security*, Vol. 17(1), Summer 1992, p.232, note 19. On the number of Scuds that actually hit populated areas, see Rubin, Uzi (2003). "Meeting the 'Depth Threat' from Iraq: The Origins of Israel's Arrow System," *Jerusalem Issue Brief*, Vol. 2(19), March 5. *http://www.jcpa.org/brief/brief2-19.htm*.

52. Yair Evron, "Deterrence Experience in the Arab-Israeli Conflict," in Aharon Klieman and Ariel Levite (eds.), *Deterrence in the Middle East: Where Theory and Practice Converge* (Tel Aviv: Jaffee Center for Strategic Studies, 1993), p.114.

53. Maoz *op. cit.*, p.57.

54. Maoz *op. cit.*, p.58.

55. Maoz *op. cit.*, p.77.

56. *Proliferation: Threat and Response*, p.34. See also Lieutenant General Patrick M. Hughes, U.S. Army, Director, Defense Intelligence Agency, "Global Threats and Challenges: The Decades Ahead," Prepared Statement before the Senate Armed Services Committee, Washington, D.C., February 2, 1999. *http://www.defenselink.mil/speeches/1999/s19990202-hughes.html*.

57. Maoz *op. cit.*, pp.69-70.

58. Asher Arian, *Israeli Public Opinion on National Security 1998*, Memorandum No. 49, July 1998 (Tel Aviv: Jaffee Center for Strategic Studies, 1998), chapter five. *http://www.tau.ac.il/jcss/memoranda/memo49chp5.html*. The question was not asked in the 2002 poll, but given that the answer regarding nuclear ambiguity changed little between 1998 and 2002, it is unlikely there would be a large difference on this question either.

59. Transcript: Munich Conference on European Security Policy, Remarks as delivered by Secretary of Defense Donald H. Rumsfeld, Munich, Germany, Saturday, February 7, 2004. *http://www.defenselink.mil/speeches/2004/sp20040207-secdef0885.html*.

19

Iran and the Problems of Non-Proliferation in West Asia

ABDUSAMAT KHAYDAROV

The current developments in West Asia have driven this macro-region to the centre of world politics. This has a growing effect on overall situation in the world and in many respects, it is one of the main elements shaping the relationship among the US, Europe, and Russia in the new century. The events subsequent to 9/11 have contributed to the importance of this macro-region, which has become the focal point of study for many security related issues like: threat of nuclear weapons proliferation, growth of international terrorism and democratisation in West Asian countries.

The globalisation of terrorism, which occurred in the recent two decades as well as the possibilities of terrorist access to WMD, has made the issue of non-proliferation assume the highest priority in the contemporary world. This is the most important reason for international concern for the West Asian countries. Particularly, the possibility of Tehran's nuclear weapon possession has alarmed the West. The main issue in the agenda is the unavailability of unambiguous assurance from Teheran that Islamic Republic of Iran (IRI) will not become a new member of nuclear club, and that it will use its nuclear facilities only for peaceful purposes and within the framework of the non-proliferation treaty (NPT) of 1968.

The basic reasons that motivate Iran to possess nuclear weapons can be divided into two categories—internal and external. At the internal level, the Iranian nuclear effort is driven by the belief that nuclear weapon is the only effective means of providing national security. The internal reasons are: to increase the prestige of the country on the international scene, to transform the country into a bigger military power, capable of resisting external threats emanating from any of the Great Powers and Israel and to preserve the existing form of government in the country. The external reasons are: to compensate for the absence of an appropriate security blanket to guarantee the continuation of the present system of authority and the belief that development of nuclear programme is the only way to deter potential aggressors. The current war in Iraq further strengthened this opinion. Other external reasons are: to increase self-reliance on its own forces to ensure its national security, to ensure balance of forces in the region and become one of the leading powers in the Middle-East and to take advantage of the weaknesses of the existing nuclear regime.

Besides these reasons, the nuclear ambitions of Tehran are based on the following three factors:

1. *Geopolitical*: IRI plays an essential role in one of the most important regions of the planet, West Asia. Iran has an exceptional military-strategic position. It has direct access to the countries of the Middle East, Caucasus, Central Asia, and Caspian Basin. It also has direct access to the Persian and Oman Gulfs and the Indian Ocean. The advantageous geopolitical position of Iran and its engagements in a region with dangerously explosive issues (i.e., ethnic, religious, military, economic, refugees, drug smuggling, terrorism, and separatism) predetermines the IRI involvement in their solution. Moreover, Iran is one of the major suppliers of oil and its territory is used for the transportation of these products.
2. *Military-political*: Today, Iran, with a population of 70 million, has one of the largest armies in the world (about 800 thousand people) and certainly the largest army in the region. The presence of such a large army is indicative of Tehran's threat perceptions. From Teheran's point of view, it does not have a friendly neighbourhood. Teheran considers the military presence of the US in the region as a potential threat to its security. The US has military presence in Iraq, Afghanistan, Persian and Oman Gulfs and after the attack on Iraq, the Iranian sense of insecurity has grown further. In the neighbourhood, Turkey as a NATO member cannot be considered an ally for Iran. The kingdoms of the Persian Gulf and Saudi Arabia, which are overwhelmingly Sunni, are also viewed by Teheran with great deal of alertness. Moreover, Israel that plays significant role in the Middle-East has no diplomatic relations with Iran at all and considers Iran as the number one enemy in view of its alleged links with Palestinian terrorist organisations.

3. *National and Psychological:* IRI considers itself as the successor of one of the earliest civilisations and main powers of the ancient world; the Great Persian Empire. At the same time, during almost last six centuries, Iran became the centre of Shi'ia Islam. Historically, in clear contrast with the Sunni neighbourhoods, the Shi'i-Iranian psyche took shape during these years. The Iranian leadership now seeks to pursue and defend the Shi'ite interests. This national psychology has affected its international political behaviour as well. The search for nuclear power is part of the Iranian effort to consolidate its position as a pre-eminent power in the region and re-establish its old glory.

The Iranian quest for power has the ability to influence the course of regional as well as international politics. In spring 1990, Iran together with the former Soviet Union took the first step forward to re-launch the nuclear programme, which had begun with the help of the western countries before the Iranian Revolution in 1978-79. On 6 March 1990, the former Soviet Union and Iran signed their first protocol on the project, stipulating that Moscow would complete the Bushehr plant and build additional two VVER-440 reactors in Iran.[1] Later in 1993, the Russian Ministry of Atomic Energy (Minatom) and the Atomic Energy Organization of Iran (AEOI) signed a contract for the construction of two VVER-440 reactors at Bushehr.[2] Both contracts never took effect due to financial difficulties.

In January 1995, Iran secured a contract with the Russian Minatom to finish the reactors at Bushehr.[3] Prior to 2005, Iran and Russia did not have any agreement concerning the spent nuclear fuel, which caused concern in the international community. This issue was one of main subjects negotiated between Iran and Russia until the appropriate document was signed on February 2005 in Bushehr.[4]

The latest crisis came to the fore on January 9, 2005, when Iran decided to resume its uranium enrichment programme. Subsequently, three European countries, UK, France, and Germany announced the termination of their negotiations with Tehran and joined Washington's efforts to refer Iran to the UN Security Council. Russia's Foreign Ministry also urged Iran "to revise its decisions, return to the moratorium, and maintain full and transparent cooperation with the International Atomic Energy Agency."[5] It was speculated that Russia would abstain in case of any vote at the IAEA Board of Governors upon referral of the case. However, China's position on referral to the

UN is still not clear. As it seems, Iran has the last chance to adopt Russia's proposal "to have uranium enriched for its future nuclear power plants in other countries, particularly in Russia."[6] In retrospect, it appears that Russia overlooked two issues: firstly Russia did not have an agreement with Iran concerning the spent fuel for a long time and secondly, Russia did not pay serious attention to the uranium enrichment issue, which caused recent serious concern in the world.

However, Iran has been saying that as a signatory to the Nuclear Non-Proliferation Treaty (NPT) it is entitled to a nuclear power programme and says it has broken no rule. It says it will not build a bomb but wants the technology only to make fuel for civil nuclear power. In December 2005, El Baradei, the Chief of IAEA, made his point very clear that if Teheran resumed its uranium enrichment in other plants, as threatened, it would take only "a few months" to produce a nuclear bomb. However, he warned that any attempt to resolve the crisis by isolating Iran would prove counterproductive. If "Iran would retaliate," he said "you have to go back to the negotiating table to find the solution." The reports of the IAEA chief, since the beginning of the year, suggest that Iran has not suspended its enrichment-related activities but it has been providing the agency with access to nuclear materials and facilities. The effort to resolve the nuclear efforts of Iran through dialogue is still on. The UN Security Council has demanded in the meanwhile that Iran will have to suspend its enrichment activities while talks take place about its long-term plans.

The policy of Iran in the nuclear sphere, in recent years, has resulted in decelerating the negotiation process. Moreover, during this period of time, Iran has continued its efforts in creating the production base capable of developing nuclear weapons. Some Russian experts consider that "even if there is direct threat of invading Iran's territory, Iran will not stop and will continue work in that direction and will be a country with a complete nuclear cycle."[7]

Despite the fact that the final political decision on developing nuclear weapon in Iran is not adopted, it is most likely that there is a unity of opinion in the country that the scientific and technical production capability for processing of uranium in the country must be established. Such tactics, implemented by IRI government, obtain considerable public support in the country. Therefore, it should be expected that any government in Iran, regardless of its political orientation, will pursue the nuclear programme as one of the main national priorities.[8]

A Nuclear Iran!

If Iran acquires nuclear weapons, it will lead to the following consequences in the West Asian region:

- Nuclear status of Iran will contribute to further weakening of the nuclear non-proliferation regime and create additional obstacles for advancement of the idea of establishing a nuclear weapons free zone in the Middle East. It will also have negative impact on creating a nuclear weapons free zone in Central Asia.
- Some countries of region, (most likely of these), i.e., Egypt and Saudi Arabia, will also undertake measures for developing their own nuclear programmes.
- Similar nuclear weapon development programmes may be initiated outside the region, in various countries of Latin America and Far East.
- The potential for conflicts and volatility of security situation will increase in the Middle East and the Persian Gulf region.
- It may strengthen a nuclear arms race and give further boost to the proliferation 'black' market.
- It may have impact on oil prices as well. There may be a possible increase in the world oil prices because of the escalation of tensions in the Middle East and its impact on the politics of the states in the Persian Gulf region.

Furthermore, if Iran acquires complete nuclear cycle it can accelerate competition for influence in the region between the Arab states and Iran. At the same time, it can speed up the race for acquiring weapon-delivery systems in West Asian countries, particularly in the Middle East. For the time being, Iran possesses one of the largest missile inventories in the Middle East and has acquired complete missile systems and developed an infrastructure to build missiles indigenously. Iran recently flight-tested the 1,300 km-range Shehab-3, which is based on the North Korean Nodong.[9] The Shehab-3 is capable of reaching any of Middle East countries. At present, Iran's capabilities in missile production comply with its national doctrine of protection from regional threats. Acquiring of WMDs and their delivery systems by Iran may be seen as the way to strengthen national security on the part of other West Asian countries.

On the other hand, growth of Iran's military capabilities could lead to some strengthening of solidarity among Arab nations. It is most likely that six Arab states of the Persian Gulf will increase their military cooperation within the framework of Cooperation Council of the Arab States of the Persian Gulf. It is likely that Iraq will join GCC as a full member of that organisation in the near future.

Iran's nuclear programme is under strict watch by Israel, which considers Tehran as a direct threat to its existence. The statements of the Iranian leadership that "Israel should be wiped off from the map of the world"[10] has caused great concern in Israel and forced Israel to take up this issue with US, EU, and Russia. But the most important will be fixation of the US military presence in Persian Gulf zone to safeguard oil transportation. In this case, the regional system of security will be under strong US influence.

As it appears, the erosion of the non-proliferation regime in the world and in West Asia, in particular, has become possible mainly due to the following reasons:

- Shortcomings and limitations of NPT and the entire non-proliferation system.
- Lack of coordination in efforts among nuclear powers, notably, US and Russia, in retaining the non-proliferation regime.
- The nuclear tests by India and Pakistan.
- Ineffective use of economic potentials by EU, Japan, and other influential countries to restrain so-called threshold states from developing nuclear weapons.
- Weakening the role of the UN and its institutions in preservation of non-proliferation regime.
- Divergence of views of the West Asian states toward Iranian nuclear programme.

Moreover, the non-proliferation issue in West Asia depends on certain other factors, including the Middle East conflict. Lack of initiatives focused on reducing tension in that part of the world may inspire attempts to acquire WMD, or to deny signing the documents, which assure the WMD non-proliferation regime. For example, Egypt refuses to sign or ratify the Chemical Weapons Convention (CWC).[11]

Another factor is the attempt of some countries, multinational corporations, or individuals to sell WMD components to others. Recent developments in Pakistan (case of Abdul Qadeer Khan) demonstrate flaws in the system. In fact, applying pressure on supplier states or corporations has not worked sufficiently.

The lack of organisations or forums in West Asia negatively affects the global system of security and non-proliferation in that region. The Middle East Arms Control and Regional Security (ACRS) working group is the only body engaged in non-proliferation discussions in the area. All this reflects the lack of regional cooperation in Western Asia and encourages states in the region to search for ways of improving their security by seeking WMD capabilities.

The solution for the non-proliferation issues in general, and Iranian nuclear crisis, in particular, lies in coordination of efforts among permanent members of the UNSC, EU, and the group of influential countries such as Japan, India, and Pakistan. The effort should be focused on improvement of NPT and development of effective measures to keep the threshold countries within the framework of non-proliferation regime through negotiations. At the same time this approach seems difficult to achieve, mainly because of the existing political and ideological differences among the main actors, the US, Russia, and China.

Russia's recent proposal made during Euro-Asian Economic Cooperation Summit in St.Petersburg to create a system of international nuclear centres that would operate under IAEA supervision is a feasible way to resolve this issue. Such an approach would sustain the nuclear non-proliferation regime and would allow countries lacking the capability to process nuclear fuel to create peaceful nuclear energy programmes. This approach is one that could be supported by the US and Europe.

Notes and References

1. "'Iran's Nuke Plant Deal with Russia Raises Fears," *The Washington Times*, 10 January 1995, p.A13.
2. Romanenkova, Veronika (1995). "Itar-Tass", in FBIS-SOV-95-179, September 15.
3. Koch, Andrew and Jeanette Wolf (1998). "Iran's Nuclear Facilities: A Profile", Center for Nonproliferation Studies. *http://cns.miis.edu/pubs/reports/pdfs/iranrpt.pdf*
4. *http://www.minatom.ru/News/Main/view?id=l 5283*
5. *http://en.rian.ru/russia/20060113Z42993309.html*
6. Ibid.
7. *http://iran.ru/rus/int_news.php?news_id=:36659* HpaH 6yaeT oGjiaflaTb TexHOJiornaMH nojiHoro HflepHoro
8. See *http://www.nti.org/e_research/profiles/Iran/index.html*
9. Ibid.
10. Mr Ahmadinejad, who was elected President in 2005, while addressing a conference entitled "The World without Zionism" told some 3,000 students in Tehran in October 2005, quoting Ayatollah Khomeini: "As the Imam said, Israel must be wiped off the map".
11. *http://www.nti.org/e research/e3 24a.html*

20

US-Iran Relations

Confrontation or Rapprochement

BASANT REPSWAL

Never before had post-revolutionary Iran been so willing to help the United States as it had in Afghanistan and Iraq. The Iranian government strongly condemned the 9/11 attacks and there were spontaneous candlelight vigils in Tehran—probably the only ones in the entire Middle East.[1] Iran responded positively to co-option in the war on terror. Under auspices of the UN 'Geneva Contact Group' was created, a vehicle for United States and Iran to meet and discuss Afghanistan. Staging facilities for American aerial transport, search and rescue, transit of humanitarian cargos, persuasive motivation of the Northern Alliance and at times even elimination of Al Qaeda cadres was executed by Iran. Cooperation at such scale was beyond Khatami. The force calling the shots behind the scene was the Khamenei-Rafsanjani combine.[2]

Though the Geneva Group lived through 18 months, with initial American successes, the nurtured distrust and suspicions re-surfaced. Iran was accused of spreading undue influence in Afghanistan. Iran's potential for undertaking subversive activities in Iraq, i.e., allowing foreign fighters to cross its borders into Iraq, support to mahdi militia, collusion with Al Qaeda and opposition to Middle East peace process were bandied. Denied any formal reconstructive role in Iraq, Iran was accused of holding annual conclaves for terrorist leaders.

In American perception, through Hezbollah, Iran had direct influence in Croatia and mounted extensive operations in Bosnia. In Latin America, Hezbollah is believed to be expanding its substantial drug trafficking and smuggling activities to Argentina, Brazil and Paraguay. Cells have also been established in Columbia and Venezuela. In North Africa, Hezbollah presence is visible in Sudan and supports Abu Nidal organisation, the Algerian Armed Islamic Group and Egypt's Gama al-Islamiya. With the credentials of having been implicated in June 1995 assassination attempt on Mubarak, Iranian terror machine went on to gain foothold in South Africa, Zimbabwe, Tanzania, Uganda and Kenya. Accredited with close supervision of Hamas, Palestinian Islamic *Jehad* and Peoples Front for Liberation of Palestine, Tehran currently continuously abets in Hezbollah's efforts of acquiring WMD's. Accused of direct guidance to Abu Musab al Zarqawi, propping of Ansar ul Islam and of harbouring 300 senior al Qaeda activists, Iran and its Hezbollah far exceeds Al Qaeda. Without doubt it should receive the full attention of the GWOT the soonest.[3]

Iran as a supporter of terrorists and terrorism, is cited so often, regularly, and repetitively that it has begun to be accepted as the universal truth. Ominous and convincing it may seem, none, is verifiable. Hezbollah, over the years, has definitely evolved itself into an entity which imparts stability to the political divisions within Lebanon. Hamas, too wants to emulate Hezbollah's success, provided they are allowed to enter the political mainstream.

While Clinton administration was breaking some ground in engaging the Iranians, Saudi Arabia in the spring of 1999, produced evidence which implicated the Iranians in the Khobar Towers bombing, where 19 US soldiers had lost their lives. Evidence collected by Saudi Arabia, pointed towards some disgruntled Shia segments.[4] Al Khobar Bombings of 1995, were recently duplicated in Riyadh, on May 12, 2003, which were Al Qaeda machinations. With nothing in common ideologically, this incident was once again encashed to link Al Qaeda and the Islamic Republic of Iran.

Karine A incident, could be termed most specious. January 3, 2002 accusation involved Karine A, a ship purchased in Lebanon by Palestinians, was intercepted in the Red Sea with eighty crates of weapons, loaded at the Iranian freeport in the Persian Gulf, the island of Kish. In this international arena, everything could be procured for the right price. This does not demonstrate state participation. During this time, Iranian-US relations, in the post 9/11 environment, for better cooperation in Afghanistan were warming up. To counter this, the intercept by the Israelis in the Red Sea, was blown out of proportion, immediately followed by Bush's State of the Union address, where Iran was included in the "Axis of Evil", trio.[5] Khamenei fired back with commensurate vitriol, "the Islamic Republic is proud to be the target of the hate and anger of the world's greatest evil: we never seek to be praised by the American officials."[6]

Leaving aside 'dual containment' and economic sanctions, US unfailingly characterises Iran in the worst possible terms for US culture. Besides the hostage crisis of 1979-1981, Iran is unceasingly demonised over, support of terrorism, treatment of minorities, empowerment of women, lack of representative polity and development of nuclear weapons. Efforts at regime change with assistance of a listed terrorist group, *Mujahideen-e Khalq*, attempts at restoration of Reza Pahlavi or support of Mahmud Chehregani, leader of an Azerbaijan separatist movement calling for a federated Iran,[7] have not helped. On being surprised by Ahmedinejad, promptly an old photograph depicting him as a direct participant in the 'hostage crisis' (though not authenticated) was floated. This incessant viciousness goes on and on.

During the cockier days of mid-2003, the banter among hot blooded American officials of younger profile stationed in Iraq was, "Baghdad is passé; everybody who is anybody is headed for Tehran." However, with frequent visitations of death and gore (the machinations of 'Al Qaeda in the land of two rivers'), such utterances have given way to pragmatism. Kenneth Pollack, in his musings states, "...I don't think the United states 'needs' Iran; we have been isolated from Iran for the past twenty five years and, in that time, have experienced the most extraordinary economic prosperity in our history, coupled with strategic developments that have made the United States the most powerful nation the world has ever seen."[8]

Today, with the centrifuges churning in Natanz, Iran looms large on the US policy agenda. While America is seeking to replicate important aspects of itself abroad, Iran can be neither ignored nor invaded

Why Engage Iran

The US defeat of Iraq has left the Islamic Republic as the strongest regional actor by default. With a population of around 70 million, Iran dwarfs its Arab neighbours to the west and south. Armed with the second largest gas reserves in the world and third largest oil reserves, Iran occupies a central position and straddles the route for the world's principal energy resources, buffer between the nations of former Soviet Union and oil-producing Gulf Sheikhdoms, and a bridge between Turkey and nations of Asia.

Iran has the lowest population growth rate in the Middle East-estimated at 1.2 per cent.[9] In 1976, literacy rates for men and women in Iran were 59 per cent and 35.5 per cent but this has increased to 89 and 82 per cent respectively.[10] Women account for 60 per cent of enrolees in universities.[11] Women parliamentarians in Iran command more representation than in the Senate in the US. Recruited by the *Pasdaran* and *Basij*, women have also contributed to the war effort. Even the *hijab* and other attire has acquired softer profiles with improvisations such as the manteau. Under the current Islamic regime, all professions are open to women including 'religious judges' and 146 women now serve in the courts.[12]

Iran is pluralistic, with people of different ethnicity, religiosity, and language, living in harmony. Article 64 of the Constitution also authorises representation in the *majlis* for Jews, Zoroastrians, Christians and Armenians. Jews in Iran comprise the largest non-Israeli Jewish community in the Middle East.[13]

Though the liberal press and media structures that proliferated during Khatami period have been purged to a large extent, electronic communication revolution, the generation of blogs and cell phone usage have witnessed phenomenal increase. Internet usage between 2000-2004 has recorded 1,820 per cent increase.[14] With 8,000 NGO's in Iran and 75,000 bloggers, a vibrant Iranian diaspora especially in Europe, effective population control measures, growing literary levels, the dream to follow the 'China Model', are indicators that the people of Iran are exercising themselves in meeting their aspirations.[15]

Iran today has the most developed computer industry in the Middle East, and besides its nuclear ambitions, Iranians are presently pursuing development of heavy industry such as automobiles and aircraft.

Nearly every neighbour of Iran today sits on a territory that once belonged to Iran. Iran reached demarcation agreements with Kuwait in 1965, Saudi Arabia in 1968, Qatar in 1969, Bahrain in 1970 and Iraq in 1975. However, almost all of the Arab neighbours supported Iraq in the war of 1980-1988.[16] Notwithstanding this, Iran has established constructive relations with its neighbours and has expanded international trade links. Pragmatic approach in the Caucasus and Central Asia and its cultivation of close ties with a range of regional actors, including Russia, China, Japan, India and European Union largely offsets its antagonism towards US.

Irrespective of the recent bellicose emanations from Ahmedinejad, Khatami's vice president in February 1998, in an interview to the Israeli newspaper Yediot Aharanot had said, "I support a dialogue between Iranians and Israelis, but it is too early to speak of a political dialogue between Iran and Israel."[17]

Political Thought

The Shia political thought, in general, is seized by two issues: Who should govern the community until re-emergence of the hidden Imam? And how the community should organise its affairs in the meantime?[18] The Fundamental laws of 1906-07 provided the foundation for the Persian Constitution. Amongst the General Dispositions, Article 1 of the Constitution proclaims Islam as the official religion. Article 2 of the Constitution stated categorically "...it is for the learned doctors of theology...to determine whether such laws, as may be proposed, are or are not conformable to the principles of Islam; and it is, therefore, officially enacted that there shall, at all times, exist a committee composed of not less than five *mujtahids*. The decision of this ecclesiastical committee shall be followed and obeyed, and this Article will continue unchanged until the appearance of His Holiness."[19] The Quran and Hadith do not mention 'Islamic Republic' and neither do they speak of elections, parliaments and constitutions. But the Islamic republic of Iran encompasses all three vital constituents.[20]

In 1917, Sediqeh Daulatabadi (1882-1961), significant amongst women, established a journal *Zaban-e Zanan*. She consistently advocated the importance of education. During 1941-1946, Ahmed Kasravi, encouraged by the prevailing political freedom, opened an organisation, *Azadegan* (The Free). Through this, Kasravi propagated his ideology, main essence of which was promotion of democracy, economic welfare, education, social and personal life.[21] Ayatollahs Mahmud Talegani and Shariat Madari (the chief cleric residing in Iran, at the time of revolution), vehemently opposed, rule of the *Faqih*. In *Islam-shenasi,* Ali Shariati, views Islam as compatible with reason, democracy, social and economic equality. He dominated the Iranian intellectual panorama during the 1970s. In his formulation on "modernism and civilisation", he demonstrates the heartless manner in which imperialism and, essentially European industrialised nations, have sold the mindless production of goods and consumerism as 'civilisation'.[22]

At the time of the Revolution, anti-Shah forces included secularists, religious nationalists, Marxists and Islamic Marxists, all who desired downfall of the Pahalvi dynasty. However, each group had its own peculiar vision of post-revolutionary Iran, amongst which, supporters of Khomeini's Islamic theocracy finally won.

The Parliament, the President and the Assembly of Experts are directly elected by the public. The cabinet is chosen by the president and approved by the parliament, which can also impeach them. Amongst these, the Assembly of Experts consists of 86 clerics, whose candidature is screened by the Guardian Council. It has the responsibility of appointing the *faqih*, monitor his performance and remove him if need be. In bulk of the Middle East, leadership is for life, whereas in Iran it calls for competence.

The problem lies with the unelected bodies of Guardian Council, Expediency Council and concentration of power in hands of the 'Supreme Leader'. This is a Western construct, in Iranian terms he is referred to as the *faqih* or *rahbar*. Articles 91 through 99, authorise the Guardian Council to "review all laws passed by the *majlis,* to check their Islamic credentials, and has a right of veto in interpreting the Constitution and supervising the elections of the Assembly of Experts, the President, the *majlis* and referendums."[23]

The Iranian intelligentsia over the past, its reformers, leftists, and moderates and even clerics have understood the merits of democracy, have not always approved of revolution by a minority, rejected the absolutism of the proletariat, and have appreciated the dichotomy between economic growth and social justice.

The electorate in Iran is very participative. In parliamentary elections of 2000, turnout was 70 per cent, followed by 80 per cent in the presidential elections of 2001. Parlamentary elections of 2004, saw comparatively poorer response of 50 per cent, followed by 60 per cent in the presidential elections of June 2005.[24] In the presidential elections, no fewer than five candidates won more than four million votes apiece, with Rafsanjani in the lead.[25]

Students, *mullahs*, and intellectuals alike have expressed their distaste for the absolute, undemocratic nature of the *velayat-e-faqih*. Abdollah Nouri. A young reformist cleric, former interior minister under President Khatami, has questioned, "How is it possible to give priority to the opinion of one person [the supreme leader] or a few personas or a small social group." Grand Ayatollah Ali Hossein

Montazeri, while disparaging Ali Khamenei, has argued that, "the notion of Islamic government has been perverted at the expense of popular sovereignty." Mohsen Kadivar, a modernist theologian, has disparaged the role of the *faqih* for being antithetical to Shia Islam as well as for being anti-democratic.[26] The age profile of the new cabinet averages at forty six. Though short on practical experience, their ideological perceptions are in line with that of the leadership.[27] Irrespective of the factional struggles between conservatives and reformers, Iran's foreign policy is governed by fixed principles that are shared by all its political elites.

Areas of Convergence

Contrary to the conventional wisdom, the conservative forces in Iran, and not the moderates, are in a stronger position to lead a rapprochement with the Americans. Interests of both United States and Iran converge in Afghanistan and Iraq. Both want stability, peace and moderate governments. Each wants influence in these countries, but a zero sum desire complicates issues. Iranian influence in Iraq to stabilise that country in a new regional security system with partial inclusion of Turkey and Saudi Arabia can be productive. After all, Iran shares with these two states, and the United States, not only a common interest in creating the basis for a pragmatic Shi'ite leadership that would be able to reach a power-sharing agreement with the Sunni minority but also in preventing Iraqi Kurds from trying to promote a Greater Kurdistan. Washington eventually must have a dialogue with Iran and achieve its cooperation through backdoor channels or other viable interlocutors, in case it desires a long-term stable Iraq.[28]

More lasting impact of the Iranian revolution has not been the promotion of Islamist activism, but deep divisions between Shias and Sunnis, a sectarian discourse of power, visible in West Asia, Pakistan and Afghanistan. While hardening of attitudes between Shia and Sunni divisions is not desirable, a unison of 'extremist activist' elements of these sects can create havoc, especially in present day Iraq. With no desire for accentuation of the Shia-Sunni cleavage, there is a need to recognise that Iran today does have a natural constituency. Iraq, Bahrain, Azerbaijan are Shia majority states. Lebanese Shias are the majority in its fractured ethnicity. Sudan and Oman too have significant Shia populations. Thus Iran facilitates balancing one against the other, which does serve US interests.

Given the marriage of democracy and theocracy, visible in Iran, it effectively demonstrates that it is possible to adopt western institutions without sacrificing your cultural and religious identity. Though, growing subordination of republicanism to increased theocratic exercise of power is undesirable, governance structures, tampered with religiosity will be much preferred by the Christian West to agnostics or atheists. Regarding Middle East, Iran would not stand in the way of a two state solution. With strong activism of both Egypt and Saudi Arabia, Iran too will acquiesce.

Iran is not a global threat. Mr Ahmadinejad supports constructive relations with the "Islamic world, the Persian Gulf region, the Caspian Sea region, Central Asia, the Pacific Area and Europe."[29] Given this, and Iran's power potential, a respectable place in the regional security framework, aegis United States, besides re-assuring Iran will also render Persian Gulf more secure.

Nuclear Ambitions

Iran kept activities at Natanz and Arak a secret for a long time and denied the IAEA, ready information and access. Such violations of NPT fail to brook requisite confidence and bring in gory spectre of the Shi'ite bomb, followed by an Arab version. Matters are not helped by President Ahmedinejad's threats to wipe Israel off the map, or his denial that the holocaust ever occurred. The West looked the other way while friendly regimes in Israel and Pakistan went about acquiring their nuclear arsenals. With nuclear powered Pakistan on its East, unstable Iraq on the West and American presence on either flank, not too predictable governments of Azerbaijan and Turkmenistan to its North, competition for regional influence with Saudi Arabia and the volatile Persian Gulf to its south, Iran is well justified in harbouring serious apprehensions regarding its own security. Issuance of regular veiled threats, further elevate this as an issue of preserving national prestige and sovereignty. History also is not too re-assuring. Iraq's decision to launch a war was welcomed by all Arab states of the Gulf. They supported Baghdad with approximately $ 40 billion. Kuwait and Saudi Arabia provided trans-shipment of goods and oil, while Jordan and Egypt contributed militarily.[30] In June 2004, US "Exercise Eagle Resolve-2004," was conducted by US in cooperation with the six states. Focus was on strengthening regional ballistic missile defence against any threats emanating from East or southeast[31]—not too reassuring for Iran. Creation of a formal American security infrastructure over the region with simultaneous isolation of the Islamic Republic, serves to undermine Iranian influence in the Gulf-whether it promotes peace, remains moot. 'Iran's desire for nuclear weapons is not irrational,' it has to be self-sufficient; it has to be self-reliant. Iran needs security; its nuclear calculus is not an instrument of aggressive diplomacy but a viable deterrent against the United States. When anxieties diminish, its weaponisation motivations will be amenable to reason.

US Options

United States involvement in the Iranian affairs has provoked the most extreme emotions from Iranians. In absence of an Iranian Lech Walesa, *Mujahideen-e Khalq* perceived as traitors in perception of general public and fierce loyalty of the military leadership to the current regime, a regime change is not on the cards. Recent report by the Council of Foreign Relations states, "Iran is not on the verge of another revolution. Those forces that are committed to preserving Iran's current system remain firmly in control and currently represent the country's only authoritative interlocutors."[32] The current strategic overreach of United States, world opinion, antagonism that it is likely to generate in the Muslim populations all over the world, and inherent strengths of Iran, a case for invading Iran, too cannot be entertained. An Osiraq on Iran whether logistically feasible or otherwise could bring all hell loose, besides sounding death knell for Iran's reform movement. Even the Arab street would revolt and it will provide viciously combustible fuel for Islamic terrorism.

Containment and Isolation has been the key policy so far. However, Iran's involvement with its neighbours and the global and political financial order will significantly dilute American efforts at isolation. Besides, due to lack of economic integration, Iran continues to be massively dependent upon oil. US sanctions thus, attract no leverage and thereby fail in creating any political pressure. Sanctions, denial of appropriate progressive technologies and healthcare, on the contrary inflict more penalties on the population in general and are counterproductive.

Sorting out differences in a cooperative framework, rather than a confrontational mode is indeed ideal. Some preliminaries need to be addressed first. Iran craves for respect, which is denied by the coinage of 'axis of evil', worst state sponsor of terrorism, reign of unelected leaders and so on. Iran thus, today desires, though unacceptable, to gain respect through 'nuclearisation'. It demands inclusion and integration. Accorded its rightful place in Iraq's reconstruction activities, involvement in a Persian Gulf security framework and inclusion in WTO, will facilitate raising stakes for penalising Iran, in case it fails to cooperate. Opening of its economy in such frameworks calls for reform, which in turn is likely to dilute the hold of *Bonyad-e Mustazafin,* and the likes, through which the clergy exercises its financial clout. Solution lies in collective action by all leading nations of the world, to which Washington has generally been averse. Iran would prefer the international community to lead the 'war on terror'. While, Washington espouses democracy and its supposed dividends, Iran would much prefer a state where governance is guided by religion, even to the extent of rendering republicanism subservient, as it vindicates its own regime. While, Iraq was phase 'two' of the US war on terror, with initial ambitions to 'tame the Middle East', Iran would like this to be the final phase.

Analysis

Year 2010 may be the watershed. With time, with passing away of Khamenei, Khomeneism will certainly evaporate.[33] By 2010 majority of the electorate would have no memory of either Shah or Khomeini as also no personal knowledge of the revolution. The growing clout of the 'quietist' school such as Grand Ayatollah Sistani the concept of *velayet-e faqih* is likely to loose prominence. The liberals/ leftists/ moderate conservatives/ reformers, whatever we may coin them, they will come to power sooner than later. The American neoconservatives desire to destroy/ change/ threaten the current regime and the self-evolved system of governance will only further serve to entrench the fundamentalists besides gifting the hardliners assured longevity. Somebody needs to rein them in—the neoconservatives. The Iranians as demonstrated by Khatami, have no desire for internal conflict. They are successfully able to operate through their supposedly flawed governmental structures. In August 2005, the *Washington Post* reported that a new US National Intelligence Estimate projects that Iran is about 10 years from having the capability of producing nuclear weapons.[34] Allow them time. Attention over Iran's nuclear choices too sharply (which only facilitates demonisation), needs to be kept out of focus for a while. Too much media glare, rhetoric and Israel's keenness (to do Osiraq-more of a fantasy), have escalated stakes so high that retreat by either will tantamount to loosing honour or moral authority. There is an urgent need to differentiate between public diplomacy and private dialogue. It is high time to acknowledge that Iran has a vital role in the post-war reconstruction of both Iraq and Afghanistan. It is increasingly becoming an indispensable player in the world economy. Dialogue need not wait absolute harmony. Mutual opposition of each other is prevalent at the governmental level and not shared by majority if their populations. Expansion of Geneva track discussions that were conducted with Tehran for 18 months after 9/11 may help. Solution lies in enhancing the current forum of EU-3, with inclusion of Russia, China and India. A direct dialogue and establishment of diplomatic relations between Iran and US through this 'Six plus Two' forum may facilitate a basic statement of principles like the 'Shanghai Communiqué of 1972'. Confidence building will eventually serve to address relevant domestic constituencies on both sides. Ted Koppel's remark, "What the Iranians have still not figured out is that Iran is just as irrelevant to the American people today as it was twenty five years ago,"[35] only serves to harden attitudes. Rapprochement with Washington will brighten the prospects of persuading Iran to ease on its weaponisation programme. It is an idea whose time has come, otherwise learn to live with a nuclear Iran. To quote Winston Churchill, "Americans always try to do the right thing–after they've tried everything else."[36]

Notes and References

1. Pollack, Kenneth M. (2004). *The Persian Puzzle*, Random House, New York, p.346.
2. Ibid. pp.346-348.
3. Berman, Ilan (2005). *Tehran Rising*, Rowman & Littlefield, USA, pp.13-29.
4. Pollack *op. cit.*, p.283.
5. Beeman, William O. (2005). *The Great Satan vs the Mad Mullahs*, Praeger, London, pp.142-145.
6. Fathi, Nazila (2002). "Bush's 'Evil' Label Rejected by the Iranians Leaders", *The New York Times*, February 1. cited in Pollack *op. cit.*, p.352.
7. Beeman *op. cit.*, p.135.
8. Pollack *op. cit.*, p.xx.
9. Beeman *op. cit.*, p.117.
10. Singh, Subha (2005). "Iran's Uneasy Agenda," *Frontline*, Vol 2(12), August 26, pp.55-58.
11. "Statistics of Sadreddin Beladi Musavi", Statistics Centre of Iran, Iran Daily 2005, cited by Beeman, William O. (2005). *The Great Satan vs. the Mad Mullahs*, Praeger, London, p.150.
12. Thiebaut, Azadeh Kian (2005) "Women's Movement in Post Revolutionary Iran", in M. Hamid Ansari (ed.), *Iran Today*, Rupa & Co, Observer Research Foundation, p.325.
13. Beeman *op cit.*, pp.152-154.
14. *www.internetworldstats.com/stats5.htm* cited by Beeman *op. cit.*, p.185.
15. Milani, Abbas (2005). "US Foreign Policy and the Future of Democracy in Iran," *The Washington Quarterly*, Summer, Volume 28(3), pp.41-47.
16. Mokhtari, Fariborz (2005). "Iranian Security Perceptions in Historical Context", *The Middle East Journal*, Volume 59(2), Spring, p.211.
17. "Israeli Paper Publishes Interview with vice President Ebtekar," BBC Summary of World Broadcasts, February 3, 1998. cited by Pollack *op. cit.*, p.318.
18. Gelvin, James L (2005). *The Modern Middle East*, Oxford University Press, New York, pp.20-21.
19. Ibid. p.165.
20. Ibid. pp.287.
21. Kasrawi, Ahmad (2005). "The Detrimental Consequences of Islam", Lloyd Ridgeon (ed.), *Religion and Politics in Iran*, I.B. Tauris, London, pp.58-69.
22. Shariati, Ali (2005). "Civilization and Modernization", Lloyd Ridgeon (ed.), *Religion and Politics in Iran*, I.B. Tauris, London, pp.176-195.
23. Moslem, Mehdi (2002). *Factional Politics in Post Khomeini Iran*, Syracuse, New York, p.30.
24. Beeman *op. cit.*, p.195.
25. Ballaigue, Christopher De (2005). "New Man in Iran", *New York Times Review*, Vol. 52(13), August, 11.
26. Vakil, Sanam(2003). "Iran's Fragile Faultlines", *Policy Review*, October/November.
27. Ray Takeyh, "Iran's New President 'Largely Indifferent' to What Outside World Thinks', at *http://www.cfr.org/publication/9199/takeyh.html.*
28. ICG, *Middle East Briefing* No. 18, "Iran: What Does Ahmedi Nejad's Victory Mean?", August 04, 2005.
29. Ray Takeyh, "Why Iran isn't a Global Threat" at *http://www.csmonitor.com/2005/0929/p09s02-coop.html.*

30. Chubin, Shahram (2005). "The Islamic Republic at Twenty Five", in M. Hamid Ansari (ed.), *Iran Today*, Rupa & Co., Observer Research Foundation, p.233.
31. Berman *op. cit.*, p.122.
32. Report of an Independent Task Force sponsored by the council on Foreign Relations, "Iran: Time for a New Approach", co-chaired by Zbigniew Brzezinski and Robert M Gates, p.1.
33. Berman *op. cit.*, p.85.
34. Bourseton, Jack and Charles D. Ferguson (2005). "Keep Your Enemy Closer", *Bulletin of Atomic Scientists*, November/December, Vol 61(6), pp.25, 76.
35. Ted Koppel's remark to Kenneth Pollack, in conversation, on July 11, 2004. cited at Kenneth M Pollack, The Persian Puzzle, Random House, New York, 2004. p180.
36. "Internet Humour", *The Times of India*, New Delhi, January 14, 2005, p.25.

21

Dynamics of Global-Regional Concerns on Proliferation

The Case of Iran

NOBUMASA AKIYAMA

The international nuclear non-proliferation regime is now on the edge. The Fifth NPT (Nuclear Non-Proliferation Treaty) Review Conference in May 2005, failed to demonstrate an international consensus for decisively tackling with nuclear non-proliferation and disarmament. The international community has not been able to take effective measures to stop North Korean ambition of going nuclear as well.

It is strongly suspected that Iran may follow suit. Iran has been claiming that its nuclear activities including the development of enrichment programme are exclusively for peaceful purposes. However, circumstantial evidences indicate that Iran would seek more beyond mere civilian programmes.

Now EU-3 (Britain, France, and Germany) are engaged in the negotiation with Iran for a solution of the problem. While the United States has limited role to play in the negotiation, it remains the most important player in various dimensions of the whole game of stopping Iran from going nuclear. Russia is also an important player, as it might provide actual assistance to Iran in the case of realisation of the international cooperation in nuclear programmes in Iran. So far, the international community has not been successful in persuading Iran to halt its enrichment programme. Conversely, Iran seems to have escalated its posture against international pressures. If Iran would acquire nuclear weapon capabilities, its implication is enormous.

Especially, implication of nuclear Iran for West Asia, is far greater than the case of North Korea in East Asia. The stability of the Gulf region is vital to the global economy, as it is a major source of oil supply to many countries. Inter-state relationships are pretty much complicated in the region, and the emergence of Iran as nuclear power, will destabilise the region and disturb the delicate power balance among regional players. Stimulated by Iran, other regional powers might be tempted to seek nuclear options, as well.

When we look into diverse and profound consequences of Iran's pursuit of nuclear capability, it is necessary to view it in three dimensional ways: implications for the global nuclear non-proliferation regime, consequences to and influence from regional security environment and political dynamics, and domestic political calculations in favour of nuclear might, including the prestige and esteem that supposedly flow from it along with the belief that it deters aggression from potential enemies. And it is important because most of the countries believe that in view of the international concern over non-proliferation, Iran ought to renounce proliferation-sensitive enrichment programmes, and this would provide Iran with a chance to retain nuclear option. There are other issues too like who can play a role in stopping Iran and how can we prevent the undesirable consequences, if Iran goes nuclear?

This paper seeks to have an overview of the implications of Iran's nuclear adventure for the global non-proliferation regime, and its impact on the regional security and political dynamics. I will try to argue that non-nuclear Iran is in everybody's interest and any irresponsible behaviour by any actor in the international community might bring about unwanted outcome in the world. Each member of the international community should realise their responsibilities, which could be far greater than they now realise, and they should take such responsibilities in order to avoid the consequences.

Crisis of Global Non-Proliferation Regime

NPT—Losing Credibility?

Article IV of the NPT acknowledges the 'inalienable' right of use of atomic energy for peaceful purposes. It is the legal basis, on which the Iranian claim of legitimacy of its nuclear programme rests. When President Ahmadinejad insists that Iran has a legitimate right, based on the NPT, to conduct research on enrichment and eventually to have the ability to produce enriched uranium on its own, he is making a perfectly legal claim.

Under the existing regime, it is certainly possible, in theory, for a state to acquire any kind of nuclear technology and materials necessary for producing nuclear energy without violating NPT and IAEA (the International Atomic Energy Agency) safeguards. The existing non-proliferation regime does not have any mechanism for preventing and sanctioning, if it happens, the conversion of civilian nuclear materials and activities into military ones. Although IAEA with its safeguards, as well as the United Nations, could be responsible for enforcing the provisions of the NPT, it does not have actual power to enforce them.

The distinction between civilian and military nuclear activities is so thin that the existing regime heavily relies on the 'will' of the state parties for not using their civilian nuclear facilities for military purposes. Once a state acquired nuclear weapon capability (without actually producing weapons), it could become a *de facto* nuclear weapon state, by withdrawing from the NPT. That's why the 'intention' does matter. But in the existing regime, there is no way to check the 'intention' of a state that is undertaking activities aimed at building up its capacity for making nuclear weapons. The international community could only infer the intention of a state concerned from its behaviours as well as words and promises to some extent.

Since there is no mechanism to verify the intention, Iran, as a signatory to the NPT, has moral and political obligation to demonstrate with words and deeds, that it does not have any intention to convert its civilian nuclear activities into military ones. But how can one believe Iran when the President told that Israel should be wiped from the face of the earth, or when there is increasing evidence that Iran had cheated the IAEA and undermined its international obligation by purchasing, without declaring, enriched uranium from China, and some components of centrifuge facility from A.Q. Khan's black market, and even acquired some information on the design of nuclear warhead from the same network? Once it is proved that Iran has not been sincere about its commitments and has sought to hoodwink the international community by refusing to act according to its obligations as a signatory to the NPT regime, it becomes difficult to buy the argument that it will be safe to leave such an irresponsible state with a potential of nuclear option.

Iran claims that it seeks what Japan already has. It is true that Japan has been pursuing its own nuclear fuel cycle programme, and it has come close to full operation. However, one must note three elements that Iran does not deliberately mention, or simply forgets to mention. First, Japan has a very good record of compliance with IAEA safeguards, while IAEA has dedicated substantial portion of its safeguards resources to inspection of the Japanese facilities. With such extensive efforts, Japanese facilities were accredited to the integrated safeguards. Second, Japan's nuclear fuel cycle was not a complete one. Japan does not have its own uranium mine. It does not have the conversion facility. Without cooperation with foreign countries, Japan cannot pursue its nuclear fuel cycle. Therefore, the Japanese nuclear community regards international cooperation and transparency as important elements in pursuing its fuel cycle programme. Third, the pursuit of fuel cycle activities is subject to the consent by the United States, Australia, France, and the UK, all of which are providing either materials or technology to Japan. With the United States, for example, Japan has concluded an agreement for nuclear cooperation, which includes the safeguards much stricter than those of IAEA. So the Japanese fuel cycle cannot work in isolation.

If Iran seeks to do the same as Japan, Iran also has a responsibility to do what Japan has done for safeguards and transparency in last 30 years for gaining the confidence of IAEA, and even surrender some parts of nuclear fuel cycle to the international community as a way to convince the international community about its intention of not going nuclear.

Iran was recognised by the meeting of the board of governors of IAEA as a case of 'non-compliance' with the IAEA safeguards agreement. If the sanction/enforcement mechanism would not work as it was designed, it would probably undermine the credibility of the norms that the regime provides. The consequence would be; first, other states would be more tempted to act selfishly by ignoring international norms, and second, it would trigger more brunt foreign policy including the use of force and coercion to ensure that Iran does not go nuclear.

China is reportedly cautious about bringing the Iranian case into the UN Security Council. If it is true, has China tried to persuade Iran to give up nuclear ambition? It is a case where the international community will fail to utilise the most important and credible collective security mechanism for a case that everyone is concerned about; the international community would soon face more serious challenges to international law. An anarchical international society would not be desirable for small and weak states in particular as the powerful ones could dominate in every situation.

Trend II: Is Restricting the 'Inalienable' Right Possible?

When NPT was created in 1970, Article IV (the inalienable right of peaceful use of nuclear energy and the right to receive international cooperation) was supposed to function as an 'incentive' for states to participate in this global non-proliferation regime. In order to gain wider support of the international community, the NPT regime had provided a kind of 'benefit' to developing countries in joining the treaty. But now this 'inalienable' right has come to be a focal point of tightening control and it may accompany *de facto* change of the give-and-take equilibrium in the provisions of NPT, which may lead to some kind of transformation of the nature of the regime. And such a trend might have a strong impact on whether the effectiveness and credibility of the treaty would be maintained.

After the Gulf War in 1991, the international community was surprised that Iraq had secretly acquired nuclear capability, fairly close to the capability for making nuclear bomb. It was important to note that it was easier for Iraq, under NPT, to exploit the available international cooperation in peaceful use of nuclear energy for hidden military purposes. It triggered the strengthening of IAEA's safeguards system, but this newly introduced Additional Protocol has not acquired international approval yet and the safeguards system is still suffering from the previous limitations.

For a few decades, North Korea also exploited technical cooperation and assistance from Russia in establishing its nuclear programme, and once it has acquired the capability and its clandestine nuclear weapon programmes were exposed, it withdrew from the NPT. The NPT regime was not able to properly respond to that crisis.

Now, if Iran were successful in developing its own nuclear capability despite the dissuasion by the international community, it would bring about the loss of the credibility of the enforcement mechanism of the international non-proliferation regime as well.

Under such circumstances, the international community would tend to shift its emphasis on concrete policy actions such as further tightening of control over technology transfer as well as provision of materials. For example, IAEA's Secretary General, Dr. El Baradei proposed a moratorium on construction of new fuel cycle facility and a multilateral nuclear approach to put fuel cycle activities under the multilateral control. The United States also proposed a kind of nuclear safety net, or a nuclear fuel bank plan that would guarantee supply of nuclear fuel, in the form of long-term leasing, for states that give up their independent pursuit of national fuel cycle activities. With tightening export control and technology transfer in the field of nuclear fuel cycle, it intends to squeeze up the supply of sensitive materials such as enriched uranium and plutonium since the acquisition of such materials would be the biggest obstacle and if such supply is to be controlled, the risk of nuclear proliferation would be substantially reduced.

Some developing countries expressed their concerns about such proposals. They thought such proposals might harm their rights provided by NPT under Article IV. Iran is unquestionably included among these countries. But the problem is that Iran was not the only one, others such as Brazil and Australia (to some degree) expressed their concerns that such ideas might hurt the spirit of NPT by nullifying Article IV.

It is true that such arrangements might create another, deeper cleavage between 'nuclear haves' and 'nuclear have-nots' even in civilian nuclear energy. Considering that all countries, having acquired nuclear fuel cycle capabilities, except Japan, are nuclear weapon states (including non-NPT states), a nuclear fuel bank arrangement might be seen as simply providing nuclear weapons states with more leverage in international power politics (Brazil, Argentina, Germany, and Holland host enrichment facilities only). If the supply of fuel is to be controlled by such 'nuclear haves', where enrichment/reprocessing facilities are located, how can other countries (have-nots) feel secure in the fuel supply? They might have to worry whether suppliers such as the United States or France might stop the supply because of political reasons rather than violation of rules of nuclear non-proliferation.

The United States, IAEA and other states supporting this arrangement of assurance of supply need to address such a legitimate concerns of nuclear 'have-nots'. They also have to be aware that they will take further serious and even tough responsibility to be consistent in their nuclear policy.

Trend III: More Tolerance to Non-NPT Nuclear Weapon States?

Another important element in the changing tide of the global non-proliferation efforts has been more tolerance to two non-NPT nuclear weapon states, i.e., India and Pakistan. As far as Pakistan is concerned, currently, the government seemingly contains proliferation activities, which were derived from Pakistan's nuclear establishment. It was due to strong pressure from the international community. And at the same time, since the United States needs Pakistan as an indispensable partner in fighting the war on terrorism in Afghanistan, it takes an accommodative approach to Pakistan, expressing its satisfaction about the improvement in Pakistan's non-proliferation efforts. However, it is not certain whether the nuclear black-market has really ceased to function. We have also not known how his underground network was linked with clandestine nuclear activities of Libya, North Korea, and Iran. Also, political instability with the fragile power base of the current government, corruptions, and widely spread sympathy towards the Islamic fundamentalists still indicate a potential resumption of proliferation activities by certain groups in Pakistan.

In the case of India, many countries are competing for attracting the attention of the Indian government in order to establish strong strategic partnership. The United States has concluded with India a bilateral agreement on cooperation in civilian nuclear programmes. It was a major shift in the US policy. The United States stresses the benefits of such an arrangement with India, as it will bring India into a pro-non-proliferation camp as an active partner, which will contribute to global security.

The accreditation by the US of India's nuclear weapons as well as tolerance of Pakistan's misconduct in proliferation activities may as well be called policy of double standards by the US. If Iran would have learnt some lessons from the case of India, they were; 'nuclear prestige would turn eyes of major powers to the country', and 'time will solve the problem'.

Nuclear weapon is often referred to as a symbol of great power in international politics, and access to nuclear energy technology as a symbol of technological competence of a country. Are they true? As we look into the details of India's case, we will find the myth of national prestige and esteem in nuclear technology.

Some claim that India's example demonstrates that nuclear weapons of India turned the attention of the world to India. In fact, many countries including the United States, Russia, Europe, China and Japan now seek strategic partnership with India. And this phenomenon could be interpreted as a symbol of increasing significance of India's status in the international community. But it is useful to ask the following questions: Has India gained recognition as a great power in international relations owing to its nuclear weapons? Probably, not. If it were the case, India would have been recognised much earlier, in the late 1990s till early 2000s. India is now rising mostly because its economic growth as well as outstanding expertise and human resources in the information technology industry, which have become indispensable for global economic growth.

What about nuclear power as a symbol of technological advancement? Although there are still some unexplored frontiers in nuclear science (e.g., nature of plutonium in half-life, which is more than 14,000 years, while plutonium was found only 50 years ago), nuclear technology including enrichment and reprocessing is no longer categorised as high technology. Since building a nuclear power plant involves huge capital and labour, it could be a national-pride project. But despite such extensive investment of capital and labour, the return may be not so high as many expect.

Rather, among nuclear-related technological areas, most advanced, 'high-tech' areas are safeguarding, physical protection and material accounting. Ironically, these are the technologies required for non-proliferation. Especially, development of built-in safeguards systems and more proliferation-resistant technology attract the best and brightest in the Japanese nuclear scientists and engineers (Of course, technologies of weaponisation and miniaturisation of warheads are high-tech. But state parties to NPT cannot legitimately pursue such technologies).

Security Dynamics in the Middle East

Israel's Overwhelming Nuclear Capability

Israel never officially accepted that it possessed nuclear weapons. But it is common knowledge now that it has its secret nuclear arsenal. It publicly claims that it will not be the 'second' nuclear power in the region. Certainly, Israel should legitimately address the threats to its security. However, addressing its security concern should not be excessive, so as to threaten others in the region.

Israel's current military capability, either conventional or nuclear ones, far exceeds that of the Arab community. The military expenditure of Israel is three times that of Egypt in 2003. It also received annual military assistance worth approximately 2 billion US dollars from the United States. Israel is the only country in the Middle East, which holds the 'triad' (i.e., ballistic missiles, submarine-based cruise missiles, and aircrafts, which can carry nuclear warheads). Arab states are worried about this asymmetrical strategic balance, which is vividly in favour of Israel. Here is a classic case of a security dilemma. This asymmetrical strategic balance may bring a different type of asymmetrical threat to Israel. Even if the balance of military in the conventional warfare is in favour of Israel, it may be counterbalanced by means of unconventional warfare, i.e., terrorism.

As we have seen in the US-USSR relations, arms control has served as a very effective instrument to build confidence between two hostile superpowers. Now this bilateral relationship is managed fairly well after the collapse of the Soviet Union and the emergence of democratised Russia, but we must note that such a relationship was possible because of the firm foundation for mutual trust built through the process of arms control through transparency and efficient verification measures. In Middle East, if two sides do not wish to fight a war, arms control could be a useful instrument for confidence building and the relaxation of tensions.

Now Israel maintains its superiority in conventional military as well as WMD capabilities. It has concluded a peace pact with Egypt, and Saddam's Iraq, its most vital threat, is now gone. Thus, the security environment in the region is far more favourable to Israel than ever before. Israel may want to take advantage of this environment to reduce the threats to both Israel and the Arabs. Israel claims that the peace agreement should come first while the Arab community insists that disarmament is essential to eliminate threats to the Arabs.

If Israel would initiate the arms control talks by eliminating the excessive part of its WMD arsenal, it may become a major breakthrough in the peace talks. And it would give a high credit to Israel as a country responsible for global as well as regional security and stability. It is also essential that the Arab community also needs to take a pragmatic approach from the long-term strategic perspective.

Also, like India which has become active in getting itself involved in international non-proliferation manoeuvers while being out of the NPT, Israel could also consider ways of being associated with non-proliferation mechanisms, starting with increasing the transparency about its nuclear capability. It is certainly an important confidence-building measure, which will contribute to reduction of the chance of armed conflict in the region. That would serve Israel's long-term security interests.

Nuclear Dominos in Middle East?

How Iran would behave and how the international community would respond to Iran's behaviour are the questions analysts seek answers to. There is an apprehension that there could be a dominos effect and a nuclear test by Iran could prompt its neighbours to consider whether they should pursue nuclear option. Saudi Arabia, Egypt, Syria, the United Arab Emirates, and Turkey are sometimes referred to as countries who could follow Iran and begin their nuclear ventures. Obviously, these countries may not pose immediate concerns. But all of them have once been suspected of acquiring nuclear weapon capabilities. For example, Saudi officials mentioned in early 2004, that they were interested in acquiring nuclear capability on lease from China or Pakistan. The United Arab Emirates also announced in September 2004, that they were planning civilian nuclear power programmes. This announcement was perceived by observers that it would carry deeper implication than simple interests in energy security.

Once any regional country could succeed in possessing nuclear weapons, they might feel tempted or even 'urged' to go nuclear. Two reasons: First, to ensure regional security balance. A nuclear Iran means the emergence of the single regional superpower, and it might harm strategic balance among regional powers, especially *vis-à-vis* the Arab countries. It is no wonder Arab countries may be tempted to 'correct' this strategic imbalance against a Persian country, by pursuing their nuclear capabilities.

Second, successful 'overcome' by Iran of institutional 'obstacles' for acquiring nuclear weapons, which are currently posed by the NPT regime and export control mechanisms, might lower the threshold for going nuclear. Even if these states would not attempt to possess nuclear weapons, more citizens, companies, other organisations could become involved in proliferation activities by associating themselves with terrorist organisations or black-market network.

Third, the emergence of nuclear relationships in the region with improper (or incomplete) nuclear deterrence will sharply heighten the probability of regional armed conflict, which could lead to a nuclear catastrophe. The deterrence relationship between the United States and the Soviet Union during the Cold War was gradually and carefully crafted to avoid accidental exchange of nuclear forces, based on a mutual understanding of the Mutual Assured Destruction (MAD) as the rule of the game, series of arms control agreements such as Intermediate-Range Nuclear Forces (INF), Strategic Arms Reduction Treaty (START), and Strategic Offensive Reductions Treaty (SORT), and confidence-building measures. Without such extensive measures to avoid a nuclear war, the security situation with multiple nuclear forces in the region would be far more dangerous. If Iran ever possesses nuclear weapons, it would be difficult to establish a stable nuclear strategic relationship with its neighbouring Arab rivals as well as Israel.

Conclusion

As examined above, Iran's going nuclear will bring about tremendous negative impact on global non-proliferation regime and regional stability in the West Asia or the Middle East. Furthermore, it will not suit Iran's own interests, as a nuclear option would create a 'prisoners' dilemma' situation in the region, which will make Iran's security policy even more complicated and unstable. Evidently, 'nuclear Iran' serves no one's interests. First and foremost, it is vital for the international community, the region and also Iran to bring about a halt to Iran's enrichment programme through an effective deal. Iran should be seriously engaged in the negotiation with EU-3. It must be aware that correcting the wrongdoings in terms of compliance with international rules under the IAEA is essential if Iran seeks any 'rewards' from halting its covert nuclear activities. In the meantime, the United States may consider changing its hostile attitude towards Iran in exchange of Iran's refraining from supporting anti-US terrorist activities.

The international community should also take necessary measures to prevent a nuclear dominos effect in the Middle East, if Iran ever goes nuclear. Israel should be encouraged to take a lead in arms control talks in the Middle East, by making its WMD capability more transparent, and unilaterally, first, starting reduction of excessive arms, which could eventually lead to a Middle Eastern zone free of WMD. Such an initiative will also help assuring Israeli security in return. The US support and encouragement for Israel will be essential for this initiative. Other countries in the Middle East have also important role to play in preventing a nuclear dominos. They should refrain from playing a dangerous game of nuclear dominos, which would surely be rewarding in terms of their security and economic interests.

The credibility and nearly universal adherence to the global non-proliferation regime should be maintained. The legitimate concerns of Iran on the nuclear front based on the provision of NPT must be addressed. Even though the international community is inclined towards more restrictive approach to the 'inalienable right' of peaceful use of nuclear energy, neglecting Iran's legitimate claim base on international law would not be helpful in solving this situation. And if Iran insists that its nuclear programmes are only for civilian purposes, and it has no interest in converting them into military purposes, Iran must show it not by words, but by deeds such as accepting strict safeguards, inspections, and increasing more transparency and accountability in its programme. Now that Iran has failed to do so in the past, it would need extraordinary efforts to get back the international confidence.

On the side of the international community, probably, after Iran gives up its enrichment programme and successfully demonstrates its good faiths to international norms of non-proliferation, and provide reliable evidence of its energy security needs, the international community (including the US, Europe, Russia, and Japan) could provide Iran with more proliferation-resistant nuclear systems, which have built-in proliferation-resistant safeguards technologies. In fact, it is the most cutting-edge technology.

Needless to say, Iran itself should show its good faith to the international community by favourably clearing away, with concrete actions, international concerns regarding the possibility of the emergence of a new nuclear weapon state. As has been argued above, the halting and renouncing of fuel cycle programmes by Iran is in everyone's interest, and Iran could gain better from shedding its nuclear option by bargaining for western commitments to address its security concerns. Thus, to prevent such a situation from happening, international community should come together and seek to address the issue in a peaceful manner.

Section V

Political Reforms and Popular Participation: Regional and Extra-regional Approaches

22

Democratising the Middle East

Democracy Promotion as a Means of Achieving International Security?

DIETRICH JUNG

"The East [is] a career."[1] Put into the present, this quote from Benjamin Disraeli's novel, *Tancred or The New Crusade,* nicely captures the current burgeoning interest of politicians, journalists and academics in Middle Eastern affairs. Since the terror attacks of September 11, not only has the Middle East returned to being an "all-consuming passion" for young Westerners, as Edward Said put it in his introduction to *Orientalism,* but engagement with this region also seems to offer a variety of opportunities for pursuing individual professional careers. The universities are being confronted with an avalanche of students who want to enrol in Arabic language courses and/or in the field of Islamic studies. National science foundations are allocating extra money for researching the Middle East. Political and governmental institutions are hiring so-called Middle Eastern experts, who have become a mushrooming species in various fields. And, last but not the least, representatives of a multiplicity of NGOs are prepared to penetrate the region similarly to the Christian missionaries whose presence on Ottoman territories was one of the major characteristics of regional history in the 19th century. Yet, while the latter came in the name of God, the spread of international NGOs is taking place in the name of democracy.

Democracy promotion has advanced to become a major policy instrument for a new alliance of states, international, and trans-national organisations that claim to be fostering global security through political and economic liberalisation. Roland Paris called the central ideology of this admittedly heterogeneous global alliance "Liberal Internationalism," an ideology using the normative legal framework and the institutional setting of consolidated democratic states as tools to support state- and nation-building processes in the former Third World (cf. Paris, 2002). The discursive dominance of this civilising mission of Liberal Internationalism is particularly pertinent with regard to the Middle East. Associating the origin of contemporary Islamist terrorism with the authoritarian political structures in the region makes democratising the Middle East seem a handy means for fighting transnational terrorism, enhancing global security, and achieving more political stability in the region and beyond.

For the time being, however, this Western involvement in the Middle East is not looking all too promising. In particular, the two attempts of "coercive democratisation" in Afghanistan and Iraq are increasingly looking flawed.[2] For a majority of observers, the two cases are well on their way to turning into examples of failed Western foreign policies. Indeed, apart from electoral processes that took place under rather dubious conditions, the political structures in neither Afghanistan nor Iraq resemble anything close to those in democratic states. In both countries, the incumbent governments have no control over their territories or their peoples. The nascent political institutions and their representatives are protected by foreign troops, and many state functions have been taken care of by largely in transparent collectives of foreign, international, national, and non-governmental organisations. While in these two cases, foreign intervention is struggling with building modern state structures from scratch, the more general picture in the Middle East is that of the resilience of authoritarian rule. In the light of this deeply entrenched authoritarianism in the region, a smooth and sudden shift to liberal democracy does not seem very likely. Rather, the assumption of Western foreign policy makers that democratisation could be the remedy for decades-old political ills is the result of a particular kind of wishful thinking with respect to the political mess into which post-colonial state building and international relations have led in the Middle East.

Against this background, this paper will question some of the central assumptions behind this current strategy of democracy promotion as a means of achieving international security. Firstly, it will analyse the state of the state in the region, i.e., the essential political structure that is the institutional target of these policies. In a second step, we will look at the impact of policies of democratisation and democratic political systems on peace and stability, posing the question as to what extent democratisation in the Middle East might serve Western security interests. Finally, the paper will discuss the choice of appropriate regional partners for Western initiatives to promote democracy and market economy in the social environment of the Middle East.

The State of the State in the Middle East

In recent times, the Middle East has been known as one of the most war-torn regions in the world. Since the end of the Second World War, Hamburg University's "Working Group on Causes of War" at has counted 48 wars in the region, which corroborate the violent image of Middle Eastern politics.[3] However, given the inherent relationship between war-making and state-making, this conflict prone political history of the Middle East is not without precedent. Obsessed by the inflated rhetoric about state failure and state decay, many Western observers tend to forget the violent history of European state-formation while looking at Middle Eastern politics. In Europe, the formation of modern nation-states was also accompanied by numerous wars, and the democratic state, based on the rule of law, was only a very late result of European state-building processes. The establishment of modern states, that is to say the formation of the state monopolies of physical force and taxation, is a process that almost by its very nature causes violent conflicts among competing social forces, and the violent trajectory of European history has known the intertwined emergence and disappearance of states. Consolidated democracies and the firm establishment of the rule of law built on the successful results of this conflict ridden process of the formation of state structures.

In conceptual terms, it was Max Weber who defined the central feature of modern statehood: "the monopoly of the legitimate use of physical force within a given territory" (Weber, 1991: 78). In referring to the inner justification of systems of domination, Weber precisely distinguishes political authority from mere power relations by the category of legitimacy. Both the state monopoly of physical force and the political order that has been established on this central feature of modern statehood need legitimacy, that is to say, they have to be anchored in the normative and symbolic structures of society at large. Focusing on the symbolic side of systems of domination (*Herrschaft*), Weber distinguished three ideal types of legitimate political authority: firstly, legal or rational authority, which rests on the belief "in the legality of enacted rules and the right of those elevated to authority under such rules to issue commands"; secondly, traditional authority, which rests on the "established belief in the sanctity of immemorial traditions and the legitimacy of those exercising authority under them"; and finally, charismatic authority, which rests on the "devotion to the exceptional sanctity, heroism or exemplary character of an individual person" (Weber, 1968: 215). In the light of these concepts, we can interpret the processes of modern state formation as a gradual shift from forms of traditional authority to the rational/legal type—from absolutist states to constitutional monarchies, republican democracies, and welfare states (Jung, 2001: 458-63)—whereas elements of the charismatic type are most likely to occur in the foundational phase of a newly emerging state.

With regard to the European experience, Norbert Elias (1994) distinguished two phases of modern state-building. In the first phase, state makers establish a factual monopoly of physical force over a territory and a population against various competitors. In Europe, this phase led to the formation of the authoritarian absolutist state under the rule of dukes and kings. In the second phase, and only after the firm establishment of the state monopoly of violence, the administration of this monopoly was nationalised and democratised through an institutional set of checks and balances that we now know as representational democracy. It is this process, in which the control over firmly established state monopolies moves from the hands of autocratic rulers into the hands of society as a whole, that we call democratisation. Historically, however, these phases are often intertwined and both phases can fail. The gradual democratisation of states and societies has not been a peaceful process in itself, and, as a consequence, many states have ultimately disappeared in the course of European state formation (cf. Tilly, 1990).

Applying the above concepts of historical sociology, we can observe four general patterns that have characterised the conflictive picture of Middle Eastern state formation since decolonisation. However, we should be aware that this does not mean simply reducing the complex social structures and historical developments behind this picture to some general and abstract root causes of conflict. The complexities of each of these very different paths of state formation deserve more thorough examination. There is no one-size-fits-all explanation of Middle Eastern politics. Instead, we have to analyse the region's historical particularities in the light of abstract categories. It is for this analytical task that the following generalisations might help:

1) In several states of the region, viable monopolies of physical force have not yet been established. In particular, the civil wars in Afghanistan, Lebanon, northern Iraq, Tajikistan and Yemen showed the precarious nature of Middle Eastern states with regard to this cornerstone of modern statehood. In these states, the rulers' claim to the monopoly of physical force remains contested in both its factual establishment and its legitimacy.
2) The population of many Middle Eastern states is fragmented along tribal, ethnic and/or religious lines. This fragmentation is reflected in competing political loyalties that have been major impediments for regional nation-building processes and for the consolidation of legitimate statehood. In their claim to power, ruling elites often cannot relate to an ethno-national or to a civic form of nationhood. Processes of nation-building, i.e., often very violent and forceful integration of a population into the majoritarian mainstream, as it took place in Europe, is still under way.
3) State-society relations in the Middle East are characterised by a huge gap between state elites and their populations. The mostly authoritarian regimes are detached from society and try to control the political and economic resources of the country. If a factual monopoly of physical force has been established, as is the case, for instance, in Egypt, Iran, Jordan, Saudi Arabia or Syria, this is used to defend the privileges of the ruling elite against domestic competitors. In these cases, it is the second phase of modern state-building, the nationalisation and democratisation of established state monopolies, that characterise the internal conflicts of the countries.
4) This unfinished process of state formation takes place in the context of a deep social transformation, which we try to comprehend with the help of concepts such as rural migration, urbanisation, mass unemployment, social marginalisation and the decline of traditional values. It is apparent that the institutional capacity of Middle Eastern states falls short of handling these societal conflicts, thus enhancing the legitimacy deficit of current regimes.

The *Arab Human Development Report 2002*, characterised Middle Eastern states by a "lack of accountability, transparency and integrity, along with ineffectiveness, inefficiency and unresponsiveness to the demands of peoples and development." The following report, the *Arab Human Development Report 2004*, reiterated these structural deficits and added the peoples' longing for freedom and justice in the region.[4] Taking these findings together with the precarious nature of the monopoly of physical force of Middle Eastern states, it comes as no surprise that, in post-colonial times, so many regional and local conflicts have escalated into wars. However, neither Arab narratives about colonial legacies and foreign interventions nor Western stereotypes about Middle Eastern or Islamic culture provide a sufficient explanation for the violent nature of Middle Eastern politics. The conflicts of the region have been closely related to the historical course of Middle Eastern state formation, and they are conditioned by both external and internal forces.

Since the end of the Second World War, a regional system of great power clientelism has emerged in which Middle Eastern states have acted as political rent seekers on the international level while pursuing relatively independent regional and domestic goals. Within this clientelistic arrangement, regional and international confrontations have become blurred, tying the Palestine conflict and the security situation in the Persian Gulf tightly together with the East-West conflict. This identification of international with regional perspectives has facilitated the extreme militarisation of Middle Eastern states, whose regimes use the thus-acquired means of force against both domestic and regional foes. In terms of international politics, this structural setting has been visible in the military escalation in Afghanistan, Palestine and the subsequent Gulf wars, as well as in the military and financial support that utterly authoritarian states such as Egypt, Iran, Iraq or Saudi Arabia have received from their international partners. Domestically, these resources contribute to stabilising authoritarian rule and to maintaining the gap between state and society. Negotiation processes between rulers and society, as we know them from the second phase of European state formation, have been thus essentially obstructed (cf. Krause, 1996).

Democratisation: A Means of Achieving International Security?

To a certain extent, the Bush Administration seems to have acknowledged the problematic character of the state in the Middle East and the way in which Western foreign policies have contributed to the resilience of authoritarianism in the region. In a speech at the

American University in Cairo, US Secretary of State, Condoleezza Rice stated that the US pursued stability at the expense of democracy in the Middle East. In their declaration of the "Partnership for Progress and a Common Future," the G-8 leaders confirmed this shift from stability to democracy at their summit on Sea Island on June 7, 2004. Not only Western politicians, but also many Western scholars have participated in this turn. Robert Blecher demonstrated how 'thinkers and pundits' such as Bernard Lewis, Daniel Pipes or Thomas Friedman have reoriented their policy prescriptions. At the time of the Gulf War in 1991, these shapers of US public opinion declared the Middle East immune to democratisation, whereas today they are joining the ranks of the proponents of democracy promotion in the region (Blecher, 2003). Democratising the Middle East has advanced to the status of a buzz theme in academia and journalism alike.

However well this rhetoric of democracy promotion might resonate in Western societies, it undoubtedly evokes reminiscences of colonial hegemony in the Middle East. Given the long struggle for independence in the Arab world, regional actors listen to this Western rhetoric of democratic reform with marked apprehension and do not view the current Western policies of peacemaking and state-building as less mercenary than European colonialism was. Rather, Middle Eastern audiences interpret them as an historical extension of the colonial encounter. In a recent public opinion poll in six Arab states, for instance, the overwhelming majority of the interviewees rejected outright the idea of present US foreign policies as being driven by democratic values and principles. Instead, they perceived them as totally in line with the economic and strategic self-interest of the United States.[5]

The elevation of democracy promotion to a policy implement to ensure international security signals a shift in the theoretical underpinning of Western foreign policies. In the Middle East, American foreign policy apparently has relied on assumptions drawn from classical balance of power theories, focusing on the military strength of regional states. The current idea of democracy promotion as a means of achieving peace and stability in the region, instead, relies on some popularised assumptions of the theoretical school of 'democratic' or 'liberal peace'. Putting its focus on regime types, this school suggests that domestic and international peace originates in the democratic nature of political systems (cf. Brown *et al.*, 1996; Doyle, 1986; Hegre *et al.*, 2001; Kim and Rousseau, 2005; Maoz, 1997;

Oneal and Russet, 1999, 2001; Russet, 1993; Williams, 2001). Looked at from a scholarly perspective, however, at least four qualifications should be made with regard to this theory of democratic peace.

First of all, there is no mutually accepted proof of the intrinsic relationship between the democratic structure of states and their peaceful interactions with each other. To a large extent this school in International Relations is based on the rather debatable statistical evidence that democratic states do not fight each other. Its basic assumption is that political institutions such as democratic liberties, the rule of law, electoral processes, and representative governments dramatically reduce the legitimacy of militaristic foreign policies, in particular between democracies based on market economies that recognise each other as being akin. Although it would be wrong to completely disregard the findings of this school, which can also claim roots in liberal political philosophy, it is equally flawed to take its assumptions for granted. Severe methodological problems concerning the applied definitions of democracies and the statistical viability of their samples have not yet been sorted out. So far, no academic consensus about the relationship between the democratic nature of and the peaceful interaction between states in the international system exists.

Secondly, while democratic institutions and procedures might be core elements in the establishment of lasting peaceful societal relations in the domestic realm, these relations build on consolidated structures not only in the political, but also in the legal and economic domains. Furthermore, democracies do not circumvent but rather engender conflicts. However, these societal conflicts are normally institutionalised and processed through non-violent means. Therefore, functioning democracies rely on consolidated monopolies of physical force, on a variety of non-political conditions, and on a minimal consensus of the population in the legitimacy of its formal procedures. It was against this background of institutionalised conflicts that Roland Paris questioned the idea of Liberal Internationalists to apply the competitive institutions of democracies and market economies in solving the problems of post-conflict societies (Paris, 1997).

His critique directly leads to my third, even more important qualification, the question of whether policies of democratisation will foster peace and stability in the environment of resilient authoritarianism or post-conflict societies. With regard to this question, most scholars agree on the conflict prone nature of democratisation processes. In particular societies under transformation from authoritarian to democratic rule show a high propensity for civil warfare. While democratic revolutions might result in peace and stability, their course has largely been characterised by outbursts of violence. From a theoretical point of view, the formation and nationalisation of state monopolies, still an ongoing process in Middle Eastern state formation, has a high propensity to engender instability and political violence. With regard to the more narrow aim of short-term stability in the Middle East, Oliver Schlumberger even pessimistically suggests it might be wiser to disengage from the idea of democracy promotion at all. Moreover, he reminds us of the failure of previous liberalisation policies. They have neither enhanced personal liberties nor led to more democracy on the macro-level. According to Schlumberger's analysis, the Western proponents of democracy promotion lack coherent political strategies that would allow donors to follow a clear path of engagement under the conditions of resilient authoritarianism that have characterised Middle Eastern politics so far (Schlumberger, 2006). In a similar way, this qualification applies to post-conflict societies. The successful implementation of electoral processes in previously war-torn societies needs first and foremost the building-up of state capacities and human capital. Yet, as Barnett Rubin has shown for Afghanistan, the doctrines and actions of donor states, international organisations and NGOs that are engaged in current state-building processes often rather tend to contradict their own goal: the building of viable and accountable state structures (Rubin, 2006).

Finally, it is flawed, if not even dangerous, to perceive policies of democracy promotion as the right means of fighting terrorism. On the one hand, we should recall that liberal democracies have not been free of terrorism themselves. In the 1970s and 1980s, a number of European states were confronted with terrorist organisations whose members had been recruited from the post-Second World War generations. These political extremists did not suffer under authoritarian regimes and social deprivation but in fact enjoyed the liberal rights and the social security of the democratic welfare state. They are a good example of the fact that a democratic system cannot completely prevent societal actors from pursuing their aims by violent means. Moreover, while democracy is a central element in the peaceful architecture of statehood, it also contains the seeds for conflict and even violent destabilisation (Gromes, 2005: 359). On the other hand, in the current poisoned political atmosphere between the West and the Arab world, the combination of coercive policies with an inflated discourse on democracy and human rights might even enhance the propensity among younger generations in the Middle East to pursue their political goals by violent means. The ongoing military struggles in Afghanistan and Iraq, as well as human rights violations in Abu Ghraib and Guantánamo, have clearly confirmed the basis, in fact of the accusation of Western double-standards in politics so popular among the Arab public.[6] Both cases aptly show the enormous discrepancies which can occur between the political rhetoric of democratisation and the realities of military and other coercive actions. Therefore, it is more than advisable not to justify applying coercive means in the fight against 'rogue states' and terrorist groups with the argument of promoting democracy in the region.

On closer analysis, the relationship between international security, regional stability, and democracy promotion is highly ambivalent, to say the least. While established democracies might be more peaceful in their interactions with each other, the very process of democratisation is prone to engender phases of instability and outbursts of collective violence. In this sense, democracy promotion seems indeed a questionable tool for enhancing security and stability in the short-term. In the long-run, however, a politically and economically reformed Middle East might be an asset for international security. In addition, Western foreign policy initiatives in the Middle East have to overcome the deep mistrust on the side of the region's political opposition. Whether Islamists, nationalists, or liberals, political movements in the region share the suspicion that the Western discourse on democracy is only a means of achieving rather different hegemonic political ends (Rougier, 2005: 89). Against this background, the support for "the democratic aspirations of all people" in the region has to be a genuine goal of Western foreign policy. Therefore, democracy promotion should not be presented as a means of ensuring international security, but as an end in itself.

Partners for Democracy Promotion: Rulers, Jurists and the Civil Society

Robert Dahl, the grand *seigneur* among American scholars on democracy, pointed out that we have to understand democracy as a process, in particular as a process for making binding decisions (Dahl, 1989). Yet this process rests in itself on historical conditions that are not the results of democratic decision-making. This applies in particular to the liberal, democratic nation-state that serves as a kind of institutional blueprint for the proponents of Liberal Internationalism. This model connects the idea of "rule by the people" with the institution of the modern state, based on its legitimate monopoly of physical force that the state administration exerts over its people and its territory. In comparison to this ideal type, Middle Eastern states lack essential conditions for a democratic polity and they often exist only as internationally accepted territories. In Iraq and Afghanistan, not only is it difficult to define a people who eventually should rule themselves, but the establishment and legitimisation of a monopoly of physical force has not yet been accomplished. In authoritarian states with a higher level of national integration such as Egypt or Iran, state-society relations have been characterised by the detachment of the state authorities from their people, with the former making political decisions according to their own particular interests.

In providing military security, state administration, economic prosperity, individual freedoms and the rule of law at the same time, current peace building efforts resemble attempts to square the circle. The current examples of 'shared sovereignty' (Krasner, 2004) are far from being state-like forms of political authority; rather they appear to be a kind of 'controlled anarchy' in which the interests and competencies of international, transnational, regional, national, and local actors compete with each other (Schlichte, 2003). In Barnett Rubin's words: "Electing officials to preside over a non-functional pseudo-state that can provide neither security nor services does not constitute democracy" (Rubin, 2006: 184). Democratising the Middle East first and foremost means developing the preconditions for a democratic process. Without a functioning state, a people, and a social order based on a certain level of economic and legal security, democracy degenerates into an electoral smokescreen.

This becomes apparent in previous attempts at democracy promotion by international donors which have rather increased the comparative level of illiberalism in the Arab world (Schlumberger, 2006). Yet it would be wrong to equate this resilience of authoritarian rule with a 'lack of modernisation'. Modernity and democracy are by no means synonyms, as is also proved by European totalitarianism. Applying the concept of "multiple modernities" (Eisenstadt, 2000), Middle Eastern authoritarianism should, rather, be interpreted as the specific form in which political modernisation in the region has evolved. Moreover, it is flawed to present the recent history of the Middle East as a mere period of political and social stagnation. Despite the continuation of authoritarian political structures, Middle Eastern societies have undergone dramatic changes. In order to promote democracy in the region, it is necessary to understand this interplay between continuity and change. In this process, Western foreign policies have to follow specifically designed strategies that take the different social and international environments into account under which individual Middle Eastern polities function. In the light of this critique, this paper will conclude with some tentative policy suggestions regarding three important groups of actors in the region: the regimes, the judiciary, and the civil society.

The Egyptian example shows that, although paying lip service to political and economic reforms, Middle Eastern regimes are not necessarily interested in genuine liberalisation policies. Maye Kassem (2006), for instance, analysed the recent reforms of the legislative elections, the restructuring of President Mubarak's National Democratic Party (NDP), the establishment of a National Council for Human Rights, and the Amendment of Article 76 of the Egyptian Constitution in order to allow direct multi-candidate presidential elections as a series of liberal democratic reforms, the only serious purpose of which is to reinforce and stabilise authoritarian rule. Policies of liberalisation may thus be initiated as part of the survival strategies of incumbent regimes. In this way, Western attempts to promote democracy and market economy in the region rather contradict the political aims of Middle Eastern state elites.

The liberalisation policies of the Mubarak regime camouflage a very different daily practice that is based on extended networks of patronage. These networks, however, are carefully designed adaptations to modern social contexts rather than persisting forms of tradition. They are knitted into modern politics and the particular survival strategies of the respective regimes and the distributive coalitions that have emerged around them. In principle, the current initiatives to promote liberal democracy and market economy in the region contradict these established social practices. It is therefore necessary to engage in a multilateral negotiation process that takes into consideration the different interests of state, society and individuals. It is important to strike a balance between these often-competing interests and to work towards bridging the gap between state and society. In this respect, Western initiatives should perhaps focus on supporting the development of legal accountability, of formal rules of law that can govern the hitherto almost entirely power-based state-society relations.

Instrumental for the development, implementation and maintenance of the rule of law is the judiciary. In some cases, the jurists have been co-opted by the regime; in other cases, however, only the judiciary has been able to confront the state elite in public. In Turkey and Iran, for example, the judiciary has been a core institution of the regime, strongly defending the existing authoritarian state structures and their respective Kemalist and Islamist ideologies against democratic reforms.[7] In other cases the jurists play a much more contributive role. Maye Kassem's (2006) analysis points out this different attitude of the judiciary in Egypt, the only public institution that has been able to openly criticise President Mubarak and his ruling elite (Kienle, 2001: 10).

In the current atmosphere of international legalisation, it is advisable to replace liberal by juridical internationalism (cf. Jacobsen, 1996 and Keohane *et al.*, 2000). The gradual establishment of independent judiciaries, applying a formal legal code as a standard for both sociopolitical conduct and their own professional ethics, would help not only to legalise state-society relations, but also to develop a political culture that is able to detach itself from the patrimonial heritage of the region. In contrast to the current prioritisation given to electoral mechanisms in democracy promotion, Western initiatives might consider shifting their focus to the implementation of legally grounded forms of public spaces and mutually accepted mechanisms of political accountability.

Given the enormous relevance of religious scholars of law in the Middle East, it would be advisable to include them in this process. However, this demands a certain revision of the predominantly negative image that the field of Islamic law holds in Western perceptions. We should be aware that very often only symbolical call to base state jurisdiction on the *Sharia* might have a double face. On the one side, it has served regimes and political movements as a means of both gaining religious legitimacy and forcefully suppressing internal opposition (Peters, 1994). On the other side, the authority of Islamic law has been put forward as a normative fundament for the promotion of legal accountability. From the latter perspective, the call for the *Sharia* has also to be read as protest against the predominant conditions of arbitrary rule. In morally binding both rulers and ruled, Islamic law and religious jurists could contribute to facilitating the move toward more political accountability (cf. Arabi, 2001).

By the same token, if Islamic culture in the Middle East is to be taken seriously, it is important to closely involve Islamic civil society associations in policies that attempt to democratise the Middle East. The overwhelming majority of grassroots organisations and non-state actors in the region have a religious background. To a large extent, the civil society of the region consists of Islamic organisations, and it is counterproductive to address Middle Eastern civil societies from the vantage point of a wrongly understood concept of Western secularism. Furthermore, in many Middle Eastern countries, Islamic organisations perform public services that in the West have been taken over by the welfare state. They often transcend the ethnic, social, or tribal fragmentation of society and provide social networks without which large parts of the population could hardly organise their lives. Whether we like it or not, currently it is the Islamic discourse that dominates Middle Eastern societies, and it is within this public discourse that Western policies have to resonate.

Therefore, any form of democracy promotion in the region has to take its Islamic culture seriously without turning religion into the sole explicative variable for the social and political developments in the Muslim world—a theoretical position which Maxime Rodinson once labelled theologocentrism, i.e., the attempt to explain all empirical phenomena in the Muslim world by references to Islam (Rodinson, 1988: 104). It goes without saying that this inclusion of Muslim and non-militant Islamist organisations in strategies of democracy promotion must go hand in hand with Western support for non-religious and secular-minded actors in civil society. As Bassma Kodmani has argued for the Egyptian case, it is important to cultivate societal diversity and to build new alliances within Middle Eastern societies in order to get rid of the legacies of deeply entrenched authoritarian rule (Kodmani, 2005: 21).

In addition to this acceptance of the predominantly Islamic character of Middle Eastern civil societies, economic policies have to address the fatal economic disintegration that characterises the region. In economic terms, it is thus essential to support those parts of civil society that produce for local and regional markets, instead of envisioning large free-trading zones in which the labour-intensive, small-scale entrepreneurs of the region might not be able to compete. The study by Oliver Schlumberger (2006) indicates that previous structural adjustment policies have not unleashed politically bounded market forces, but rather tied the commercial interests of the regional bourgeoisie more closely to the self-interest of the regimes. In this way, the economic top segment of regional civil societies is also benefiting from the authoritarian *status quo*. It also is this particular coalition between regimes and big businesses that hampers democratic reform and suffocates economic development. In supporting smaller enterprises, economic strategies also would bypass the widespread structures of crony capitalism that have accommodated the interest of big business with the will to power of the regimes.

In conclusion, one-size-fits-all solutions for democratic and economic reform in the Middle East do not exist. Promoting democracy in the region is not a simple task. Breaking up the entrenched structures of authoritarianism while attempting to build up a state, as is the case in Afghanistan and Iraq, entails high political risk and requires meticulously calibrated strategies which address various regional and local actors. There is no blueprint for reform and no guarantee that democratisation will automatically come along with security and stability. However, the negative balance of decades of state-centred security policies in the region begs taking the risk. Yet this shift in foreign policy only makes sense if democracy promotion becomes an end in itself. In the light of the atmosphere of suspicion and mistrust that so far has characterised the political climate between Western countries and the Middle East, it would be disastrous to view democracy promotion as a mere means of expedience for implementing Western security interests.

Notes

1. Quoted after Said (1978: 5).
2. Regarding the history and recent developments of these two cases, see: Bellin (2004-05); Dodge (2003, 2005, 2006); Dorronsoro (2005); Papagianni (2005); Rubin (1995, 2006); Synnott (2005).
3. The research results of the AKUF are accessible on *www.akuf.de.*
4. The Arab Human Development Reports are accessible via: *http://www.rbas.undp.org/ahdr.cfm.*
5. In a public opinion poll conducted in May 2004 by Zogby International and the University of Maryland, democracy scored lowest as motivation for the US intervention in Iraq. Most respondents viewed the intervention as being motivated by the desire to control Iraqi oil and to protect Israel, as well as by the intention to dominate and weaken the Muslim world ("Arab Attitudes Towards Political and Social Issues, Foreign Policy and the Media," accessed on August 1, 2005, *www.bsos.umd.edu/SADAT/pub/*).
6. There is no doubt that future Western interventions in post-colonial state-building processes will also have to face these double-standard accusations. From a scholarly perspective, Stephen Krasner was certainly right to perceive the international norm of state sovereignty as a form of "organized hypocrisy" (1999). Yet his policy advice to legitimize Western interventions in state-building processes by different forms of "shared sovereignty" or new kinds of international "trusteeships" (Krasner, 2004: 119).

   seems to be rather questionable. Even if badly governed or collapsed states were indeed the major problem of international security, something that seems to be more the result of an inflated rhetoric than of empirical findings (cf. Jung, 2005, Schlichte, 2005), openly abandoning the established norms of sovereignty is not only a very sensitive issue that does not resonate very well in the Third World, but also might have a very negative impact on international security in general. Moreover, the label of shared sovereignty hides the fact that, in the end, international power relations rather than the norms of "good governance" will decide who is qualified to claim sovereignty and who not.
7. For a comparison of state-society relations of Turkey and Iran with regard to the religious field, see Jung, 2004: 69-75.

References

Arabi, Oussama (2001). "The Dawning of the Third Millennium on Shari´a: Egypt's Law No. 1 of 2000, or Women May Divorce at Will", *Arab Law Quarterly,* Vol. 16(1), pp.2-21.

Beilin, Eva (2004-05). "The Iraqi Intervention and Democracy in Comparative Historical Perspective," *Political Science Quarterly,* Vol. 119(4), pp.595-608.

Blechner, Robert (2003). "Free People Will Set the Course of History'. Intellectuals, Democracy and American Empire, *Middle East Report Online,* March, accessed in October 2005.

Brown, M.E., S.M. Lynn-Jones, and S.E. Miller (eds.) (1996). *Debating the Democratic Peace,* MIT Press, Cambridge, Mass.

Dahl, Robert A. (1989). *Democracy and Its Critics,* Yale University Press, New Haven and London.

Dodge, Toby (2003). *Inventing Iraq: The Failure of Nation Building and a History Denied,* Columbia University Press, New York.

———. (2005). *Iraq's Future: The Aftermath of Regime-Change,* Adelphi Paper 372, The International Institute for Strategic Studies, London.

———. (2006). "Iraq: The Contradictions of Exogenous State-Building in Historical Perspective", *Third World Quarterly,* Vol. 27(1), pp.187-200.

Doronsoro, Gilles (2005). *Revolution Unending-Afghanistan: 1979 to the Present,* translated from the French by John King, Hurst & Company, London.

Doyle, Michael (1986). "Democracy and Management of International Conflict", *Journal of Conflict Resolution,* Vol. 37, pp.42-68.

Eisenstadt, S.N. (2000). "Multiple Modernities", *Daedalus,* Vol. 129(1), pp.1-29.

Elias Norbert (1994). *The Civilizing Process. The History of Manners and State Formation and Civilization,* Blackwell, Oxford.

Gromes, Thorsten (2005). "Innerstaatliche Konflikte und die Wirkung von Demokratie", in Egbert Jahn, Sabine Fischer and Astrid Sahm (eds.), *Die Zukunft des Friedens. Band 2: Die Friedens-und Konfliktforschung aus der Perspektive der jüngeren Generation,* Wiesbaden, VS Verlag für Sozialwissenschaften.

Hegre, Håvard, Tanja Ellingsen, Scott Gates, and Nils Petter Gleditsch (2001). "Toward a 'Democratic Civil Peace'? Democracy, Political Change, and Civil War. 1816-1992", *American Political Science Review,* Vol. 1 (March), pp.33-48.

Henry, Clement M. and Robert Springborg (2001). *Globalization and the Politics of Development in the Middle East,* Cambridge University Press, Cambridge.

Hulsman, John C. (2005). "Lawrence of Arabia and the Perils of State Building", *Heritage Lectures,* 900, 2005, *www.heritage.org/research/middleeast/Iraq/hl900.cfm.*

Jacobson, David (1996). *Rights Across Borders. Immigration and the Decline of Citizenship,* Johns Hopkins University Press, Baltimore and London.

Jung, Dietrich (2001). "The Political Sociology of World Society", *European Journal of International Relations,* Vol. 7(4), pp.443-474.

———. (2004). "Globalization, State Formation and Religion in the Middle East: Is Islam Incompatible with Democracy?", *Distinktion: Scandinavian Journal of Social Theory,* Vol. 8, pp.61-78.

———. (2005). "New Wars, Old Warriors and Transnational Crime. Reflections on the Transformation of War", *Cooperation & Conflict,* Vol. 40(4), pp.423-34.

Jung, Dietrich, with Wolfango Piccoli (2001). *Turkey at the Crossroads. Ottoman Legacies and a Greater Middle East,* ZED Books, London.

Kassem, Maye (2006) "Democratization Reforms as a Means of Stabilizing Authoritarian Rule in Contemporary Egypt", in Dietrich Jung (ed.), *Democracy and Development: New Political Strategies for the Middle East?* New York: Palgrave.

Keohane, Robert O., Andrew Moravcsik and Anne-Marie Slaughter (2000). "Legalized Dispute Resolution: Interstate and Transnational", *International Organization,* Vol. 54(3), pp.457-488.

Kienle, Eberhard (2001). *A Grand Delusion. Democracy and Economic Reform in Egypt,* I.B. Tauris, London.

Kim, Hyung Min and David L. Rousseau (2005). "The Classical Liberals Were Half Right (or Half Wrong): New Tests of the 'Liberal Peace,' 1960-88", *Journal of Peace Research,* Vol. 42(5), pp.523-543.

Kodmani, Bassma (2005). "The Dangers of Political Exclusion: Egypt's Islamist Problem", *Carnegie Papers: Middle East Series,* Number 63, October, *www.CarnegieEndowment.org/pubs.*

Krasner, David (1999). *Sovereignty: Organized Hypocrisy,* Princeton University Press, Princeton.

———. (2004). "Sharing Sovereignty. New Institutions for Collapsed and Failing States", *International Security,* Vol. 29(2), pp.85-120.

Krause, Keith (1996). 'Insecurity and State Formation in the Global Military Order: The Middle Eastern Case', *European Journal of International Relations,* Vol. 2(3), pp.319-354.

Maoz, Zeev (1997). "The Controversy Over the Democratic Peace: Rearguard Action or Cracks in the Wall?", *International Security,* Vol. 22(1), pp.162-198.

Oneal, John R. and Bruce Russet (1999). "Assessing the Liberal Peace with Alternative Specifications: Trade Still Reduces Conflict", *Journal of Peace Research,* Vol. 36(4), pp.423-42.

———. (2001). *Triangulating Peace. Democracy Interdependence and International Organizations*, W. W. Norton & Company, New York.

Papagianni, Katia (2005). "Transitional Politics in Afghanistan and Iraq: Inclusion, Consultation and Public Participation", *Development in Practice*, Vol. 15(6), pp.747-759.

Paris, Roland (1997). "Peacebuilding and the Limits of Liberal Internationalism", *International Security*, Vol. 22(2), pp.54-89.

———. (2002). "International Peacebuilding and the Mission Civilisatrice", *Review of International Studies*, Vol. 28, pp.637-56.

Peters, Rudolph (1994). "The Islamization of Criminal Law: A Comparative Analysis", *Die Welt des Islams*, Vol.34, pp.246-73.

Rodinson, Maxime (1988). *Europe and the Mystique of Islam*, translated by Roger Veinus (French 1980), I.B. Tauris, London.

Rougier, Bernard (2005). "Le Grand Moyen-Orient: un moment d'utopie internationale?" *Critique Internationale*, Vol. 26 (janvier), pp.79-94.

Rubin, Barnett R. (1995). *The Fragmentation of Afghanistan: State Formation and Collapse in the International System*, Yale University Press, New Haven.

———. (2006). "Peace Building and State-Building in Afghanistan: Constructing Sovereignty for Whose Security?," *Third World Quarterly*, Vol. 27(1), pp.175-185.

Russet, Bruce (1993). *Grasping the Democratic Peace: Principles for a Post-Cold-War World*, Princeton University Press, Princeton.

Said, Edward W. (1978). *Orientalism*, Vintage, New York.

Schlichte, Klaus (2003). "State Formation and the Economy of Intra-State Wars," in Dietrich Jung (ed.): *Shadow Globalization, Ethnic Conflicts and New Wars: A Political Economy of Intra-State War*, Routledge, London and New York.

———. (2005). "Gibt es überhaupt 'Staatszerfall'? Anmerkungen zu einer ausufernden Debatte", *Berliner Debatte Initial*, Vol. 16(4), 74-84.

Schlumberger, Oliver (2006). "Dancing With Wolves: Dilemmas of Democracy Promotion in Authoritarian Contexts", in Dietrich Jung (ed.) *Democracy and Development. New Political Strategies for the Middle East?*, Palgrave, New York.

Synnott, Hilary (2005). "State-building in Southern Iraq", *Survival*, Vol. 47(2), 33-56.

Tilly, Charles (1990). *Coercion, Capital and European States, AD 990-1990*, Basil Blackwell, Cambridge.

Weber, Max (1968). *Economy and Society. An Outline of Interpretive Sociology*, edited by Guenther Roth and Claus Wittich, Volume I, Bedminister Press, New York.

———. (1991). *From Max Weber: Essays in Sociology*, edited, with an introduction by H.H. Gerth and C.Wright Mills, with a new preface by Bryan S. Turner, Routledge, London.

Williams, Michael (2001). "The Discipline of the Democratic Peace: Kant, Liberalism and the Social Construction of Security Communities", *European Journal of International Relations*, Vol. 7(4), pp.525-553.

23

The Political Reforms in West Asia after 9/11

SAHAR A. AL-MAJALI

Arab countries of West Asia differ significantly in size, population, and levels of economic and social development. Besides, Arab societies are complex, with religious, tribal and ethnic affiliations often exerting more influence on people's thinking and actions than the concept of nationality. Such diversity throws up unique challenges when it comes to promoting good governance. Imbalances in power and influence—political, economic, religious or military—have contributed to the development of extremism, the denial of human rights, and civil conflict in some of the societies. International terrorism, and some governments' attempts to counter it, has wielded adverse impact on the process of granting democratic freedoms as the fight against terror has proved instrumental in slowing down the process of political reforms.[1]

Impact of 9/11

The terrorist attacks of 9/11 have spawned deliberations on accelerating the pace of promotion of political reforms in the countries of West Asia. The ongoing developments in Iraq have further energised this transatlantic debate. This scenario has brought the United States and Europe into a closer exploration of each other's actual and intended approaches to promotion of political reforms in the West Asian region. Viewed in a broad perspective, these debates have centred on the agenda of Group of Eight (G-8)'s June 2004 Summit, EU–US Summits and the United States' proposed Greater Middle East Initiative. Despite initial reservations about each other's proposals, the EU and the United States seem to have reconciled to the ground realities by acknowledging the EU's more pervasive presence in much of West Asia and the EU is engaged with new US initiatives that appear to address the EU's own pleas for a focus on the root causes of instability.

In the aftermath of the attacks of September 11, Europeans have exhibited striking commonality with US policy reassessments. European ministers and policy makers have frequently asserted a link between terrorism and political repression. Tony Blair has argued on many occasions that security can best be achieved by "spreading our values." European Commissioner Chris Patten has expressed the opinion that, "fostering human rights should become an integral part of the fight against terrorism."[2] "The best protection of our security is a world of well-governed democratic states"[3] is the new message emanating from the new European security strategy, agreed upon in December 2003. Recently, German Foreign Minister Joschka Fischer was reported to have said that, "security is a broader concept in this fight against terrorism: social and cultural modernisation, as well as democracy...are of almost greater importance [than traditional security issues]."[4] The former French Foreign Minister Dominique de Villepin, also struck a similar logic, in acknowledging that security thinking could "no longer be explained by a series of alliances."[5]

The EU's enthusiasm for promoting political reforms in West Asia can be gauged from various new initiatives introduced by European governments recently. A "Task Force for Dialogue with the Islamic World" was set up in Germany with one million euro 'Anti-Terror Package' identified support for governance reform as a top priority. France devoted an increased 474 million francs to governance work in 2002–03, basically through an expanded *zone de solidarité prioritaire* that included Lebanon, the Palestinian Territories, Yemen, Algeria, Morocco, and Tunisia. The UK launched a new Arab Reform strategy by making available an additional amount of £7 million under a new conflict prevention fund, with governance, security sector reform, and 'engagement with political movements' outlined as priority areas. The Danish government has introduced a similar initiative on Arab reform, with €15 million a year available for new projects.

In the post-Cold War era characterised by the processes of globalisation, liberal economic reforms and privatisation to attract the FDI investments, many countries of West Asia are gradually opening up to the possibility of political change. There seems a new willingness emerging on the part of many Arab governments to admit the essentiality of some sort of political change. "Even the most conservative among them are willing to say that Arab countries are bound to change politically, although in their own fashion and in their own time."[6] Arab intellectuals are articulating their views about the need for change more openly than before. The Arab press is crowded with debates about reform and democracy. The civil society organisations, and business groups in some West Asian countries are usually issuing democratic manifestos which are creating a new and confusing array of declarations. Arab organisations issued the Sanaa Declaration, the Alexandria Declaration, and the Beirut Declaration and these have been joined by innumerable statements about democracy from the Arab League, foreign ministers meeting with representatives of the G-8 in Rabat, and others.[7]

The initial enthusiasm for 'reform trends' has spurred optimistic predictions that the region is, at length, responding to the global trend toward democracy. It has been argued by Saad Eddin Ibrahim, Egyptian sociologist, that the prospects for liberal democracy in the region have never been so bright.[8] American journalist concluded after a visit to West Asia, Fareed Zakaria, an American journalist, concluded that, "everywhere in the Arab world, people are talking about reform...the wind is behind those who advocate free-market, modern, Western-style reforms."[9] President George W. Bush spoke of Arab democratisation as a certainty and in a June 2004 speech he declared, "voices in that region are increasingly demanding reform and democratic change...now freedom is stirring in the Middle East and no one should bet against it."[10]

Broadly speaking, the definite steps adhered to by Arab governments to reform their political systems fall short of the much-hyped rhetoric, although some reform is taking place. The 'eloquent declarations' and the debate over democracy in the media in most of the Arab countries is within the limits of the law of the land so as not to incur the wrath of 'still intolerant' regimes. In other words, substantial political change is still a far cry. However, optimism looms large in the slowly evolving penchant for reforms due to domestic craving and external pressures.

Agenda for Political Reforms

What should constitute the 'Agenda for Political Reforms' in the Arab countries of West Asia is a controversial issue that lacks unanimity both among Arab as well as Western scholars. A segment of Arabs seems wary of imposition of Western style of democracy. It is seen as an alien Western construct designed to provide the West greater clout in economic and political affairs of the region. It is further argued that Western style of democracy is detrimental to the cultural and social mores of Arab society that could spell the breakdown of social cohesion and order. Reformers call for an Arab 'way forward' which is outward looking and embraces ideas from the wider world.

The system of checks and balances between the executive and legislature in some Arab countries is either lacking or not adequately defined. Parliaments exist mainly to enforce decisions taken by the ruling elite, and lack teeth to investigate or challenge decisions. This scenario is unlikely to undergo substantial transformation in the immediate near future because 'undemocratic governments' are tolerated, and supported, by the West because they provide a stability that suits Western politico-economic and strategic interests.

The growth of a wide gamut of policy-based political parties and efficient civil society organisations is considered as *sine qua non* for the process of democratisation. The feasibility of development of Western-type secular political parties, open to all irrespective of religion, is ruled out in the Arab scenario where over 95 per cent of the population is adherent of Sunni sect of Islam. Islam is a central element in education, the law and in people's everyday lives. Similarly, the influence of Islam on politics is pervasive. This factor can be utilised to cement the bonds of brotherhood, equity, compassion and goodwill between the ruler and the ruled in accordance with the tenets of Islam. At the same time, there is also a need for encouraging women's political participation by special transitional measures such as quotas for women's representation. Undoubtedly, women's participation in politics is making headlines in countries like Kuwait, Jordan and other countries but more needs to be done for their increased participation.

Unlike the West, a complete separation of politics and religion is not possible in the Arab context. The strong constituency for 'political Islam' deserves a political voice. The 'Islamists', a not-so-well defined term for the proponents of 'extreme manifestations' of Islam, are required to be involved into the process of political reforms because their talent, energy and clout with the common masses can be utilised for the benefit, sociopolitical cohesion and peace and stability of the society rather than outright rejection. Moderate Islamists can be helpful in stemming the tide of extremism and helping the others of their ilk to join the national mainstream. Engaging the Islamists in dialogue is better than rejecting them because their sway over the masses in streets and rural areas is more than other leaders or political activists. Thus, the need of the hour is to strengthen the forces of peace and stability so that the process of political reforms—a long-term goal—can be reached by gradually involving all the segments of the population, particularly the women.

Rationally organised and adequately resourced civil society can undoubtedly serve as an effective instrument for change in any country. The evolution of non-governmental organisations (NGOs) is still at a nascent stage in most of the countries of West Asia. The development and influence of existing NGOs and pressure groups is spasmodic in this part of globe. The fear of NGOs being suppressed or being taken over by the governments themselves always looms large.

The civil society organisations (CSOs) and movements regarding human rights, the promotion of women's interests, and the defence of media freedom have witnessed incremental growth in some Arab countries where they have come to wield increasing influence. The *bona fide* civil society organisations should be encouraged to develop positive and viable relations with the local government in order to serve as a genuine partner to achieve common objectives and contribute to policy. It devolves on the CSOs to broaden their membership and resource bases, develop regional networks for the exchange of good practice and engage openly and constructively with the governments of the countries in which they operate. In this manner, NGOs/CSOs can ward off unnecessary pressure of strong governments and can play a positive and constructive role in the successful promotion of reform.

Role of Media

Media, both print and electronic, has come a long way in carving out a niche of its own in the Arab world while confronting the odds of severe restrictions and even strict censorship. Media is the most potent means of enlightening the public opinion about the right perspective on impending political reforms. Undoubtedly, the Arab media is vibrant and diverse, it needs to be further strengthened to play its designated role of 'inform and reform' the people. There is a need to improve laws safeguarding freedom of speech. The advent of the Internet and satellite TV has opened new vistas for communicating with ease. This new medium has eased and broadened uncensored access to information.

The Doha-based Arab satellite news channel Al-Jazzier, launched in 1996, has come to wield a powerful effect on local politics. The network through its countless news articles, reports, and commentaries, has been giving CNN, BBC, CNBC and other networks a good run for their money in the region. There are other networks like Saudi-owned Middle East Broadcasting Center (MBC) operating from Dubai's 'Media City', Abu Dhabi Television, and other state-owned television networks. However, Al-Jazeera has redefined the format and content of Arab news networks. Throughout the region, opponents of Arab regimes across the political spectrum, Israelis, Americans, and Europeans have had an opportunity to offer their views on this channel. There have been attempts at banning Al-Jazeera or close its offices in different countries but it continues to survive the odds.

West Asian countries lag behind other regions in Internet connectivity. There prevails censorship in some Arab countries, prominently Saudi Arabia, Syria, and Tunisia, on what can be seen on websites in these countries or monitor what people are looking at on the Web. The Internet is a valuable tool for the people to reach beyond the borders of the state-controlled press. Web logs or 'blogs' are finding favour with the Arab world for their unvarnished commentary on important issues of the day. For instance, Bahrainis dissatisfied with the way the traditional press covered the 2002 scandal involving the national pension fund, GOSI, could visit the 'Bahraini blogsite' or 'Mahmood's Den' on the Internet for a more trenchant analysis of that episode.[11]

Apart from the revolutionary impact being wielded by 'new media,' newspapers and magazines in the Arab World are also undergoing substantial transformation both in content and form. They are exploring new avenues to engage in debates that were hitherto off-limits. *Cairo Times,* the English-language weekly has set a standard for outspoken criticism of the Egyptian government. *Al-Masri al-Yom,* launched in mid-2004, in Egypt, is setting new standards of journalism. Jordan's privately owned Arabic daily, *Al-Ghad*, Saudi Arabia's *al-Watan al-Hayat*, have published articles openly questioning a variety of previously taboo subjects including the role of women in society, extremism, and the intersection of religion and politics. These are positive developments that are indicative of a significant opening in Arab societies. The emergence of the new Arab media is a positive development which can be further encouraged to be the harbinger of political reforms by becoming a link between the people and the government to ward off any outside intervention.

Role of Education and Information Technology

Education is an essential ingredient to encourage participatory governance. Educated people are better equipped to discern and appreciate the working and governance of their society. They are better placed to make informed political and economic decisions as well as articulate their needs. Levels of education differ in countries of West Asia, particularly for families in the rural areas. Women are best agents of social change and development. Hence, priority must be accorded to female education. Illiteracy rates, particularly amongst women and girls, are among the highest in the world. Special attention needs to be focused on providing primary or secondary education. Besides, specific measures are to be designed so as to ensure access to education for the less well-off. Socioreligious 'obstacles' to formal education are required to be warded off.

Many countries in West Asia, particularly the Gulf countries, are dependent on imported labour for skilled and semi-skilled jobs thereby resulting in the flight of valuable foreign exchange. This trend can be reversed by establishing more technical institutions and Information Technology can also be harnessed to keep pace with global trends. There are strong internal pressures for wholesale reform of education systems, not least because of the need for Arab countries to keep up with a rapidly globalising world. Extensive use of information technology and ITES-related services will augur well for countries like Egypt, Jordan, Syria and other countries of the region. This will go a long way in making the region as a business and technology hub of the Arab world while catering to the requirements of Europe in this regard. This will also help reduce unemployment and afford an opportunity to engage women in more gainful employment.

Role of the External Actors

The European Union and the United States are two main external actors that wield considerable impact on the developments in West Asia. There is a widespread feeling among some segments of population in some countries of the region that the US presence in West Asia, notably in Iraq and the perennial ongoing Arab-Israel conflict have prevented reform, not least because they generate extremism. However, others have argued that the process of democratic reform is too important to be delayed or sidelined until these issues are settled. Yet robust US engagement to resolve the Israeli-Palestinian dispute would, many believe, much ameliorate Arab mistrust of US intentions in the region and gain the US greater credibility as an advocate of democracy.

The United States intends to promote democracy in the Arab world and to help remove the conditions that encourage extremism and terrorism in the region. While some question US motives, doubts are also expressed about the *bona fides* of this policy and its durability. The US should act in tandem with the European Union to raise, with all countries of this region, principles of human rights, political representation, tolerance, the rule of law, women's rights and transparency in decision making. The issue of promoting political reforms is a long-term process that needs to be sustained as a matter of policy. Undoubtedly, Iraq provides a major test for US policy. The establishment of security and of a properly functioning democracy in that country will constitute a model worth emulating for other countries of the region. The current realities in Iraq have created more room for doubts and intentions of Washington than raising optimism about positive hopes for others.

The US Government works through bilateral and multilateral governmental initiatives and provides assistance through such initiatives as the G-8's Broader Middle East and North Africa Initiative and the bilateral Middle East Partnership Initiative (MEPI). The G-8's Forum for the Future is regarded to have made a promising start, but needs to be developed into a meeting point where governments, the private sector and NGOs are involved on an equal footing.

There is widespread agreement that reform cannot be imposed on the Middle East but has to be generated by the governments and people of the region. External actors can endeavour to assist. To do so, there is a need to develop a country-by-country strategy, contextually sensitive, and broaden contact with actors in the region beyond a relatively narrow constituency of liberal and secular circles. Arab countries and the European Union (EU) have strong economic and trade relations as well as geographical proximity and cultural affinity. Some question whether the EU has sufficiently used its assets to promote political reform, despite its longstanding funding of NGOs in the region. Recent policy initiatives are intended to remedy this by providing benchmarks and incentives. There is a dire need to maintain consistency in implementing this new approach.

Initiative for reforms should emanate from the countries of the region and while doing so, the regional actors should work out details in this regard to involve the non-state actors, particularly the civil society organisations as well as credible NGOs to embark on the process of reforms designed to work for the benefit of the native population. The factor of involving all segments of society, including the Islamists, should be involved so that a unanimous decision is arrived at. The sentiments of the people, particularly their attitude towards the United States as well as the European Union, should be taken into consideration before embarking on any cooperative endeavour with either of them.

Conclusion

The success of reforms depends on the mindset of the people for whom the reforms are addressed. Thus, it is essential to prepare ground for generating awareness among the people about the objectives for which the proposed reforms are intended. Media, both print and electronic, can play pivotal role in this regard. Besides, the authorities in the concerned countries are called upon to introduce incremental doses of reforms keeping in consonance with the local needs. Reforms can not be imposed overnight from the above by transplanting the Western model. Any model designed to be applied in the Arab region should be based on genuine needs and aspirations of the Arab people imbibing Islamic and Arab values. Western ideas of freedom, equality, rule of the law, etc. are to be tailored in accordance with the local and regional mores, customs and traditions in order to make them acceptable to the people.

Despite country-specific differences, the conditions required for reform to succeed *inter alia* are concerted popular demand for change: citizens should know their rights and be able to pursue them, which requires respect for fundamental freedoms such as freedom of expression and assembly, etc. It is equally important to have strong, stable, internally democratic and visionary institutions that will help guide society through the process of change, even if it is slow and painful. Adequate and well-concerted steps are called for to prepare leadership cadres at the grassroots level by involving youth through democratic means so that they get properly trained and gain sufficient experience to represent their constituencies in the provincial and national legislatures to formulate policies. The judiciary, civil society organisations, religious leaders, professional or labour groups, and a responsible and professional media are significant players in the process. Specific attention is required to be paid to involve women and youth as effective agents in promoting reform.

The recipient countries accepting external help, either from the United States or European Union, in the form of support for reforms, including financial and other development assistance, have to be cautious about the real intentions of the aid provider. In order to help mitigate the negative effects of change, and to train, for example, security and defence forces in their role as guarantors of democracy and security, there is also need to exercise caution that such assistance contributes to national and regional peace and security and does not exacerbate tension or jeopardise the ongoing peace process, say the Palestinian-Israel dialogue.

The mutual spirit of developing true partnerships and promoting reconciliation can go a long way in providing meaningful manifestations to genuine aspirations for reforms. Far more can be achieved once all actors in society try to resolve animosities and tensions through peaceful means and preferably by peaceful negotiations and shun recourse to violence. It is equally important that all actors involved in the process of reforms should share common objectives and interests. In the international arena, there has to be more transparency and inclusiveness. Western governments must be prepared to listen to the views of all representative groups, not just those that tell them what they want to hear. The modalities for harnessing the assistance programmes must be worked out and owned by the countries to which they are directed and not designed in western capitals to further their own political or economic agendas.

Notes and References

1. Mattinson, Stephan (2005). "Building Participatory Governance and Representative Institutions in the Broader Middle East and North Africa", *Conference Report on Wilton Park Conference* 779, London, March 7–10, p.1.

2. Patten, Chris (2003). Special Seminar with NGOs (speech, Brussels, July 14, 2003), available at *http://europa.eu.int/comm/external _relations/w29/2.htm*, accessed on November 12, 2004.

3. "A Secure Europe in a Better World", European Security Strategy, adopted by the heads of State and Government at the European Council (Brussels, December 2003) p.10, available at *http://www.iss-eu.org/solana/solanae.pdf*, accessed on November12, 2004.

4. Fischer, Joschka (2005). Speech to the 40th Munich conference on Security Policy, February 7, 2004, available at *www.securityconference.de/*, accessed November 17, 2005.

5. de Villepin, Dominique Law,Force and Justice (speech, International Institute for Strategic Studies London,March 27, 2003), available at www.ambafrance-us.org/news/statmnts/2003/villepin_iraq032703.asp, accessed on November 22, 2004.

6. Ottaway, Marina (2005). "Evaluating Middle East Reform: How Do We Know When It Is Significant?", *Carnegie Papers, Middle East Series*, Number 56, February, p.4.

7. Ibid.

8. For details see, Ibrahim, Saad Eddin (2004). "An Open Door", *Wilson Quarterly*, Vol. 28: 2, Spring, pp.36–46.

9. Zakaria, Fareed (2004). "The Good, the Bad, the Ugly", *Newsweek*, May 31, 2004.

10. Remarks by George W. Bush, at the Air Force Academy Graduation, Colorado Springs, Colorado, June 2, 2004, available at *http://www.whitehouse.gov/news/releases/2004/06/20040602.html*, accessed on October 29, 2005.

11. Council on Foreign Relations (2005). "In Support of Arab Democracy: Why and How", *Report of an Independent Task Force*, Washington, DC, p.31, available at *www.cfr.org.*, accessed on November 21, 2005.

24

Political Reforms in Saudi Arabia

The Prospects

GULSHAN DIETL

Since the mid-1990s, in a relatively short span, there were six regime changes in the Arab world. In Qatar, King Khalifah was overthrown and King Hussein of Jordan, King Hassan of Morocco, King Isa of Bahrain, President Hafez al-Asad of Syria and President Yasser Arafat of Palestine died. In the recent past, the death of King Fahd in Saudi Arabia has been followed by the deaths of King Maktoum of Dubai and King Jaber of Kuwait.

The first wave of fresh faces at the helm had set off a flurry of expectations. The 'Arab Spring' was welcomed among a cacophony of voices for reform, elections, constitutional changes, women empowerment, an end of cronyism and corruption, and so on. However, things did not change much after that. Even a pretence of changes was quickly abandoned. The reaction the second time around is muted by comparison.

This paper proposes to scrutinise political reforms in Saudi Arabia in the above context. It seeks to bring out that there have been no major departures from the normal political evolution and makes a case that similar trends may not replicate themselves elsewhere.

The Saudi Royal Family

The Sauds were tribal rulers in a small principality of Diriyya near Riyadh. It was Abdul Aziz who laid the foundation of the present Saudi state and expanded it to its present boundaries. Simultaneously, he upgraded his own designation. Thus, he took the title of Emir in 1902 on the capture of Riyadh, Sultan in 1921, and Malik in 1926. In October 1986, King Fahd officially adopted the title Khadim al-Haramain (Servant of the Two Holy Shrines); emphasising the religious aspect of his authority.

After Abdul Aziz, the succession in the family has been horizontal; the Crown passing from brother to brother. Behind this (yet another unique feature of the country!) form of succession lies an accident of history. Abdul Aziz's first son died young; Saud was the second son and Feisal the third. Abdul Aziz spotted his rightful successor in Feisal and roped him into the Saudi government system and diplomatic missions from the early age of 14. Saud, however, could not be bypassed and, therefore, when he was declared Wali al-Ahd or the Crown Prince, a simultaneous declaration was made that Feisal was to succeed him.

The Saudi royal family is not just the sons of Abdul Aziz, but also their sons and grandsons, and the descendants of the allied branches of the extended clan. The clan is estimated to number roughly 5,000 male members out of whom nearly 200 to 300 are active in the government. The Abdul Aziz family is the strongest within the clan; in fact, the line of succession does not extend beyond this family. It is made up of 30-plus sons, 100-plus grandsons and 500-plus great grandsons of the founder king.

Within the Abdul Aziz family, there are further divisions. King Fahd headed the most formidable of these—the Sudairis. They are popularly called the "Sudairi Seven," the full-brothers whose mother came from the Sudair tribe. Fahd was the eldest of them. Sultan is now the senior most Sudairi and the Deputy Prime Minister and Minister of Defence. Nayef is the Minister of Interior, Ahmad is his deputy, Salman is the Governor of Riyadh. Among the second generation of Sudairis, Bandar bin Sultan is the most prominent. After a long stint as the Kingdom's ambassador to the United States, he has returned to head the newly constituted National Security Council. The Sudairis, though united against the others, have not been free of petty feuds among themselves.

King Abdullah, the Prime Minister and the Chief of National Guard, is considered a counterweight to the Sudairis. His mother belonged to the Shammar tribe in the north of the country, where he retains a strong political base. He has no brother.

The third major group is the Bani-Feisal group. It consists of the eight sons of King Feisal, who ruled from 1964 to 1975. Prince Saud al-Feisal, the Kingdom's Foreign Minister, is the most prominent among them. Turki al-Feisal is the Saudi ambassador in the US.

Each of the factions within the family competes for superiority and a bigger largesse on the basis of its political weight, numerical strength and closeness to the ruling monarch.

The Saudi Royal Council

In Saudi parlance, the term "Ahl al-Aqd wal-Hall" refers to the high and mighty. Literally translated, the term would mean "the people who bind and loosen." A more accurate translation would perhaps be "the decision makers." It is a term used for the Royal Council, which consists of a select group of princes and leading *ulama.*

The exact composition of the Royal Council is never divulged; in fact, it could vary according to the seriousness and urgency of the issue at hand. Its powers and functions are still shrouded in a denser mist. It has so far been mentioned only in times of grave crises. This would suggest that the Council is a purely an ad hoc body, called only for temporary fire-fighting functions.

There is too much at stake for the princes to operate without a functioning hierarchy of their own. The Saudi princes and princesses get annual royal pensions. The amount is fixed on the basis of seniority, political influence, and the fact of a prince being in favour or otherwise at a given time. Then, there is the royal representation in the Council of Ministers. How many should be appointed as ministers, who should be given what portfolios, etc. are decisions that cannot be taken by the King alone. Similarly, the governorships of the provinces and a number of other political and administrative posts are held by the Saudi princes. All these things need continuous deliberation. To that extent, the Royal Council is the rulers' power-base and also acts as a check on his power.

In 2000, the then Crown Prince Abdullah finally set up a publicly announced Council of the Royal Family. It was to be headed by him personally. The purpose was not so much to manage the affairs of the royal family, but to appease public opinion by divorcing family finances from the government outlays. According to opposition sources, nearly 40 per cent of the state revenue is reportedly spent on the royal family.

King Abdullah

King Fahd ascended the throne in 1982 on the death of King Khalid. The latter was a retiring person by nature; and did not keep well during the last years of his rule. Fahd had been looking after the royal responsibilities during Khalid's illness and was no novice to the job. His long reign continued till 2005. This was the most turbulent period in the Kingdom's history. There were economic setbacks, terrorist incidents and four wars in the neighbourhood. The Kingdom was an active participant, the major contributor and the sole assembling ground and launching pad during the Gulf War.

And if Fahd ruled while Khalid reigned, the present King Abdullah has officially served as an 'Acting King'. On January 1, 1996, a royal decree asked him to take over duties of kingship, as Fahd needed to "spend some time resting and recuperating." Within a few months, Fahd decided to return. The effective governance of the country remained with Abdullah nonetheless. To that extent, he has been the *de facto* ruler for a decade already.

Upon taking over as the king, Abdullah has made certain highly demonstrative gestures in a quick succession. He has met the Shias, signalling a more inclusive religious fraternity in the country. He has pardoned the democratic activists, signalling a more tolerant political milieu. He has visited poor neighbourhoods in the country; putting poverty on the agenda rather than under the carpet. He has raised the salaries of the government employees by 15 per cent for the first time in over 20 years. He has earmarked $20 billion for housing loans, education and welfare.

On atmospherics, he has dispensed with the royal protocol of having his subjects kiss his hand.

Promising a Parliament: A History

The *Sharia* makes it obligatory for the ruler to consult his people on public matters. In the Saudi context, it is translated to mean talking to select individuals and groups on an ad hoc basis at an informal level. An institutional channel of communication between the ruler and the ruled has remained an elusive dream so far.

It was way back in 1928, that King Abdul Aziz, the founder of the Kingdom, issued a royal decree announcing the formation of a legislative council for the provinces of Nejd and Hijaz. It is to be noted that the state itself came into existence a few years thereafter. The decree was never implemented. Since then, every single Saudi ruler has felt compelled to make promises regarding the Majlis every time he has been threatened with a legitimacy crisis. Thus, in 1960,

King Saud used the bait in an attempt to checkmate his younger brother Feisal. He lost, nonetheless. Feisal, in his turn, promised to fulfil the aspirations of the people in this regard. After Feisal's assassination in 1975, King Khalid rededicated himself to Feisal's ideals and made ritualistic statements on the subject.

The overthrow of the Shah in Iran in 1979, had an extremely unnerving effect on the rest of the Gulf monarchies. In Saudi Arabia, there was a dramatic seizure of the Mecca Mosque and spontaneous popular uprisings among the Shias in the eastern province. Both these developments were perceived to have been inspired and assisted by the revolutionary regime in Iran. The talk of a Majlis was revived once again. A committee was formed under the leadership of the Interior Minister Prince Nayef to draft a constitution for the Kingdom with a provision for the Majlis.

The deliberations of the committee would have gone on forever if the country had not been shaken to its very foundations by the Gulf crisis in 1990-91.

The Basic Law

In 1990, the Iraqi armies attacked, occupied and annexed the neighbouring state of Kuwait. The Iraqi President Saddam Hussein came in for a universal denunciation for his foreign misadventures and domestic oppression.

The crisis impacted the Saudi political discourse in two ways; one, demonisation of Saddam Hussein did not carry much conviction, as the Gulf monarchies were by no stretch of imagination thriving democracies themselves. Two, the exiled Kuwaiti ruler held a mega event in Jeddah in October 1990. Called the Kuwaiti People's Conference, it brought the Kuwaiti ruler and some 1,200 Kuwaiti citizens together for two days' deliberations, centred around the strategy to liberate Kuwait and to resurrect parliamentary democracy in the liberated Kuwait. If the Kuwaitis could demand democracy from their rulers in the very heart of Saudi Kingdom, so could surely the Saudis themselves!

King Fahd made statements to reform the country's political system and establish central and provincial consultative councils. On March 1, 1992, he proclaimed the Basic Law. A committee of 10 officials, headed by Prince Nayef, the Saudi Interior Minister, drafted the document. For all intents and purposes, it is the Constitution. In the Saudi context, however, calling anything else but the Quran the Constitution would be a sacrilege. In fact, the first article of the Basic Law defines the *Sharia* to be the Constitution of the country.

This document consists of 82 articles, laying down the political structure of the country. More particularly, it provides for a consultative council. It also spells out the fundamental rights of the Saudi citizens.

The Basic Law puts the King at the top of the political structure. He rules according to and supervises the implementation of the Sharia. He is the 'final resort' of the legislative, executive and judiciary authorities; he is the commander-in-chief of the armed forces, with absolute power to defend the country from an external force, declare a state of national emergency and suspend the Basic Law in times of war.

As president of the Council of Ministers, he appoints and dismisses the ministers, and approves the Council decisions. It is the King, who appoints the members to the Consultative Council; approves, amends or vetoes the Council by-laws; and controls its budget.

Additionally, it is the King who chooses the Crown Prince and relieves him of his duties. The provision, thus, sets aside the traditional practice of choosing the ruler through a consensus within the family and of elevating the Crown Prince to the position of the ruler as a natural successor.

The Nominated Parliament

In early March 1993, the formation of a 60-member consultative council was announced to mark the 10th anniversary of Fahd's rule. It was to be a body representing the business, professional, academic and religious elite. Appointed by the King for a period of four years, it would have the right to put questions to the Council of Ministers, make suggestions on the laws-in-the-making and submit them to the King, and consider the international agreements. Neither in the mode of is formation nor in the scope of its authority was it a great leap forward.

The royal decree establishing it said, "it would operate in compliance with the existing system of government and in adherence to Sharia." The government has made it abundantly clear that it is not a pale version of Western-style parliament. Instead, it is projected as an organic development of consultative process that goes back to the Prophet.

Since its formation, it has been expanded in number if not in the sphere of activities; thus, in 1997 it went up to 90, in 2001 to 120 and in 2005 to 150. Various ministers have appeared before it to appraise it of their ministries' working.

Political Ferment on the Ground

Till recently, articulating political issues was taboo in Saudi Arabia. The political debate and demands, as a result, originated either from across the state borders or anonymously from underground. The ones who chose to speak out openly either lost their freedoms or their jobs. However, from the Gulf War onwards, many have chosen to do so in an increasing number and with greater intensity and efficacy.

As the US assembled an international coalition of 28 states in Saudi Arabia to repel the Iraqi aggression and liberate Kuwait, the religious radicals' disgust at the presence of the US troops on Saudi territory assumed a more strident tone. Thousands of cassettes in which the speakers identified themselves started spreading the message of bitter anger. Mosques and religious educational institutions turned into effective platforms to denounce the 'heresies'.

Sheikh Safar al-Hawali and Salman bin Ouda deviated from their predecessors among the dominant Wahhabi *ulama* in that they expressed their opinions on issues beyond theology, morality and proper Islamic conduct. Their regular Friday sermons included their views on internal politics and regional and international affairs.

In 1993, a group of six prominent academics, lawyers and clerics announced the formation of the Committee for the Defence of Legitimate Rights (CDLR), seeking to eliminate injustice and support the oppressed. In terms of the Saudi Basic Law, the CDLR should have been a perfectly legitimate group. The reality on the ground was entirely different. The members of the CDLR were quickly dismissed from their jobs, its office was closed and its spokesman, Muhammad al-Masari, was arrested. Fourteen more arrests of the supporters followed; and an activist, Abdullah al-Hudhaif, was beheaded after a secret trial. The Council of Senior Islamic Scholars issued a statement condemning the establishment of such an organisation as unnecessary since, according to them, the rights of the people were well protected in the country.

Masari was later released and escaped to London. From there, he has continued to pursue his campaign for human rights. In 1996, internal divisions spawned within the CDLR and a rival group was born. Headed by Saad al-Faqih, it calls itself the Movement for Islamic Reform.

In addition to the vocal dissent, there were instances of violent opposition. In November 1995, a car bomb exploded outside the offices of Saudi National Guard killing seven, most of them American security personnel training the National Guard. In June, the next year, there was a powerful explosion outside a military housing complex in al-Khobar near Dhahran. It killed 19 American security personnel and injured nearly 400. The responsibility for the attacks was taken by Islamic Movement for Change, which had warned of more attacks if the American military did not withdraw.

The war against Afghanistan sharpened the tone and tenor of critical voices. *Fatwas* were issued, threatening to excommunicate the royal family if it supported the US offensive.

Osama bin-Laden transited to a career in terrorism in these circumstances. He refuses to refer to his country as Saudi Arabia. To him, it is Bilad al-Haramain or the country with two sacred precincts. The United States, similarly, is al-Wilayat al-Mulhida al-Muttahida or the United States of Atheists. The American military presence is likened to the later-day crusaders.

During and since the war against Iraq, the open defiance has continued; sometimes winning small concessions, sometimes yielding ground. Thus, the conservatives were successful in mounting pressure to get Jamal Khashoggi, the editor of *al-Watan* dismissed from his job for critiquing Hanbali theology. In quick succession, Hussein Shobokshi, a popular columnist, was banned from writing after he published an article that fantasised his daughter driving him to a polling booth. The piece ruffled a few feathers as the Saudi women could not drive and the men could not vote.

On the other hand, Sheikh Safar al-Hawali, who was one of the very first to defy the rulers in early 90s, has volunteered to endeavour to bring people from the *dairat al-unf* (circle of violence) to the *dairat al-miyul al-fikriyya* (circle of intellectual deviation), and then to reform them. Sheikh Salman al-Oudah has reconciled himself to supervise the website *www.islamonline.net*, which deals with theological issues and shuns politics. Nearly a thousand clerics have been suspended from their jobs and retrained to expunge strident militancy from their sermons.

In May 2003, there was a series of suicide bombings in Riyadh, killing at least 35. The employees of US security firms, training the Saudi National Guard, were targeted. Since then, there are intermittent government claims of having unearthed and destroyed the terrorist cells followed by more explosions to prove the government wrong.

By October 2003, politics of dissent had manifested itself overground, when hundreds demonstrated in Riyadh and Jeddah demanding economic and political reforms. More than 150 of them were arrested. In march 2004, Matrouk Faleh, Ali Demaini and Abdullah Hamid were arrested for "issuing statements and collecting signatures on petitions" calling for adopting a "constitutional monarchy and using Western terminology in demanding political reform." They were handed down long prison sentences after a closed-door trial. Pardoning the three was almost the first royal act of King Abdullah when he took over in August 2005.

Politics of Petitions

Petition has emerged as a unique tool in Saudi political engineering in recent years. It was in December 1990, when the first petition was sent to the King; or more precisely, when the King acknowledged the first petition that was sent to him. Titled "Petition for Change" its 43 liberal conservative petitioners requested for a consultative council, the revival of municipal councils, independence of judiciary, equality of all citizens irrespective of their ethnic, tribal, sectarian or social origins, freedom of the media, reform of the religious police, throwing open numerous public fields for the participation of women and comprehensive reform of the educational system.

In February 1991, "Memorandum of Advice" followed. The signatories were religious leaders, judges and preachers. Sheikh Abdul Aziz bin-Baz, Sheikh Safar al-Hawali and Sheikh Salman al-Auda were the most prominent of the petitioners. The thrust of its requests was distinctly different from the earlier petition. It also urged the formation of a consultative council but the one that would work on the basis of the *Sharia*. Additionally, it requested a competent committee to repeal laws and regulations that were not in conformity with the *Sharia*, encouragement of the Islamic banking, freedom of the media to work within the confines of the *Sharia*, an Islamic foreign policy that avoided alliances against the *Sharia* and upheld the Muslim causes, and rights for the individuals and society within the safeguards of the *Sharia*. Finally, it requested that religious and proselytising institutions be developed and strengthened with financial and human resources. This last point went directly against the earlier petition that had sought a curb on their powers.

It was more than a decade that a new spurt in petitions happened. In January 2003, 104 intellectuals and academics began the new series, as it were. It was titled "National Reform Document." The petition grouped their appeals in four axes. The first called for national and regional legislative assemblies on the basis of direct elections, an independent judiciary, freedom of expression, assembly and human rights, establishment of civil society institutions including labour unions. The second insisted on fair distribution of wealth, fighting corruption, bribery and expropriation of public lands, strengthening and empowering the oversight and accountability institutions and making them accountable to the legislative council, allocating a part of the national income for future generations, and diversifying economy. The third called for a greater interaction between the society and the leadership through encouraging national unity, ensuring minimum needs of each citizen as regards employment, housing, education, healthcare, fair trial, minimum wage, retirement benefits, and rights to the women according to Islam. The fourth called for positive initiatives from the government to infuse public trust and patriotism—for instance, announcing general amnesty for political prisoners or giving them an open and fair trial; reinstating expelled judges and university professors; and lifting all restrictions such as travel ban, threat of imprisonment, expulsion from work, forced statements to stop writing or expressing opinion. The fifth axis appealed for a National Dialogue Convention.

In April 2003, 450 Shias, including 24 women, presented a petition titled "Partners in One Nation" to the then Crown Prince Abdullah. It enumerated their grievances: exclusion from employment in military, security and diplomatic services, bias in the curricula and media, absence of freedom to set up cultural institutions or religious centres, arbitrary arrests, interrogation, prevention from travelling, personal searches and so on. They sought a redressal of the situation on these counts. More specifically, the petition called for an unequivocal announcement of the Kingdom's respect for all religious sects, appointment of a national committee with the participation of Shias to deal with the question of sectarian discrimination, treating sectarian hatred as a crime, creation of an official body of Shia Islamic scholars under the guidance of the Ministry of Endowments and Islamic Affairs to arrange their religious and cultural affairs, removal of restrictions on Shia religious ceremonies, printing and importing of Shia publications, and referral of royal decrees to the Shia Sharia courts.

In September, 306, including 51 women, sent a "Petition in Defence of the Nation." It referred to the "National Reform Document" that was presented in January, reiterated its arguments and gently nudged the authorities to move from the stage of adopting the reform agenda to the stage of action and implementation according to an explicit schedule.

Politics of the Patriarch

National Dialogue

The leadership showed a willingness to engage with the petitioners and listen to their opinions. In June 2003, a "National Gathering for Intellectual Dialogue" was held with invited participants in Jeddah. In August, King Abdul Aziz Centre for National Dialogue was formally established in Riyadh to make it an ongoing project and to provide it an institutional structure. In December, a second dialogue was held in Mecca under the theme "Extremism and Moderation: A Comprehensive Approach." The third Dialogue was held in June 2004 in Medina and was devoted to the "Women's Rights, Duties and Education." The fourth was in December in Dhahran with the theme being "Youth Issues: Reality and Expectations." The fifth, and the last to date, concluded in December last year in Abha to deliberate on "Us and Others." This Dialogue was telecasted live on the Saudi television.

It is difficult to assess the Saudi people's involvement in these Dialogue sessions; and, consequently, their impact on Saudi society. That the venue is moved from city to city ensures a greater number of local participants. The dialogues end with recommendations that are general in nature. Whether they provide an input into the official policies remains to be seen. In the meanwhile, at least once, the recommendation to reform academic curricula created an instant hostile reaction. An open letter signed by a long list of signatories that included judges and university professors warned against such a move.

Elections to the Municipalities

In October 2003, an announcement was made for elections to half the seats in the municipal councils. An elaborate exercise in voter registration was undertaken; a Saudi male citizen of 21 years and above was eligible to vote. Political parties were not allowed. A staggered schedule was worked out for the elections between February and April 2005. The voting procedure was free and fair, although the voter turnout was low. Five hundred and ninety two (592) counsellors were elected for 178 municipal councils. The elected representatives will constitute only half of the municipal councils; the rest are to be nominated. And the municipal councils are not envisaged to be the pulsating centres of power sharing. They are administrative bodies with strictly local mandates. It was only in the late December that the councils were called to session.

Empowering Women

The women's issue has been a major area of attention and substantive accomplishment. In September 2005, the Cabinet approved a new Labour Law that replaces the legislation issued 37 seven years ago. The Law raises the rate of Saudi employees in the private companies to 75 per cent; urges the private sector to provide Saudi workers with adequate training to equip them to take up various jobs and create a suitable working atmosphere for them. And most importantly, it allows women to work in any field that suit them. The interpretation of the Law and its implementation will determine its actual impact on the women in the long run.

Last year, women were allowed to contest for seats on the board of the Jeddah Chamber of Commerce and Industry. The results were stunning; Lama al-Suleiman and Nashwa Taher were elected on a 12-member board. Since then, Nadia Bakhurji has won a seat on the 10-member board of the Saudi Engineers Council in its first-ever election. They have now staked their claims to contest for seats on the board of the Saudi Journalists Society. There are reports that they may also be inducted into the judiciary and rule on cases involving women. Recently, Adila bin-Abdullah, the King's daughter, held a weekend workshop on "Women and the Millennium Development Goals" and forwarded its recommendations to the Parliament for consideration.

Abdullah has also raised hopes of women, who are rearing to get behind the wheels and drive away. He promised in very first interview as the King that the Saudi women will "one day be able to drive." Young girls, since then, have taken to simulated driving lessons readying for that day.

Lastly, a dent has been made in an unbelievably restrictive legal code that seeks to set standards of social behaviour. The legal code bans dating, cinemas, theatres, concert halls, clubs, discotheques and so on. In November 2005, after a gap of nearly 20 years, cartoon films were screened in a hotel on an experimental basis. The shows were open only to children and women. It may or may not lead to cinemas being made open to the citizens, as is generally understood.

Managing Extremism

Since 9/11, the state has sought to rein in the Wahhabi *ulama* as also their puritanical ideology. Sheikh Saleh bin Abd al-Aziz Aal As-Sheikh, the Minister for Islamic Affairs, Endowment, Guidance and Call, has launched an Anti-Extremism Project. It is to be implemented in a series of campaigns over a period of many years. The project implementation would utilise the services of *ulama*, mosque *imams*, preachers, scholars and Dawa workers, in addition to writers, poets, playwrights, historians and journalists. The 'reformed' militants are projected widely to promote the project. Thus, Mishari al-Zaidi and Mansour al-Noqaidan, the former militants well versed in Wahhabism, are now calling for a fight against extremism.

The government constructs and repairs the mosques, pays salaries to the Imams and defines qualifications for the positions to be filled. There are estimated to be 10,000 *ulamas*. Also, there is the religious police al-Hayat al-Amr bil Maruf wa Nahiyan al-Munkar (Committee to Encourage Virtue and Prevent Vice), which is authorised to ensure that the women observe the dress code and the men attend the prayers.

In August 2003, some 12,000 people were questioned on terror-related suspicions. Abdullah ruled out any room for neutrality or hesitancy in a conflict between what he termed "power of good and of evil." The Council of Senior Islamic Scholars echoed his opinion. Since then, many militants have surrendered themselves under the amnesty announced by the state; and many more have been released from detention. A number of prominent scholars, intellectuals, psychiatrists and parliamentarians have volunteered to take part in counselling programmes for the detainees. A television serial in which repentant radical Islamists reveal what they call the "Deceit in the Name of Jihad" is being aired in prime time. The offers of amnesty, shoot-outs, prolonged gun-battles, lists of wanted terrorists are the methods used in an all out effort to co-opt or crush militancy.

Human and Civil Rights

The Saudi Basic Law was the first codification of human rights under the Saudi law. Article 26 of the Law states, "The state protects human rights in accordance with the Islamic Sharia." The Law restricts the power of the religious police, stripping them of the right to arrest and harass people. It also restricts the right of the government to spy on its citizens and guarantees the privacy of post and telecommunications—they cannot be confiscated, delayed, read or listened to except in cases defined by the statute. It includes freedom from arbitrary arrest, punishment and search; and freedom to private property.

The document makes no mention of freedom of expression and information. Free association and assembly are not mentioned either. The press remains under the government control.

The state-sponsored National Society for Human Rights has been functioning since 2004 and has looked into cases related to employment, bureaucracy, judiciary, naturalisation, domestic violence, prisons, etc.

In late October last year, an 18-member Human Rights Commission was formally set up with a clearly defined mandate. It is entrusted to ensure strict implementation of rules pertaining to human rights by all concerned government departments and private bodies, detect violations and take corrective measures; to follow up the implementation of international agreements on human rights; to summon representatives of government and private bodies as part of its investigation; to visit prisons and detention centres without prior permission; and to submit reports to the King.

The state practices Quranic punishments of flogging for adulterers, death for those who act against God and his Apostle, and severing the hands and feet of thieves. These punishments remain on the statutes, but are sparingly carried out at present. The Amnesty International and the Human Rights Watch have consistently reported infringements of privacy, pervasive discrimination against women, widespread use of capital and corporeal punishment, arrests without judicial warrants, incommunicado detainment for long periods, and so on.

In the realm of civil rights, small steps have been taken towards better access to information. In January 1999, the government decided to allow local Internet Service Providers to operate, although it was careful to build a 'firewall' against the sexually explicit, anti-Islam and politically offensive sites. Till then, the country's 40,000 cyber-buffs used to dial long distance to the US, Bahrain or Cyprus to connect to the Internet. The *AwalNet* is an interesting site in the Saudi context as it has a home page with a link to a women's site with postings on fashions, child-care, cooking and so on. Its cyber cafes have an all-women staff for guiding the women customers through a virtual cyber tour, keeping them out of trouble in the chat-rooms and trouble shooting in cases of various snags and contingencies.

The press is privately owned. The Ministry of Information appoints and may remove Editors-in-Chief. It provides guidelines to the newspapers on controversial issues. The Saudi Press Agency is government-owned and expresses official government views on most subjects. The government also owns and operates the radios and television channels. The foreign books and magazines are closely scrutinised before these are released to the Saudi market—the offending articles are either blackened, or removed, or their pages glued together, or outright confiscated and prevented from circulation. The foreign journalists are thoroughly scrutinised before being granted visas and are closely monitored after they are permitted entry. Two of the most widely read newspapers are owned by the Princes; *al-Okaz* by Prince Nayef and *al-Watan* by Prince Bandar bin Khalid.

Interestingly, it was the Saudi-owned Orbit company that had entered into a contract with the British Broadcasting Corporation's (BBC) Arabic Service to produce news that eventually led to the trailblazing Arab multi-media company of Al-Jazeerah. The Orbit at some point decided to pull out of the contract and the Qataris swept in.

Every year, a million plus pilgrims come to the country to perform Hajj. There is a strict prohibition on bringing in books, pamphlets or pictures of political, propaganda or ideological orientation.

Prospects

The issue of political reforms predates the rule of King Abdullah, the US-led democracy-promotion project in the region, and even the state of Saudi Arabia itself. King Abdullah's leadership does not commence with his formal accession to Kingship either. In a sense, therefore, it is an incremental process. In the past few months, there is a perceptible quickening of pace, nonetheless. There are some highly demonstrative gestures and a few initiatives of substance in various spheres of public life. If the pace does not slacken and initiatives are not stalled; political reform will be firmly on the national agenda.

Two very serious challenges to the entire project need to be noted here. One, the convention of horizontal succession in the country may prove to be a mixed blessing. To date, it has served as an inclusive mechanism; it has managed to keep most of the family factions involved and given them all a stake in its sustenance. But it has also translated into a rule by the men of the same generation.

Abdullah is an octogenarian and his rule is bound to be short. Quick successions in turn are bound to encourage and activate the younger aspirants to the throne. Will the Saudi political trajectory repeat an earlier instance, when, for instance, short rules by octogenarian leaders eventually led to the emergence of a younger Mikhail Gorbachev in the Soviet Union? And what will that imply for Saudi state and its *raison d'etat*?

For the past 15 years, the whispering for popular participation has grown louder. The Basic Law enumerating citizens' rights and the state's obligation to protect them; a nominated parliament; and half-elected local municipalities are the net gains at the end of a persistent many-pronged battle. Not a great deal in themselves, these steps could acquire an irreversible momentum, with unpredictable direction and results.

Two, the Gulf is the most penetrated region in the world. The external powers are permanently poised there and have repeatedly resorted to violent shows of their might. At the Saudi end, the regime is well aware that the US presence in and around its territory simultaneously underpins its stability and undermines its legitimacy. The uncertainties are far too many and the consequences are unpredictable. The United States, to date supportive of the existing political structures, has become more critical, more demanding and far more intrusive than hitherto fore. There is a constant hectoring of the rulers on various counts.

As a result among the people, there is a widespread perception of being at the receiving end of the global aggression in which the rulers are seen to be helpless bystanders, or worse still, collaborators.

The most dangerous moment for a bad government usually comes when it starts to reform, said Alexis de Tocqueville. There are far too many uncertainties in the present situation to predict the course or consequences of the reforms underway. Whether the monarch will continue to rule as hitherto, or reconcile himself to becoming a constitutional king, or leave the scene altogether are open questions.

25

Turkey's Democratic Experience and Its Influence on Regional Muslim Countries in the Post-9/11 Era

SIRET HÜRSOY

Shaping the world order in the 21st Century has begun with the terrorist attacks in New York and Washington on September 9, 2001. This has also been the moment of fears that the developments could lead to a so-called 'clash of civilisations'. In such an international security environment, Turkey has been instinctively attributed a 'pivotal role' to mediate between Christian and Islamic civilisations as the successor of not just the Ottoman Empire but the numerous civilisations that rose and fell at the crossroads of Europe, Asia, Middle East and Africa. As a predominantly Muslim populated state, Turkey's historical choices for modernisation along Western lines as well as its democratic and secular system present it to the rest of the world as the best alternative to a 'Christian-Muslim confrontation'. The way Turkey's image has been perceived by its Muslim neighbours in the West Asia and the Middle East is changing for the better, mostly as a result of Turkey's successful democratic experience and its increasing constructive influence on regional Muslim countries. As a result, Turkey is widely considered as a convincing 'example' that offers a genuine appeal to the region. Therefore, Turkey's democratic experience, Western connections, Eastern characteristics and its unique geographical location all served to increase Turkey's international attraction in the aftermath of 9/11.

Turkey's 'Pivotal' Geographical Location

At the crossroads of unique geographical regions, Turkey sits astride the East-West (extending from Europe to Central Asia) and the North-South corridors (extending from the Caucasus and Black Sea regions to the Middle East and Eastern Mediterranean) so that it acts as 'a pivotal state' of the Euro-Atlantic and the Eurasia zones. The concept of 'pivotal state' is explained in the *Foreign Affairs* with reference to Turkey as, "a hot spot that could not only determine the fate of its region but also affect international stability...The classic example of a pivot state throughout the nineteenth century was Turkey, the epicentre of the so-called Eastern Question; because of Turkey's strategic position, the disintegration of the Ottoman Empire posed a perennial problem for British and Russian policymakers."[1] Not only Turkey, but India and Pakistan are also considered as 'pivotal states' in the West Asia region.

There are several criteria for being 'a pivotal state' in order not only to be a moderating element in violent conflicts of region(s), but also to affect the balance of power at the global level in the 21st Century: Firstly, possessing a geostrategic location, a potential economic and military power, and a considerable amount of population. Secondly, having the potential political power to develop multi-dimensional, multi-layered, inter-regional and trans-regional foreign policy at the crossroads of different civilisations, religions, cultures and ideas as "a multi-regional country."[2] Turkey's 'pivotal' status and characteristics is further explained in detail as: 'At a multifold crossroads between East and West, North and South, Christendom and Islam, Turkey has the potential to influence countries thousands of miles from the Bosporus...Turkey enjoys solid economic growth and middle-class prosperity...A prosperous, democratic, tolerant Turkey is a beacon for the entire region..."[3] Turkey's location is also at the crossroads of world's wealthiest oil and natural gas reserves, as well as it lies over the transportation lanes and communication networks from East to West and from North to South.

Although Turkey is at the crossroads of different regions, its face is looking adamantly towards West, it has a common geographical border with Europe and is culturally entwined with Western values, despite the pull of its Islamic creed toward the East. Turkey is a valuable strategic pivot country in a most dangerous part of the world as a regional stabiliser between East and West, as a valued member of all European institutions, as well as a respected power by Russia. As one of the Russian analysts, Aleksandr Dugin, put it: "Turkey is an East went to the West; but, a country that its centre still remained the East. At the same time, Turkey is a state of the West that has melted in itself the values of the East. The states like contemporary Russia and Turkey are established on the rubbles of Eurasian empires. The roots of Turks are in the wide geography of our continent. Their course of direction is the West. Turkey, as a melting pot of peoples and states as well as a Eurasian country that of having strong accumulation of historical and political will, is converting peoples into a new historical community."[4] In fact, even though Turkey is put into the European camp geographically since the Treaty of Paris in 1856, there has been a widespread—but unspoken—perception beneath the surface that it culturally and religiously remains outside the boundary of Europe. A clear cut response about the geographical location of Turkey and the reason for pursuing a 'multi-dimensional foreign policy' by the Turkish Ministry of Foreign Affairs' can be found in its website as, "[Turkey] reconciles the West with the East and the North with the South and is active in all continents. Turkey serves by way of its geographic disposition and close historical and cultural ties across a vast landscape as a crucial bridge for dialogue and interaction between civilisations at the heart of Eurasia."[5] The need to balance Turkey's interests between the East and the West has always been the principal challenge to Turkish foreign policy.

Since the Ottoman period, the Turks has been the geostrategic part of the Western 'security community' as a 'security provider' and resolutely contributed to the emergence of Western Europe from the Cold War as a 'pluralistic security community'. A 'pluralistic security community' is defined by Karl Deutsch[6] as a "zone of peace" where the war between Western European countries is almost unthinkable. The concept of 'security through integration' in the Cold War era, which was a byproduct of the shared perception of the Marxist-Leninist threat, became an intuitive European integrationism and transnationalism in economic, social, foreign, security and defence spheres in the post-Cold War era.[7] As a consequence of geographical integration in the sphere of security, the concept of 'security interdependence' emerged as a part of security complex in the European region as a whole.[8] The conceptualisation of security is in fact a political process. That is, in the aftermath of the collapse of the Soviet Union, the overriding political changes in the global security environment have gradually led to a greater geographical interdependence between regions by bringing politics more into 'geo-security'. This, in turn, increased the importance of the concept of 'geopolitics' in a way that 'geography overrides political processes in shaping not only security policies, but politics in general'.[9] Therefore, as a result of increasing globalisation in the post-Cold War era, the narrow definition of 'security interdependence' in a regional security complexes, which was previously confined to the Euro-Atlantic region, began to gain a strong Eurasia security dimension as a part of global web of security interdependence.

'Geo-Cultural' Security Role of Turkey

Although Turkey is defined as 'a pivotal state' between different regions in the Euro-Atlantic and Eurasia zones, its meaning simply remained as geographic and geo-strategic until September 11, 2001, terrorist attacks in the US. However, the September 11 and the subsequent following terrorist attacks in November 15, 2003, in Istanbul, have brought to fore much desired political, economic and cultural connotations of geography. This new type of threat in the form of terrorism is resulting in clashes among values, civilisations, religions, cultures and ideas. Therefore, a sheer geographical and geo-strategic description of Turkey has also gained a geo-cultural security dimension. Turkish authorities claim that secularism and democracy are 'universal values' and they should be added into peaceful resolution of conflicts during a search for reconciliation between Western and Islamic values and cultures. As Prime Minister Recep T. Erdoðan put it in his speech at the University of Oxford, "[threat] no longer rests on substantive military-nuclear balances of power or on deterrence, and is fed by such abstract notions as socio-economic welfare, conflict of civilisations and extremism...our universal values face today a threat which has no ideology, nationality, religion, geography or an address. The threat perceptions are no longer limited to our borders. They are in minds, ways of life, thoughts and beliefs of those who portray themselves as the 'other'."[10] Therefore, as a result of Turkey's efforts to utilise its cross-cultural resources after 9/11, it came to be recognised by the international community with non-military aspects of security. This geo-cultural security dimension of Turkey is a valuable factor for both Christian and Muslim worlds in commonly addressing 'universal values' that are crucial in winning battles for hearts and minds of people in addition to the military anti-terror campaign. Turkey stands out against the Christian-Muslim confrontation as a unique alternative in the Muslim world by presenting itself so that it is possible to reconcile Islam within a modern, Western-style, democratic and secular system.

The metaphor of Turkey as 'a pivot state' has different geo-strategic, geopolitical, and geo-cultural dimensions since September

11. Thus, their importance in a new era of antagonism posits Turkey as a valuable partner between the East and the West in fighting against new kind of terrorist threats. This is exactly the reason why Turkey was attacked by terrorists in November 2003, because Turkey has traditionally pursued a policy of inclusion into the Western political system and successfully continued to be a veritable harbinger as an only pro-Western, religiously secular, economically free-market and politically multiparty democratic Muslim state amongst the 55 Islamic countries of the world.[11] It is because of this dual-identity character of Turkey that it relentlessly calls upon the world to draw a distinction between Islam and terror.

The unique Turkish experience epitomise that secularism, democracy, rule of law and human rights are universal values with no geographical, cultural or religious boundaries, nor they represent values specific to particular cultures or religions.[12] An interesting paradox is that while Turkey was seen by most of the Western countries as a 'torn' country in identity crisis between the clash of Western and Islamic civilisations in the beginning of the 1990s,[13] it began to be portrayed as a 'pivot' country which possessed the quality to play the role of 'moderator' between different civilisations by standing out as a specific example of modernisation with its multi-cultural and dual-identity characteristics in the beginning of 2000. Turkey's secular, free-market and democratic ideology in a predominantly Muslim setting has been widely perceived and promoted by the West as best alternative 'model' to fill the ideological gap left by communism and to halt the spread of radical Islamic ideas into Central Asia, Middle East and North Africa. This Western understanding is refuted by the Turkish elite on grounds that the 'ideological' Iron Curtain that once led to a division between the East and the West could now be replaced by a 'cultural/religious' Iron Curtain,[14] if Turkey is tried be presented as a rigid 'model' outside the EU. Given that Turkey's successful—yet incomplete—democratisation experience is unique, it cannot be easily transformed to other Muslim countries in the region as a 'model', albeit Turkey might constitute a good 'example', for at least two main reasons:[15]

Firstly, Turkey could only become an example for the Muslim world with its unique experience in democracy and economic development. Turkey underwent significant democratising reforms such as codifying the legal system, adopting parliamentary forms of government, reconciling Islamic and Western law, and achieving education reforms. These developments are deeply rooted in two-centuries of Western orientation plans that had began in the Ottoman period and continued with revolutionary transformations in the Turkish society to develop a secular democratic identity through marginalising the anti-secular Islamic identity since the establishment of the new Turkish republic by Mustafa Kemal Atatürk in 1923. Since the multinational and the Islamic basis of the Ottoman Empire had come to an end, the ambition of the Kemalist elite in Turkey was to reach the 'contemporary level of civilisation', a formulation for Western level of standards, by establishing homogenous national identity and strict secularism as well as striving for integration into the European political order.[16] According to the Kemalist elite, the notion of an Islamic state should be removed from the political discourse by establishing secularity and rationality on the basis of a set of epistemological and normative principles so that modern industrial economy and social progress would flourish. Therefore, Turkey can not easily be taken as 'model' by other Muslim countries simply due to the fact that a democratic regime can neither be established overnight nor flourish, if it does not emerge through home-grown initiatives that are based on domestic social, political, and economic dynamics. As Brzezinski argued, "without political dignity derived from self-determination, there can be no democracy."[17] As long as domestic enthusiasm for political and economic reforms in a Muslim society is lacking, over-selling the Turkish 'model' with international pressure from outside the region would most likely cause to serious repercussions. Not to mention that the rise of Arab nationalism against the legacy of the Ottoman rule in mind, most of the Islamic countries in the Middle East and West Asia would consider the secular Turkish 'model' as acting an instrument for Western intervention into the region. Given the high levels of instability, political tensions and economic deprivation, outside powers should not expect quick changes in the region. However, Turkey's successful experience in political and economic reform processes could still be a useful example and a source of inspiration to its neighbours.

Secondly, Turkey has a long-standing institutional relationship with the West as being the member of the Council of Europe, NATO, WEU, OCSE and entered into the last—painstaking—stage of the EU membership process with the start of accession negotiations on October 3, 2005. Muslim Turkey's EU membership dilemma would clearly affect the way in which the EU is perceived throughout the

Islamic world. The significance of Turkey-EU relations for the Muslim world is manifested in *The Economist* as, "...an EU that is open to Turkey should send a message to the troubled Muslim world of today: the West does not consider Islam and democracy incompatible as long as Islam doesn't...the start of negotiations is an historic chance for Europe but also for the Muslim world to show that their two great civilisations are not fated always to clash."[18] No country in the Western Asia or in the rest of the Islamic world had similar experience or would likely to exercise a comparable relationship with the West in foreseeable future.

For these reasons, Turkey cannot be a model that would easily be copied or duplicated as such. Although Turkey cannot financially and economically afford to share its experiences effectively or to meet the expectations of the regional countries, it is still ready to contribute in their transformation efforts as much as it can. As it is stated in the website of the Turkish Ministry of Foreign Affairs, "Turkey, with her experience in democracy and economic development, and making use of the multiple ties with the vast majority of the countries in the area, has been able to take part in their transformation efforts."[19] However, any forceful attempt by Turkey to bring Western values into the region may not only cause to deepening the existing political gap between Turkey and Islamic countries, but even lead to the deterioration of the existing bitter relationship between Christian and Islamic societies. Therefore, the more hegemonic and unilateral Washington's policies in the West Asia and the Middle East become, the greater the likelihood that Turkey will find itself more closer to the multilateralist policies' of the EU. In view of the pro-EU orientation of regional civil society organisations, Muslim countries would be much more receptive to democratic and secular Turkish experiences if Turkey is backed up by the EU financially and institutionally, instead of by the pro-Israeli US.[20]

Turkey's Global Identity

As a result of Turkey's geo-strategic, geopolitical, and geo-cultural position, it acts as a global cultural/religious facilitator for reconciling the interests of the different geographical groups in the UN. Owing to Turkey's cross-civilisational identity, it takes part in the proceedings of both the Asian Group and Western Europe and Other States (WEOG), while participating only in the elections of the latter.[21] Due to this unique position of Turkey between the East and the West, the

Secretary-General of the UN, Kofi Annan, decided to put the Turkish Prime Minister Recep Tayyip Erdoðan together with the Spanish Prime Minister Jose Luis Rodriguez Zapatero in charge of the project called the 'Alliances of Civilisations' and assigned them the task of deepening and developing the political, cultural and educational relationships between the West and the Muslim world.[22] Moreover, Turkey's position and role in the UN's WEOG, Asian Group and in the Organisation of Islamic Conference (OIC) incrementally gained importance after the Central Asian Turkic Republics and the Caucasus states joined the UN with the end of the Cold War. Given that Turkey is assigned a global role by the UN for deepening and developing a dialogue and cooperation between civilisations, Turkey is accordingly declared its candidature on June 21, 2003, to one of the two non-permanent seats of the UN Security Council allocated to the WEOG for the term 2009-10, for which elections will be held during the 63rd session of the UN General Assembly in 2008.[23]

An only Muslim country that has begun accession negotiations with the EU, a member of NATO and at the same time having a seat in the OIC, Turkey is in a unique position of nurturing a culture of understanding and cooperation between civilisations. Following the 9/11 incident, Turkey took the initiative to organise the OIC-EU Joint Forum at the level of Foreign Ministers under the heading of "Civilisation and Harmony: The Political Dimension" on February 12-13, 2002, in Istanbul.[24] This Forum reiterated that harmony, rather than clash between the two civilisations, is imperative and underlined that Islam should not be associated with terrorism. As a result of Turkey's growing self-confidence and prestige in taking on the task of trustworthy advocate of democratic practices in the Islamic world, a Turk, Prof. Dr. Ekmeleddin Ihsanoglu has become the first-ever elected Secretary-General of the 57-member OIC for a four-year term on January 1, 2005.[25] In particular, democratic Turkish Parliament's decision of 'no' to Washington on March 1, 2003, on assisting the US military operation in Iraq—even some Arab countries dared not do—has also been helpful for the Turkish profile to be viewed with greater sympathy and respect by many Muslim states. Therefore, Turkey's membership in the OIC is an asset for the East and the West, when it comes to deepening and institutionalising relationships with the Muslim world.

Turkey has also been instrumental in the foundation of Economic Co-operation Organisation (ECO) as an intergovernmental regional organisation together with Iran and Pakistan in 1985. ECO is the successor organisation of Regional Co-operation for Development (RCD) in the West Asia region, which remained in existence from 1964 until the revolution in Iran in 1979. ECO gained its legal basis with the Treaty of Izmir in 1977 and, after some amendments, it became fully operational in early 1991. The main objective of the ECO is to promote mutual trade, economic development, technical and cultural collaboration. ECO's presently includes ten members: Turkey, Iran, Pakistan, Afghanistan, Azerbaijan, Kazakhstan, Kyrgyzstan, Tajikistan, Turkmenistan, and Uzbekistan. ECO is encompassing an area of 7 million square kilometres, inhibited by nearly 300 million people, and possessing rich energy resources. In an era in which world's economy has moved rapidly towards the free market economy, Turkey is widely sharing its economic and political experiences with the ECO member states and helping them in their transition into free market economy as well as introducing them democratic principles.[26]

The Turkish elite have expressed their willingness to share their own experiences with regional Muslim countries for the initiation of a democratic reform process. In the mean time, Turkey and other six predominantly Muslim states (Afghanistan, Algeria, Jordan, Bahrain, Iraq and Yemen) have been invited to participate to the "Plan of Support for Reform" of the G-8 Summit in June 2004. Accordingly, the G-8 Summit established a "Forum for the Future", which will be a new multilateral platform for the states in the region under the so-called label of the "Broader Middle East and North Africa", in order to prepare a ground for Muslim states to work together on a reform process. Within the framework of the 'Broader Middle East and North Africa Partnership Initiative' of the G-8, Turkey has also been given co-chairmanship of the "Democracy Assistance Dialogue" together with Italy and Yemen. This dialogue is including government and civil society organisations with a key objective of sharing experiences in democracy programmes in the region in order to develop them and initiate new ones, as well as to establish joint initiatives including twinning and capacity-building projects.[27] It is widely recognised commonly by the US and the EU side that a region ranging from North Africa to West and Central Asia needs political and economic transformation. In view of this, the US and the EU believe that a secular and democratising Turkey could offer important lessons to the initiation of a reform process in its broad neighbourhood. Of all the states in this broad region, Turkey is making the greatest progress in political and economic development, improving relations with its neighbours and so gaining self-confident, prestige and independent global vision. However, Turkish authorities strongly believe that, without accurately understanding the region's problems and needs, developing a common objective on paper would not be brought into practice.

Conclusion

It is evident from above that Turkey is an exclusive 'security provider' in its large neighbourhood, a catalyst for making the West more multicultural, and a harbinger of the introduction of democratic and liberal values into the Islamic world in general and into the Eurasian region in particular. In the West Asia and the Middle East, Turkey has a unique profile through maintaining a well-balanced position between various disputes and conflicts ranging from Arab-Israeli conflict, Syria, Iraq, Iran to Azeri-Armenian dispute over Nagorno-Karabakh. Nobody could expect a viable solution to these multifaceted problems without Turkey's constructive input. Therefore, the prospect of Turkey positioning itself into a place of developing further channels for an intensive dialogue and cooperation with its Muslim neighbours and between Western and Islamic civilisations would more likely to contribute to a stable region in the long-term. After all, Turkey has not only begun to be considered as an alternative against radicalism and religious intolerance, but also a bridge between the Islamic world and the Western Christian world since 9/11. The more Turkey is defined as a 'model' to its Muslim neighbours, not only Turkey will be dragged into internal political, social and religious instability but also the less it will appear to be a successful mediator between the East and the West. Therefore, the true understanding of Turkey's cross-civilisational identity and its democratic experience would result in repudiating the spectre of the 'clash of civilisations'.

Notes

1. Chase, R.S., E.B. Hill, P. Kennedy (1996). "Pivotal States and US Strategy", *Foreign Affairs,* Vol.75(1), January/February, pp.33-34; Makovsky, Alan O. (1999). "Turkey", in R.S. Chase, E.B. Hill, P. Kennedy (eds.) (1999). *The Pivotal States: A New Framework for US Policy in the Developing World,* W.W. Norton, New York and London.
2. This phrase was often used by the former Turkish Minister of Foreign Affairs Ismail Cem. See Cem, Ismail (2001). *Turkey in the New Century–Speeches and Texts presented at International Fora (1995-2001)*, Rustem Publishers, Nicosia, North Cyprus, pp.74-78.
3. Chase *op. cit.*, p.47.
4. Dugin, A. (2005). "Is it possible to unite the values of the East and the West successfully with one another?", *Trud,* January 12. Available at *http://www.byegm.gov.tr/YAYINLARIMIZ/DISBASIN/2005/01/17x01x05.htm#%206* Visited on: 25.07.2005.
5. Ministry of Foreign Affairs, Republic of Turkey. Available at *http://www.mfa.gov.tr/MFA/ForeignPolicy/Synopsis/SYNOPSIS.htm* Visited on 06.10.2005.
6. Deutsch, K. and S. Birrell (1957). *Political Community and the North Atlantic Area: International Organisation in the Light of Historical Experience,* Princeton University Press, New Jersey.
7. Hürsoy, S., *The New Security Concept and German-French Approaches to the European 'Pillar of Defence', 1990-2000,* Tectum Verlag, Marburg, Germany, p.23.
8. Barry Buzan defined the security complex as Washington Post, February 27, 2001. Speech by Prime Minister Recep Tayyip Erdoðan (2004). "Why the EU Needs Turkey", *EU-Turkey Working Papers* No.1, Centre for European Policy Studies, August, p.20.
9. Bilgin, P., "Turkey's Changing Security Discourses: The Challenge of Globalisation", *European Journal of Political Research,* Vol. 44, p.187.
10. Speech by Prime Minister Recep Tayyip Erdoðan, "Why the EU Needs Turkey", University of Oxford, May 28, 2004. Available at *www.sant.ox.ac.uk/esc/esc-lectures/Erdogan1.pdf.* Visited on 08.08.2005.
11. Speech by Hürsoy, Siret (2004). "Turkey and ESDP", during the occasion of *One-Day Analytical Seminar on Turkey and ESDP'*in cooperation between the Strategic Research Centre of the Turkish Ministry of Foreign Affairs and SIPRI, Stockholm, September 22. Seminar report is available at *http://web.sipri.org/contents/director/TURKEYESDPSUMMARY.html.* Visited on 02.02.2005.
12. Emerson, M. and N. Tocci (2004). "Turkey as a Bridgehead and Spearhead Integrating EU andTurkish Foreign Policy", *EU-Turkey Working Papers* No.1, Centre for European Policy Studies, August, p.33.
13. Huntington, S.P. (1993). "The Clash of Civilizations", *Foreign Affairs,* Vol. 72(3), Summer, pp.22-48; Huntington, S.P. (2001). *The Clash of Civilizations and the Remaking of World Order,* Simon Schuster, New York; Rosenfeld, S.S. (2001). "Turkey's Long-Running Identity Crisis", *Washington Post,* February 27.
14. Öymen, O. (1997). "Turkish Foreign Policy: Current Challenges", *Policy Watch* No. 238, Washington Institute, Special Forum Policy, Feb. 27. Available at: *www.washingtoninstitute.org/templateC05.php?CID=1117.* Visited on 22.08.2005.
15. A 'model' is defined by the Oxford English Dictionary as "design of something that is made so that it can be copied in another material"' and an 'example' as "fact, event, etc. that illustrates or represents a general rule; specimen showing the quality of others in the same group or of the same kind."
16. Heper, M. (2000). "The Ottoman Legacy and Turkish Politics", *Journal of International Affairs,* Vol. 54(1), pp.63-82.
17. Brzezinski, Z. (2004). "The Wrong Way to Sell Democracy to the Arab World", *New York Times,* March 6.
18. "Turkey belongs in Europe", *The Economist,* Vol. 365, Issue 8302, July 12, 2002, p.13.

19. Ministry of Foreign Affairs, Republic of Turkey. Available at *http://www.mfa.gov.tr/MFA/ForeignPolicy/Synopsis/SYNOPSIS.htm*. Visited on 06.10.2005.

20. Everts, S. (2004). 'An Asset but not a Model: Turkey, the EU and the Wider Middle East", *Centre for European Reform Essays*, London, October; Hürsoy, S. (2004). "EU's Long-Term Stability Strategy for the Middle East", *Turkish Policy Quarterly*, Vol.3(2), Summer, pp.185-194; Bal, Ý. (1998). "The Turkish Model and The Turkic Republics", *Perceptions*, Vol.3(3), September/November.

21. The five geographical groups within the UN are: African States, Asian States, Eastern European States, Latin American and Caribbean States, Western European and Other States.

22. Özdemir, Ý. (2005). "Zapatero'nun Doðru Yolu", *Radikal*, August 03.

23. See the leaflet advertising Turkey's candidature for one of the non-permanent seats of the UN Security Council on the following website: *http://www.mfa.gov.tr/MFA/ForeignPolicy/InternationalOrganisations/UN/Turkey_CandidatetotheUnitedNationsSecurityCouncil2009_2010.htm*. Visited on 11.07.2005.

24. For more information go to: *http://www.oic-eujointforum.org.tr/* and *http://www.byegm.gov.tr/on-sayfa/oic/oic.htm* Visited on: 20.10.2005.

25. Ünal, E. (2002). "West and East Attempt to Bridge Differences in Turkey", *Turkish Probe*, No. 473, February 10; Khan, S.S. (2004). "Turkey: Groping for Destiny?", *The Nation*, August 29, p.7.

26. For more information, go to the following web page *www.ecosecretariat.org;* Ünal, E. (2002), "West and East Attempt to Bridge Differences in Turkey", *Turkish Probe*, No.473, February 10.

27. Emerson, M. and N. Tocci (2004). "Turkey as a Bridgehead and Spearhead Integrating EU and Turkish Foreign Policy", *EU-Turkey Working Papers* No.1, Centre for European Policy Studies, August, p.20.

References

Bal, Ý. (1998). "The Turkish Model and The Turkic Republics", *Perceptions*, Vol. 3(3), September/November.

Brzezinski, Z. (2004). "The Wrong Way to Sell Democracy to the Arab World", *New York Times*, March 6.

Buzan, B., O. Wæver and J. de Wilde (1998). *Security-A New Framework for Analysis*, Lynne Rienner Publishers, UK.

Buzan, B (1991). *People, States and Fear: An Agenda for International Security Studies in the Post-Cold War Era*, Boulder, Lynne Rienner Publishers, US.

Hürsoy, S. (2004). "EU's Long-Term Stability Strategy for the Middle East", *Turkish Policy Quarterly*, Vol. 3(2), Summer.

Section VI

Non-Traditional Challenges to Security: The West Asian Experience

26

Fundamentalism and Democracy

EMMANUEL SIVAN

Let us first dispose of the never-ending argument whether Islam is compatible with democracy. Democracy, at least as a form of governance for large societies, is a modern idea, born out of the Enlightenment. The ancient Greek model was till then considered appropriate, at best, for local governance. It was Montesquieu who first argued that large, centralised states might be governed this way.

As a modern idea, democracy was bound to challenge traditional thinking, including that of historical religions. The latter had to grapple with it for a long time, with the mainstream rejecting it (remember Protestant England in the 17th century), and some dissidents accepting it. The latter justified this innovation by reference to this or that element of the religious tradition considered to be a prefiguration of democracy. Islam's problems with democracy are no different than those of other monotheistic religions. The intervention of exogenous factors accounts for some crucial differences.

A case in point is the Catholic Church which had combated democracy for the century and a half following the French Revolution. Indeed the Church saw the Revolution (especially in its Jacobin version) as a danger to the very foundations of Christian civilsation. And what were these foundations? They were the ideas that states and individuals are beholden to obey the moral values carried by Revealed Religion. The French Revolution, and its offspring democracy, was predicated upon the notion of human autonomy and human reason as sole basis for any social and political organisation. Jacobinism had in fact persecuted the Church and tried to liquidate its role as the exclusive mediator between the sovereign God and his creatures. No wonder the Church stuck to the alliance between Altar and Throne.

The Church's opposition to the "democracy heresy" arrived at its pinnacle in the Encyclicals of Pope Gregory VII (1832, 1834) which condemn democracy as an anticlerical force and reject Liberalism as "founded upon that false and mad principle, namely that every man has the right to enjoy freedom of conscience." This opposition was formally consecrated in the Syllabus of Errors, issued by Pius IX in 1864, which excommunicated the idea "that the Pope should reconcile himself to the ideas of Progress, Liberalism and modern civilization."

Dissenting voices were heard in the early 19th century through Italian, French, German and Austrian movements calling for such a reconciliation. But these harbingers of what would later be dubbed Christian Democracy, were tiny and marginal. They would soon be dealt with heavy blows by Nazism and Fascism.

It was only in the 1940s, that influential thinkers from within the mainstream, for example the French philosopher Jacques Maritain and American Jesuit John Courtney Murray, constructed a theory (based on an early medieval idea, the 'Subsidium') which bridged the chasm between Catholism and Liberal Democracy. This theory soon got the blessing of Popes Pius XI and and Paulus VI. Why the breakthrough?

First, because of the Totalitarian experience, which taught the Catholic mainstream that western democracies are a shield and a guarantee against the pagan-like notions of the right-wing dictatorships, and later against atheistic communism. Second, the fears about freedom of conscience were gradually dissipated with the decline of the anticlerical tendencies among European liberals. The *Subsidium* thus, came to stand for desacralisation of politics, with the power of the Church taken to serve as a check on the State's power, while within the State there is to be no religious compulsion, and pluralism and politics, based on social accord, reign supreme. Man has a natural right to freedom in his search for meaning, with no compulsion in matters of belief. And, third, Catholic thinking has undergone an even deeper process of 'Protestantisation' in the sense that the individual had come to be recognised as both the source and the end of society.

Further, the individual person is naturally social, and can only achieve the fullness of human development in communities ("the principle of solidarity"). The purpose of such communities, and actually of all human relationships, is to give help to individuals, as they pursue, freely, their obligation to work for their own development. State or society should not, therefore, replace or displace this individual self-responsibility. (There might be exceptions to this rule, but they apply only rarely). State and society provide the conditions for the possibility of the exercise of self-responsibility.

1

The Catholic experience provides a useful grid for understanding the uphill battle waged by democracy within the world of Islam. It is a battle where the endogenous difficulties are exacerbated by two exogenous factors. Unlike the West, where civil society expanded over the last two centuries, in Islam civil society has been in decline, under the pressure of the military-populist regimes; democratic liberties have been strictly curtailed (especially rights of assembly, association and expression), while the middle class, which had been in the West, the model of voluntary organisation and the socio-economic basis of defence of political liberties in the framework of market society—has seen its size and autonomy greatly reduced by the interventionistic State. These factors have nothing to do with Islam as a historical tradition; they represent environmental forces which tilt the balance in favour of the non-democratic tendencies of this very tradition.

The upshot is clear: the Islamic mainstream, or religious establishmant, is on the whole anti- (or at best, non-) democratic, sharing the same fears the Church had nourished a century ago. It is, moreover, beholden to a centuries-old habit of subservience to the powers-that-be, born out of economic dependency and the view that the Ruler is a bulwark against anarchy and fratricidal wars. Both dependency and perception are fortified by the experience of over 50 years of post-colonial regimes.

The most intriguing question is why didn't the Islamic establishment follow in the footsteps of its Catholic counterpart and ponder the legacy of decades of dictatorships in Islamic lands? Why did not the horrors of the Saddams, Assads and co. tilt the religious establishment towards democracy? Why didn't they learn the merits of pluralism for the very survival of religion?

Was it due to their fear that the radicals would reap the benefits of democracy? Or was it their own authoritarian bent in matters religious which makes them natural allies of authoritarian rulers ?

There are no doubt, liberal Islamic thinkers and movements who claim that Islam is a diverse, heterogenous and ever-evolving historical phenomenon. In our day and age it must, accordingly, espouse modernity so as to ensure Islam's evolution in a new world environment. Public, reasoned discussion is a precondition for thrashing out the various choices, for there's no single, uniform, one-size-fits-all solution, as the Estalishment (and Muslim radicals) propose. There are in effect a variety of options coexisting in this dynamic modern predicament. Only democracy can guarantee that such a debate takes place.

So the survival of Islam, which hangs upon its ability to adjust to change, is dependent upon the democratisation process and particularly the creation of a law-abiding State. Many of these thinkers[1] who are available now in two French-language anthologies[2] have worked out sophisticated doctrines which develop the ideas of individualism, human rights and contextual interpretation of the Law (*Sharia*), by testing out long forgotten medieval concepts (e.g., that of *Maqasid*, or original intent of the legislator, conceived by the 14th century Andalousian jurist, al-Shatibi).

Yet with the exception of Indonesia, where there does exist a mass liberal movement, the liberals are few and far between. The fault lies not with their creed which is eloquenly elaborated, but with their poor organisational and mass communicative capacities. They were and still are an elite phenomenon. Also, they are unable to woo the modern middle class, whereas the latter is held hostage to the fear that democracy would ineluctably bring about the seizure of power by the Islamist radicals (as in Iran in 1979 and as almost happened in Algeria in 1991-92). And the radicals are violently and loudly anti-democratic, the way the Catholic Church had been a century ago. They believe that there's only one way to reshape Islam, in order to preserve it from contamination by secularist modernity, i.e., their way —by application of the *Sharia*; and that too in a rigid, uncompromising manner.

Pluralism, diversity, transparency and public reasoning are anathema for them. "Islam is the solution", as their slogan goes. And there's but one way of comprehending Islam.

If the lessons of the totalitarian experience have not been drawn by the Islamic mainstream, and if liberal forces are not strong enough, committed enough (especially given the overall weakness of civil society), what about the possibility of change from top to bottom, Gorbachev style? Do not the regimes see the need for some liberalisation, all the more since the collapse of Saddam's regime happened amid the striking indifference of Iraqi citizens (except at the Sunni Triangle)? This collapse has indeed been amply commented upon in the Arab press, officious and oppositional alike, as continuing a pattern set already by the fall of the Rumanian dictator 15 years ago. For one thing, as the late Hafiz al-Asad has said, Gorbachev has brought it in two major ills, avarice and anarchy. And this is just the example NOT to follow; better do it the Chinese way: economical opening long preceding the cultural one, while keeping tight hold of politics in the hands of the ever-cautious powers-that-be.

Yet there is more in play here than learning from the experience of others. The Arab state is (and considers itself to be) in crisis during the last two decades. The measure of the crisis is set forth when one takes into account the order of priorities of the Arab (and most other Islamic) regimes.

Priority no. 1—Defend the resources base of the regime, i.e., oil and gas (if they possess them), and particularly the public sector, which is the major provider of jobs and income for the regime's stalwarts, lackeys and clients.

Priority no. 2—Take care of the needs of the elites the regime relies on.This is all the more critical in countries where the elites are recruited from ethnic and/or religious minorities, which had been property-less. (e.g., Alawites in Syria).

Priority no. 3—Maintain the broad social base of the regime, which consists of those groups who live off the public sector (civil servants, industrial workers, even peasants).

These social groups expect the regime to recompense them for their support through welfare services, subsidies for basic products (bread, flour, cooking and heating oil, cooking gas), education, jobs for the young. Should the regime find it arduous to provide the latter–because of inflation, unemployment or reduction of the state's revenues–its "executive legitimacy" (i.e., capacity to "deliver the goods") will be damaged. It would find itself all the more hard put if, concurrently, its "systemic legitimacy" (also known as ideological fundament) is likewise incapacitated. This is indeed what happened to the populist-military regimes with the decline of (and disenchantment with) Pan-Arabism and Arab Socialism from the 1970s onwards.

One can, thus, account for the grave political implications of the decline of oil prices since the mid-80s Arab oil-rich countries witnessed sharp cuts in GNP and revenues, and that too at a time of huge, unexpected expenses (loans, or rather grants, Iraq to finance the Iran-Iraq War, covering the costs of the Allies's war against Saddam in 1991). Poorer countries (*viz.* Syria, Egypt, Yemen, Jordan) were likewise hard hit: directly, due to drastic cuts in aid from Gulf states; and indirectly because of the decline of the number of their citizens employed in the Gulf who used to send substantial remittances to their families back home. Foreign aid and remittances first shrunk only moderately, but this turned into a steep decline following the 1991 war, when migrant-workers were expelled en masse (either as retribution for their home countries' support for Saddam, and/or because these workers were now deemed a security risk).

When the Soviet block disappeared, Soviet foreign aid all but disappeared as well, while exports to East European markets shrank considerably. The Arab state found itself in a corner, all the more so as it continued to incur high (and stable) expenses, whereas fertility rates remained high (with the exception of Tunisia and Egypt where family planning campaigns bore fruits). The rural exodus to towns, in search of livelihood, went on unabated, and created dangerous pressures upon the dwindling state resources (for housing, roads, etc.). If the state had in the past absorbed high school and university graduates in the civil service, it now found itself increasingly unable to do so, at least not in the former scale.

The public sector was supposed to be—in terms of Arab Socialist vision in particular—the last refuge of the state after the rente revenues had been hard hit (that is revenues due to geological or geo-political location, not to productive initiative). Yet, the productive part of the public sector had been in grave trouble. Its plants were manifestly inefficient because of their heavy bureaucratic structure and high degree of politicisation. They had always preferrred, Soviet-style, make-work and the concentration on cheap consumer products for the local market, with low value added, rather than export-oriented, competitive production or contribute to infrastructure building.

As long as the regimes enjoyed the revenues of the 'rente', they could take care of their clients, despite the low productivity of the public sector. Their bluff was called overnight by the steep decline in oil prices. One could no more pay off all social groups to the same extent (in providing 'guaranteed' jobs for graduates, doling out food susidies, health and welfare services). Drastic budgetary cuts had to be, and were made. The productive part of the public sector declind due to inadequate capital outlays for investment, especially in renovation and maintenance.

The political elites (particularly in the oil-rich states) could have reduced these cuts, had they reduced their own lavish share of the 'rente revenues'. They could have also fostered the development of the productive private sector, a potential source of revenue from profits' taxation. The first option was never tried, not even in Algeria on the the eve of an Islamist upheaval. So pradatory and egotistic, nay even 'Kleptocratic' these elites had grown to be. The second option was tried, but just in a desultory, unsystematic fashion. The reason being perhaps the fear of the political elite from the eventual rise of an entrepreneurial class which might impinge upon the nature of the regime; for one who pays taxes might demand the right to control the way these are spent.

2

In a nutshell, by a partial retreat from certain economic activities, through:

(a) Transfer of some activities to the private sector (Banking and foreign trade were partly privatised at first, then private investment was encouraged, among others) with the help of legislative measures which put strict limits on abuse of bureaucratic (and political) power. This related, however, to property rights and the law of contracts, but not, repeat not to human and civil rights.

(b) Moving the responsibility for welfare and educational services to NGO's in civil society, thus reversing a century-old trend of expanding the state at the expense of society.

(c) Some political activity was allowed, notably the founding of parties, which had been till now frowned upon as an insult to the cherished national unity.

Yet, and here is the rub, this is perhaps liberalisation, especially in economics, but by no means democratisation. In a way, if some relaxing of the reins is evident in certain fields, there's greater regimentation in others. Authoritarianism still reigns supreme.

One remarks, first and foremost, the continued, and crushing ascendancy of the executive branch and the absence of an authentic separation of powers. The autonomy of the judicial system is relative and varying, that of the legislatures even less so. Within the executive branch, democratic rules (e.g., transparancy, parliaments serving as mere *chambre d'enregistrement.*

As for human and civil rights, there are few guarantees against arbitrariness on the part of the powers-that-be. *Habeas Corpus* defence against arbitrary arrest or perquisition—is virtually unknown. Government controls on right of association (budgetary supervision, the power to disperse their managing committees) persist.

And the fact that the judges rarely enjoy a tenured position makes their intervention as checks-and-balances, for example, at the level of 'cassation', not quite reliable.

Elections are, on the whole, *faites par le pouvoir en place;* today almost as well as in the past, for the regimes, given the present crisis, don't think they can afford the luxury of really free elections. The rulers themselves tend 'to' be reelected, not just twice, but thrice and more. Transmission of power from father to son among republican presidents—to ensure stability and legitimacy in time of permanent crisis—to become the norm, not just in Syria, Saddam's Iraq, Yemen, Libya, but elsewhere too.

The state has indeed retreated in certain economic and political areas, but within the *camp retrenche* which it holds, defends its position tooth and nail. It has recourse to emergency laws, to rampant torture (and not just against Islamist subversives), to 'disappearances', to adminstrartive detention, etc. The *pre carre* they defend might be smaller then in the heyday of Nasser, Boumedienne and Asad Sr. Yet it is defended with utmost firmness and ferocity.

Mainstream Islam, and even Liberal Muslims (with certain notable, and noble exceptions) tend to accept this state of affairs, if for no other reason than for fear of the alternative; Islamist opposition winning free elections is their worst nightmare. So authoritarianism is the medication one should swallow to immunise oneself against the Islamist malady. And when one accepts use (and abuse) of emergency laws against Islamists (not just terrorists, but also Islamist civil society associations), there's a spillover effect to other areas as well. Very few Egyptian liberals (Islamic or secular) rose in the defence of sociologist Sa'd Eddin Ibrahim, held in jail for more than three years, twice convicted, and then, in a third trial disculpated. And what was his sin according to the prosecution? Trying to organise international supervision of parliamentry elections, and 'besmirching Egypt', by writing (in foreign publications) about the discrimination from which Copts have to suffer.

3

If the "democratic wave", announced by Samuel Huntinton, has passed over the Arab (and Islamic) world, this doesn't have much to do with Islam. The root causes are rather the crisis of the nation-state in the Middle East and in the Maghreb and the resulting regimentation, the structural weaknesses of Liberal Muslims, and the ever-present danger (real, imagined, sometimes grossly exaggerated) of Radical Islamism.

True enough, there are in Islam-as-a-historical-tradition—as there had been in Catholicism—factors which militate against this modern invention which is democracy. Islamic doctrine calls for subservience to authority, be it Allah, men of religion or the *de facto* ruler (who is always preferable to anarchy). The idea of *Shoura* (or consultation, especially prior to the selection of a Caliph) is not. As many apologetic writers would have it, a sort of public reasoning with representatives of the people, and, therefore, the basis for democracy.

In effect it was just about the duty to consult with Muslim jurists or with other notables. And, anyway, it was respected more in the breach than in the act. Indeed, the early Islamic history, post-Muhammad, was pretty violent and strife ridden. (No wonder that much like in 19th century France and 20th century Spain, democracy is supposed to lead to anarchy and chaos). As for the idea of the individual and his rights, it was as alien to traditional Islam, with its stress on the *Jama'a,* or collectivity—as it was for historical Catholicism.

But does the past necessarily determine the future? Enough was said here about Catholicism. And think also of the recent experience of Taiwan and South Korea. Despite being impregnated with a Confucian tradition, which primes hierarchy, authority and social cohesion, here they are, with their successful transition to democracy. The impact of culture and religion is surely important, yet one must not disregard the eventual impact of processes like economic modernisation, spread of literacy, the communications revolution, the rise of the new middle class and of civil society. Such factors transformed the two supposedly Confuscian countries. Couldn't they one day, transform certain Muslim countries? After all, in some respects Islam might be more receptive to democracy than Confuscianism. It has, for instance, a stronger sense of the rule of law, being, like Judaism, a nomocratic polity.

The present paper has highlighted factors of 'structure and conjuncture'. It is useful to remind the reader that there's a third level to unfolding history, namely the realm of human consciousness, freedom of action and volition; however limited by the other two levels, this dimension does exist. It is there that creativity (in ideas and in organisation) finds expression; and transformative leadership takes the initiative. The trouble with this level is that it is barely foreseeable. Yet human initiative, with its ability for improvisation, might sometimes make constraints into opportunities and hopes into reality.

It is this that Polish poet Arthur Medzieczki highlighted so well in the mid-80s, during Solidarity movement's darkest hour:

What does the political scientist know?
He knows the latest trends
He knows the present state of affairs
The history of ideas.
What doesn't the political scientist know?
He doesn't know the despair
He doesn't know the political game
Which consists
In forsaking the game.
And it never occurs to him
That nobody knows
When will irreversible changes occur
Like sudden cracks in an icy surface.

Notes and References

1. Muhammad Sa'id al-'Ashmawi, *Hadad al-'Aql*, Cairo, 1974; id., *Jawhar al-Islam*, Cairo, 1986; id., *Al-Islam al-Siyasi*, Cairo, 1987; id., *Ma'alim al-Islam*, Cairo, 1989; id. *Muwajahatuna Ma'a-l-Tatarruf*, Cairo, 1996; id., *Madha Jara Limisr*, Cairo, 1991; Hussein Ahmad Amin, *Fi Bayt Ahmad Amin*, Cairo, 1989; id., Dalil al-Muslim al-Hazin, Cairo, 1984; id., *Hawla Tatbiq al-Shari'a*. Cairo, 1985; id.,Mi'at al-A.zam, cairo,Risala min Tahti-l-Maa, Cairo, 1982; id., ImamFi Fadlayn, Cairo, 1980; Sayyid Mahmud al-Qimni,Dawr al-Hizb al-Hashimi, Cairo, 1990; id., Hurub Dawlat al-Rasul, cairo, Cairo, 1983; id., Al-Fashiyun wal-Watan, Cairo, 1999; Khalil. Abd al-Karim, *Al-Islam Baynal-Dawla wa-l-Din*, Cairo, 1995; Quraysh, Cairo, 1993; id., *Al-Ta'ifiya Ila Ayn*, Cairo, 1987; id., *Fikrat al-Takwin*, Cairo, 1997; id., *Al-Judhur al-Ta'rikhiyya lil-Shari'a*, Cairo, 1990, Ridwan al-Sayyid, *Al-Mas'ala al-Thaqafityya*, Beirut, 1998; id., *Al-Jama'a wa-l-Mujtama*, Beirut, 1997; id., *Siyasat al-Islam al-Mu'asir*, Beirut, 1997; id., *Azmat al-Fikr al-Siyasi al-.'Arabi*, Beirut, 2000; id., *Al-Hurriya al-.Aqa.idiyya*, Beirut, 1995; Hamid al-Ansari, *Hadith al-Intilaq*, Beirut, 1995; id., *Mabani al-Fiqh*, Beirut, 1992; id., Muhammad Talbi, *Alaqat al-Islam wa-Nafsuhu*, Tunis, 1992; Muhammad Cherfi, *Islam et Liberte*, Paris, 1988; Abd al-Majid Sharfi, *Qira'at al-Nass al-Dini*, Tunis, 1989; id., *Al-Islam wa-l-Hadatha*, Tunis, 1991; id., *Tahdith al-Fikr al-Islami*, Casablanca,1998; id., *Al-Islam Bayna-l-Risala wa-l-Ta'rikh*, Casablanca, 2001; Hamid Nasr Abu Zayd, *Al-Imam al-Shafi*, Cairo, 1992; id., *Mafhum al-Nass*, Cairo,1994; id., *Al-Khitab al-Dini*, Cairo, 1992; Tafkir Fi Zaman al-Takfir, Cairo,1995; Kamal Abu Majd, *Ru'ya Islamiyya Mu'asira*, Cairo, 1991; Muhammad Shahrur, *Kitab al-Kur'an : Qira'a Mu'asira*, Damascus1992; id., *Dirasat Mu'asira fil-Dawla wal-Mujtama*, Damascus, 1994. Cf. M. Hatina, *Studies in the Thought of Faraj Foda*, (Hebrew), Tel Aviv, 2000.
2. Abdou Filali-Ansari, *Reformer L'Islam*, Ed. La Decouverte, 2003; Rachid Benzine, *Les nouveaux penseurs de l'Islam*, Albin Michel, 2004. This modern school of political thought should be interpreted against the historical background set in Patricia Crone's monumental *God's Rule: Government and Islam*, Columbia UP, 2004.

27

Jehadist Insurgency and the Crisis of Legitimacy in the Islamic World

JOSEPH McMILLAN

Jehadist Terror as Globalised Insurgency

In the four years and more since September 11, many observers groping for an explanation of what took place that day—and anxious to rule out the possibility that it may have been related to US policy and conduct—have laid the blame for violent Islamic radicalism at the feet of the Arab governments of the Middle East. Islamic terrorism, so this argument goes, stems from a lack of democracy and related channels for political expression under the authoritarian regimes that rule throughout the region. Others would have it that the United States is coming under attack because people in the Middle East perceive an inconsistency between American rhetoric about democracy and human rights and US policies that prop up repressive dictatorships in predominantly Muslim countries.

There is clearly an association between violence carried out in the name of extremist interpretations of Islam and the internal political problems of much of the Muslim world, but the causation is more complex than is captured by the simple formula that a lack of freedom causes terrorism. Throughout history, billions of people have lived under authoritarian governments more repressive than most of those in the Muslim world without generating a significant terrorist backlash. Indeed, both inside and outside the Islamic world violent, political dissent has always been the lowest in the most repressive states—for obvious reasons. Moreover, it is difficult to see how the agenda espoused by Islamic radicals, also known as Salafis, can be construed as growing out of a lack of democracy. Even such relatively peaceful Salafi organisations as the Egyptian Muslim Brotherhood promise to dismantle whatever trappings of democracy their countries have now and institute in their place a form of Islamic totalitarianism should they ever come to power.[1]

While both terrorism and insurgency have been topics of intense attention in the last few years, only a few analysts[2] have put the two concepts together outside the confines of the Iraq problem. These analysts begin from the insight that what we are dealing within Al Qaeda and the movement associated with it is not simply terrorism in isolation but rather terrorism as one manifestation of a globalised insurgency. In other words, it is at root a political confrontation that can accurately be described in the terms with which Bard O'Neill defined "insurgency" in his 1990 book, *Insurgency and Terrorism*:

> A struggle between a non-ruling group and the ruling authorities in which the non-ruling group consciously uses political resources...and violence to destroy, reformulate, or sustain the basis of legitimacy of one or more aspects of politics.[3]

It goes without saying that violent extremist Salafis, often referred to as jehadists, use both violence and non-violent political means in the pursuit of their goals. Because most of the insurgencies of the last century relied primarily on a protracted war strategy and guerrilla tactics, some people assume that such a strategy and such tactics are defining characteristics of insurgency. But as the eminent British counter-insurgency expert, Sir Julian Paget, pointed out nearly 40 years ago, "the methods insurgents adopt may include guerrilla warfare, but insurgents may equally well resort to civil disobedience, sabotage, or terrorist tactics."[4]

Some also question whether the terrorist acts committed by jehadists can seriously be intended to gain political control and replace the existing order. What this objection fails to recognise is that, while non-Muslims in America, India, Europe, and elsewhere may be the literal targets of the violence, they are not the political audience for whom the violence is undertaken. Terrorist attacks against American office buildings, British and Spanish transport systems, and Indonesian hotels are less a device for weakening the will of the societies they strike than a means of galvanising support among Muslims through the production of high-profile martyrs, an example *par excellence* of what the 19th century anarchists called "propaganda by deed."[5] They are, thus, a complement, in classic insurgency fashion, to the 'propaganda by word' that is disseminated in sermons, tracts, tapes, lectures in radical Islamic schools, and over the Internet.

Legitimacy and the Sources of Insurgency

For our present purposes, however, the most important implication of seeing the jehadist terrorist movement as a globalised insurgency is the understanding that, like any insurgency, it is fundamentally a challenge to the legitimacy of the order that controls the lives of the population. As always, the relevant population is whatever group constitutes the insurgency's political audience. In this case, that is the worldwide Islamic *umma*. Before exploring the strategic implications of that understanding, however, it is necessary to examine what is meant by the term "legitimacy."

The concept of legitimacy was introduced into modern social science by the German sociologist Max Weber in the early 20th century. Weber described legitimacy as the attitude—shared by people subject to a particular set of social relationships—that the order within which those relationships exist has binding moral authority behind it.[6] Legitimacy may stem from a variety of sources—tradition, religious belief, logical deduction from abstract principles, conventional acceptance of a tacit social contract, and others. Whether a particular social order is legitimate does not depend on the source from which it claims to derive its legitimacy. In fact, most social orders actually derive their legitimacy from a combination of prescription, revelation, convention, and legality.[7] Nor is it something that can be judged according to abstract standards from outside the society concerned. The key to legitimacy is whether the order enjoys the consent of the governed, keeping in mind that popular elections and representative bodies are not the only or even necessarily the most important means by which consent can be expressed.[8]

Legitimacy and the International System

Ordinarily one thinks of governments as entities that are legitimate or illegitimate. A legitimate government is one that is able to induce compliance with its rules without the constant use of overt coercion. Conversely, an illegitimate government is often threatened, or feels itself threatened, with disorder, instability, and domestic turmoil. As Weber used it, however, the concept of legitimacy is relevant to all sorts of social relationships, not just political ones, and not just within the confines of the nation-state. G. John Ikenberry has pointed out that international orders may be viewed by their members as legitimate or illegitimate just as domestic ones can.[9]

Traditionally, however, only states themselves would have been considered to have standing to judge the international order as legitimate or illegitimate. But in a world shaped by the process of globalisation, in which political, economic, intellectual, and other social relationships increasingly transcend cultural and political boundaries,[10] and in which non-state-actors increasingly feel empowered to challenge the traditional state, it is only reasonable to expect that how individuals perceive the legitimacy of the wider international order will become increasingly salient in how they behave politically.

I would contend that this is precisely what we are seeing now. Most Muslims clearly do not view the US-led international order as legitimate. By large majorities, they believe it is wrong for the United States to possess unrivalled and unchecked power, and that the United States uses that power in an attempt to dominate the world.[11] Given that frame of reference, they inevitably interpret almost any use of US armed force, no matter how benign or justified, as reinforcing their sense of international illegitimacy. Even more widespread is the view that traditional cultures ought to be protected against the impact of globalisation.[12] At the same time, many Muslims, and particularly Arabs, view the international order not only globally but within their own region as being equally illegitimate. In the Arab world, this perception applies not only with respect to Israel's presence in the Middle East but also, more subtly, to the transformation of Arab power relationships that resulted from the concentration of the region's oil resources in countries that had traditionally been seen as peripheral players in Arab politics and culture.[13]

In short, many Muslims and others who feel their lives affected by forces they do not accept as having moral authority over them are lending support to groups that promise to combat what they see as an illegitimate international order. Thus, the critique of the *status quo* offered by radical Islamists and acted upon by jehadist terrorists has as much to do with the legitimacy of the international order, of which the United States is the dominant member, as it does with the domestic legitimacy of the states within which most Muslims live.[14]

Legitimacy in the Arab-Muslim States

It is not a new notion to observe that many states with majority Muslim populations, and especially Arab states, suffer from a shortage of domestic political legitimacy. The weak legitimacy of most Arab regimes has been commonplace among regional specialists ever since Michael C. Hudson in 1977 described it as "the central problem of government in the Arab world today."[15] It was to this legitimacy deficit that Hudson attributed the pervasive instability, unpredictability, and repression that characterise Arab domestic and international politics.

In broad terms, this deficit is a consequence of the disorderly collapse of the Ottoman Empire following World War I and its replacement by what David Fromkin described as "an artificial state system" that has made the Middle East "a region of countries that have not become nations even today."[16] More specifically, it stems from at least five mutually reinforcing aspects of Arab society and politics as they were constituted in the wake of World War I and then reshaped during the process of gaining independence from the colonial powers in the 1950s and 1960s:

- The lack of a sense of shared identity binding the citizens of many Arab states. Higher loyalties to the wider Arab "nation" and the universal Islamic *umma* on the one hand and lower loyalties to sub-national tribal, ethnic, sectarian, and local affinity groups on the other often overwhelm any sense of attachment to the territorial nation-state that theoretically exercises sovereignty.
- Serious internal inconsistencies in the popular political value systems of many Arab societies. Decades of progressive rhetoric extolling democracy, equality, liberation, and unity have been overlaid on centuries of religious rhetoric extolling justice, tradition, and the importance of rightly (i.e., religiously) guided government. The combination produces a mix of inflated expectations that any government would be hard pressed to fulfil.
- The historic transplantation into the Arab (and wider Muslim) context of post-colonial and anti-colonial political structures that not only do not match well with underlying social structures but were in many instances expressly designed to destroy traditional modes of social organisation.
- The influence of classical theories about the nature of the Islamic state, which generally deprecate the concept of

multiple, independent sovereigns within the community of the believers.

- The failure of the region's governments to keep up with evolving expectations, driven by the information revolution, about the role ordinary people should play in determining their own destiny.

Neither international nor domestic illegitimacy in itself is a sufficient cause for the eruption of an insurgency. An ideological appeal based merely on an abstract challenge to the legitimacy of the prevailing order may appeal to a few narrow elites, but will not in itself persuade the masses on whom the success of an insurgency depends that violence is justified to change the *status quo*. If, however, a population has a set of concrete grievances against the *status quo*, the legitimacy or illegitimacy of the underlying social order can be the critical element in determining whether these grievances are resolved peacefully or ignite into violent resistance.

Where the great majority of people accept the prevailing order as fundamentally legitimate, notwithstanding whatever particular dissatisfaction they may feel with particular policy outcomes, they will consider themselves morally obliged to help maintain that order. In that case, insurgents will find it difficult to recruit new fighters, generate financial and material support, and move with impunity among the population. When the authorities call for assistance in combating terrorists, they will get it, or at least enough to make a difference.

Conversely, if most people do not accept the existing order as legitimate, even relatively minor discontents can lead rapidly to mass alienation. A radical, violent ideology can then mobilise the most susceptible among this alienated population to take up arms. Not everyone will resort to violence; the decision to become a guerrilla or a terrorist depends on a multitude of highly individual factors. But if the existing order is not seen as legitimate, even people who do not subscribe to the agenda of violence will tend to reject cooperation with the forces of the *status quo*. They will be unwilling to take personal risks on behalf of a system to which they are not emotionally and morally committed.

To argue for the centrality of legitimacy in confronting the jehadist insurgency, therefore, is not to argue for the irrelevancy of the multitude of discontents felt by Muslims who actively or passively support the insurgency—and even by most Muslims who do not support it. Some of the discontents that make people susceptible to the jehadist message are obviously common to much of the Islamic world—unhappiness at the relative weakness of Islam *vis-à-vis* the West; concern over the plight of Palestinian, Kashmiri, Chechen, Bosnian, or Iraqi Muslims; resentment against the intrusion of alien ideas into traditional societies; a sense of powerlessness in the face of relentless globalisation. Others are peculiar to individual countries, or even to districts and provinces—high unemployment, a lack of clean water, unfair land tenure laws, poor education, dysfunctional courts, and so on.

Even if it does not explain everything, the lack of domestic and international legitimacy is a common thread that runs throughout the societies that have shown themselves most susceptible to the radical appeal. It therefore, behooves those who believe that the success of the jehadist insurgency would be disastrous both for the Muslims and for the world at large to find ways to address this legitimacy deficit as a key element of a strategy against terrorism.

Enhancing Systemic Legitimacy

Because the legitimacy deficit that facilitates insurgent success exists at both the international and domestic levels, steps to enhance systemic legitimacy must be taken at both levels as well. In taking these steps, it is important to realise that there is considerable interplay between the two levels; measures taken to reinforce the legitimacy of the international system may also strengthen the domestic legitimacy of the states within that system and *vice versa*. On the other hand, there may also be unintended and deleterious consequences flowing in both directions. Decision makers must be attuned to these interactions and prepared to adjust policies quickly when unintended consequences appear.

Enhancing the Perceived Legitimacy of the International System

At the international level, much although not all of the burden for strengthening legitimacy rests on the United States as the perceived leader of the international order. Carrying this burden will be complicated by the obvious fact that most Americans naturally do not share the perception that this US-led order is illegitimate. This conflict of perceptions, of course, immediately suggests the need for more effective explanations of US policy, but such 'propaganda by word' will not achieve its purpose if it not accompanied by non-rhetorical steps—'propaganda by deed'—to mitigate hostility to America's role in the modern international system.

One such step, one on which many European critics focus their attention, would be for the United States to pay greater heed to the United Nations' function as a validating mechanism in the eyes of foreign public opinion. While the political benefits of giving greater attention to the UN would certainly not be as dramatic as some might contend, they would not be negligible, either, particularly in regions such as Europe where supra-national organisations are widely seen as vested with legitimacy in their own right. The pay-off would be less clear in the Islamic world, where regard for the UN is quite low,[17] although the disrepute in which the body is held among Muslims may be driven by their belief that it functions mainly as an instrument of US policy. Because of this widespread Muslim distrust of the UN, it would also be useful to promote the development of more—and more effective—multilateral institutions in which Arab and Muslim interests can be articulated and represented within both regional and functional contexts. This effort should include:

- Shaping a regional security landscape in West and Central Asia and North Africa that includes multilateral institutions of varying degrees of formality.
- Creating structures by which Arab and other Islamic states could interact collectively with the G-7, along the lines of the G-8 arrangement by which Russian interests and concerns are represented.
- Developing initiatives through which predominantly Muslim states can be integrated into institutions that transcend the Islamic/non-Islamic divide.

Finally, it must be recognised that a foreign military presence (especially an American one) in any predominantly Muslim country has become an easy target for jehadist ideologues to generate sympathy for their insurgency. Given the transnational nature of the insurgency, it does not even matter if the people in the host country welcome the foreign presence; that 'infidel' troops are on Muslim soil is now enough to stoke the propaganda furnaces elsewhere regardless of the facts on the ground. Not only does such a presence reinforce the idea that the United States and its allies act illegitimately, it can also weaken the domestic legitimacy of the governments in the countries that host foreign forces.

This is neither an argument for eliminating the US forward presence in the Middle East or South Asia, certainly not until the situations in Iraq and Afghanistan are substantially more stable, nor for eschewing the use of military force as a key component in the fight against jehadist terrorists. It is, rather, an argument for lowering the visibility of these forces to the greatest possible extent,commensurate with mission accomplishment, and for more explicitly taking into account that each use of force is, whether intentionally or not, an act of 'propaganda by deed' in the struggle for the hearts and minds of the people whose support is necessary to defeat the jehadists in the long run.

Helping Build Domestic Legitimacy in Muslim States

On the domestic side, enhancing legitimacy requires action on two main tracks: (1) broad political reform to create political orders that are more congruent with the needs and expectations of the region's people, and (2) strengthening of regional states' governance and administrative capabilities.

In the first area, as Ralph Braibanti wrote of the restructuring of the Japanese political system after World War II, what is necessary to strengthen domestic legitimacy "is to establish a polity which is in consonance with the social organization and traditions of the people."[18] Such polities can then shape popular expectations over time while building a sense of national loyalty and solidarity to supplement—and perhaps supplant—the higher and lower order loyalties that now characterise many Muslim societies. At the same time, they must also find ways to accommodate public demands for participation in a way that is consistent with the society's beliefs and attitudes, as well as to change and adapt as these beliefs and attitudes evolve.[19]

Obviously, this means that a one-size-fits-all approach to political reform will be counterproductive. The notion of what constitutes legitimate governance varies with culture and even from country to country within a cultural zone. The differences within the Islamic world can be seen in the way that different populations value competing political ideals. Asked to choose whether it was more important for the government to guarantee that no one was in need or to leave people alone to pursue their own goals, Lebanese participants in the Pew Global Attitudes Project opted for non-interference by a margin of 52 to 47 per cent. Yet in neighbouring Jordan, social protection was valued over individual freedom by nearly two to one. In South Asia, Muslim Pakistan and predominantly Hindu India both opted for non-interference, while Muslim Bangladeshis lined up overwhelmingly in favour of social protection.[20]

Furthermore, public perceptions in different countries are sharply at odds with many outsiders' assumptions. For example, a 2002 Pew report found that citizens of Pakistan and Uzbekistan, two countries considered by most external observers to be dictatorships with little legitimacy, had overwhelmingly positive views of their government (72 per cent and 88 per cent respectively). Meanwhile, the government of Jordan—which Americans tend to view favourably—gets positive marks from less than half of its people.[21] Moreover, from a Western perspective, Muslim attitudes on right governance may seem internally inconsistent. As has been widely remarked, majorities in almost all Muslim countries say they want democracy. Yet in Jordan, according to the Pew research, while 68 per cent believe democracy can work in their country, only 28 per cent think it is very important to have honestly contested elections.[22]

These findings suggest at a minimum that focusing on 'democracy' as the key to defeating jehadist terror is considerably trickier than it may appear at first glance. While the word itself has widespread resonance, its underlying meaning seems to be elusive. Indeed, there may well be a tension between the seemingly related desiderata of legitimacy and democracy, at least democracy as understood in the secular West, if the citizens of Muslim states do not share the underlying assumptions about the nature of society and politics from which the logic of democracy is derived.

It is, of course, impossible to speak of what 'all' Muslims believe on these matters, but it is clear that 'most' have a fundamentally non-Western understanding of the purpose of politics.[23] They would not agree, for instance, with Harold Lasswell's famous definition that politics is the process of deciding who gets what, when, and how. In the Islamic view, politics is about establishing the law of God and representing the transcendent order within the temporal world. Beliefs about the admissibility of disagreement and dissent, the relative priority to be accorded to the individual or the community, and the processes by which political decisions should be made all divide most Muslims from most Americans and Europeans.

This does not justify the condescending conclusion that Muslims are simply not capable of democratic governance. Muslim political scientists often point to three elements of Islamic teaching—consultation, consensus, and independent judgment—that they contend would make Islam particularly hospitable to democratic principles if properly interpreted. And while the historic practices in Muslim countries has not been encouraging in this regard, neither was the historic experience in Germany or Japan before World War II, Spain and Portugal before the deaths of Franco and Salazar, or most of Latin America before the wave of democratisation in the 1980s. It also seems clear that, even as residents of the Islamic world resent the deluge of Western culture, exposure to that culture is nevertheless altering the way they look at their own governments and the expectations they have of their political and economic systems. It is inevitable that political development will occur in response to these changes, whether by evolution or revolution. However, reiterating the key point, to be sustainable the systems that emerge after these changes will still have to be founded on the shared values and assumptions of the people of the countries concerned, not on those of people with a radically different worldview.

Even if the forms of government that have succeeded in one country could be transplanted successfully to another, we would still have to face the fact that foreign models—especially American models—are viewed in the Islamic world with intense suspicion. For example, while people in most of the Arab world react positively when pollsters pose the proposition that democracy is a good system, the support falls off sharply if the question is rephrased to specify 'American-style democracy.' Any efforts outsiders make to promote political reform in the Islamic world must therefore, be as low-visibility as possible and painstakingly deferential to local popular (although not necessarily regime) sensitivities.

Efforts should therefore, focus on fostering the 'process' of political development rather than on establishing fixed benchmarks of democracy and freedom against which every country will be graded. An early order of business must be nurturing the grass roots skills necessary for sustainable participatory government: aggregating diverse individual interests into politically effective coalitions; ascertaining constituent desires and needs and converting that knowledge into effective action; managing contention without having it devolve into enmity; developing budgets that reconcile competing needs; and so on. Outside experts can advise on the development of organisations and networks that cut across existing social groups as a means of building a broader sense of identity and loyalty. Finally, the development of indigenous cadres of political and constitutional experts who can take a leading role in shaping political arrangements that satisfy the many competing needs peculiar to each country can be promoted.

Initiatives to strengthen the legitimacy of regimes that are under insurgent criticism or assault must go hand-in-glove with efforts to build their capacity to govern their national space effectively. At worst, persistent inability to exercise control over the national space is a manifestation of actual or impending state failure; at best, a government's inability to assert its authority in the ways its people expect is corrosive of its credibility. In cases where governments have consciously made long-standing arrangements with local populations to refrain from asserting state authority, it is necessary that they reevaluate this policy given the ability of outside organisations with broader agendas to take advantage of the power vacuum in these ungoverned spaces, both to coopt the local population into insurgency and to use their territory as a base for operations.

Increasing capacity for governance also means improving the ability of states to deliver the services their people expect. To some extent, the problems of uncontrolled spaces and unserviced spaces overlap; it is hard to induce construction workers, power line repairmen, teachers, and doctors to work in areas into which even the army and police are unprepared to venture. But the problem is not merely geographic. Many regimes, and not only in the Muslim world, have failed to provide such basic institutions as functioning courts of law. Citizens seeking the most modest action by local administrations are often met with prolonged delays. The same applies, only more so, to the social services—housing, healthcare, education—that 21st century governments are expected to furnish.

In the Islamic world, we have seen a clear pattern of radical organisations moving into the political vacuum created by these governmental shortcomings. As has been widely noted, Pakistan's failure to provide an adequate system of public education has been offset by the provision of *madrasas* by the major Islamic political parties and affiliated organisations.[24] In many Egyptian villages, the only functioning clinic is not run by the health ministry but by an Islamist non-governmental organisation. In the legal area, recognising that justice delayed is justice denied, people turn away from the slow, inefficient, and often corrupt official judicial system and turn instead to informal arbitration, often provided, once again, by radical Islamist groups.[25] This process weakens even further the credibility of state authorities and their ability to control the national space; they become outsiders in their own land and jehadists and other radicals occupy the void.

Finally, if the measures taken to enhance systemic legitimacy and build capacity for governance are to become self-sustaining, it will also be necessary to develop the economic and social base to support them. Without educational systems that prepare citizens to think critically for themselves, participatory political systems will be easy prey for radical demagogues. Without economies in the success of which people understand themselves to have a stake, it will be difficult to obtain the kind of collective sacrifice for the sake of the common good that successful self-government demands. This will particularly be the case as globalisation places strains on selected sectors of many economies, strains that may well create new challenges to the legitimacy of both the state concerned and the international system itself.

Addressing material discontents is a corollary to political reform, not a substitute for it, but it is a particularly important corollary considering that people in much of the Islamic world have been conditioned for 40 or more years to measure the performance of their governments in terms of material benefits delivered. This was the basis of the social contract that Gamal Abdel Nasser struck with the people of Egypt to justify his authoritarian Arab socialist state, and it was the justification by which King Faisal of Saudi Arabia secured public support to overrule the conservative religious establishment in the cause of modernisation.[26] Ultimately, this is a fragile basis on which to construct a system of government, and indeed a major objective of our effort to enhance systemic legitimacy must be to help regional states construct more enduring affective foundations for their polities. But if the evolving political systems cannot do a better job of delivering what people expect in the short term, there may be no opportunity to lay this lasting foundation in the longer term.

Conclusion

Just as the lack of legitimacy is not a sufficient explanation for support of violent Islamic movements, so reinforcing systemic legitimacy and governmental effectiveness will not be a sufficient solution. As in fighting any insurgency, success will require a combination of armed force and a broad collection of political measures aimed at separating the insurgents from the population in which they live and operate. This paper has not touched on the necessity of defusing regional flashpoints, such as Palestine and Kashmir, that are such an important motivation of popular discontent, nor the need to control cultural confrontation that plays into the jehadist message of irreconcilable conflict between the House of Islam and the House of War, nor the importance of countering radical ideology with alternative diagnoses and prescriptions for the problems facing the Islamic *umma*.

Even as we address these other sources of discontent, however, we must constantly keep in view that success in the war on terrorism hinges on being able to change the way people in the Islamic world perceive the legitimacy of the international order and the legitimacy of their own governments. It is an axiom of counterinsurgency strategy that winning or losing depends on depriving the insurgents of public support and toleration. Samora Machel, the head of the Front for the Liberation of Mozambique in the 1970s and 1980s, was once asked how he could conduct a guerrilla war in a country almost devoid of natural geographic sanctuaries, such as mountains. He replied, "Our people are our mountains." By taking the people away from the jehadists, we can take away their mountains. When that happens, the eradication of the terrorists that remain will not be long in coming. But the people can be taken away from the jehadists only if they are given something to take the place of the insurgent message. The best thing they can be given is a political order in which they can believe.

Notes and References

1. "The very first of [the Muslim Brotherhood's) fifty demands stipulates the dissolution of parties, the unification of the nation's political forces, and their direction toward a single goal." Rejwan, Nissim (1998). *Arabs Face the Modern World,* University Press of Florida, Gainesville, P67.
2. E.g. Kilcullen, David (2005). "Countering Global Insurgency," *Journal of Strategic Studies,* Vol.28, pp.597-617.
3. O'Neill, Bard (1990). *Insurgency and Terrorism* Brassey's, Dulles, VA, (emphasis in original), p.13.
4. Paget, Julian (1967). *Counter-Insurgency Operations: Techniques of Guerrilla Warfare,* Walker & Co., New York, p.14.
5. According to Hoffman, Bruce (1998). *Inside Terrorism,* Columbia University Press, New York, p.17, This concept was first articulated by the Italian anarchist Carlo Pisacane. Mikhail Bakunin explained it in "Letters to a Frenchman on the Present Crisis" (1870): "We must spread our principles, not with words but with deeds, for this is the most popular, the most potent, and the most irresistible form of propaganda." Prince Pyotr Kropotkin seems to have been the first to use the exact expression "propaganda by deed," in his pamphlet *Revolutionary Government*(1880).
6. Weber, Max (1964). "The Fundamental Concepts of Sociology (1913)", in Talcott Parsons (ed.), *The Theory of Social and Economic Organization,* Free Press, New York, p.124.
7. Ibid. 124-132.
8. Bickel, Alexander M. (1975). *The Morality of Consent,* Yale University Press, New Haven, p.16.
9. Ikenberry, G. John (2001). *After Victory: Institutions, Strategic Restraint, And the Building of Order After Major Wars,* Princeton University Pres, Princeton, p.52.
10. The U.S. Defence Science Board defines globalization as "the integration of political, economic, and cultural activities of geographically and/or nationally separated peoples." Final Report of the Defence Science Board Task Force on *Globalization and Security,* (Washington: Office of the Under Secretary of Defence for Acquisition and Technology, 1999.
11. Pew Global Attitudes Project (2004). *A Year after Iraq War,* Pew Research Center for the People and the Press, Washington, p.19.
12. Pew Global Attitudes Project (2003). *Views of a Changing World* Pew Research Center for the People and the Press, Washington, p.104.
13. Ajami, Fouad (1998). *The Dream Palace of the Arabs,* Pantheon, New York, pp.169-172.
14. Fukuyama, Francis (2004). "The Neoconservative Moment," *The National Interest,* No. 76, Summer, p.63.
15. Hudson, Michael C. (1972). *Arab Politics: The Search for Legitimacy,* Yale UP, New Haven, p.2.
16. Fromkin, David (1989). *A Peace to End All Peace,* Avon, New York, p.17.
17. Pew (2004). *op. cit.,* p.10.
18. Braibanti, Ralph (1984). "The MacArthur Shogunate in Allied Guise," in Robert Wolfe (ed.), *Americans as Proconsuls: United States Military Government in Germany and Japan, 1944-1952,* Southern Illinois UP, Carbondale, IL, pp.85-86.
19. Rejwan, Nissim (1998). *Arabs Face the Modern World: Religious, Cultural, and Political Responses to the West,* University Press of Florida, Florida, p.4.
20. Pew (2003). *op. cit.,* p.105.
21. Pew Global Attitudes Project (2002). *What the World Thinks in 2002,* Pew Research Center for the People and the Press, Washington.

22. Pew (2003). *op. cit.*, p.34-40.
23. The average adult literacy rate in majority-Muslim countries is some 57 per cent. Given that as a starting point, it is certain that well over half of the population has no significant exposure to modern Western political or social thinking, except in the context of negative propaganda.
24. Fair, C. Christine (2004). "Islam and Politics in Pakistan," in Rabasa, Angel, Cheryl Bernard, Peter Chalk, C. Christine Fair, Theodore W. Karasik, Rollie Lal, Ian O. Lesser, David E. Thaler (eds.), *The Muslim World After 9/11*, Rand Corporation, California, p.276.
25. Even in areas where governments normally function reasonably well, such groups have shown a greater nimbleness in responding to episodic lapses in regime performance, as when they were able to get aid to victims of the December 2004 Tsunami in Southeast Asia faster than the governments concerned.
26. Thaler, David (2004). "The Middle East: The Cradle of the Muslim World," in Rabasa, Angel, Cheryl Bernard, Peter Chalk, C. Christine Fair, Theodore W. Karasik, Rollie Lal, Ian O. Lesser, David E. Thaler (eds.), *The Muslim World After 9/11*, Rand Corporation, California, p.92, 104.

28

Religious Terrorism

Territorialised and De-territorialised

RYOJI TATEYAMA

Among the series of *Patterns of Global Terrorism* published by the US State Department, the 1995 version was the first one which cited the threat of religious terrorism with utmost seriousness. It said, "Terrorism by extremist individuals or groups claiming to act for religious motives continued to dominate international terrorism in 1995." The previous reports also referred to religious terrorism, but merely mentioned religion as one of the causes of global terrorism.

The year of 1995, in fact, witnessed many brutal terrorist attacks based on religious motives, such as the assassination of Yitzhak Rabin, the Israeli prime minister, who was gunned down by a Jewish extremist, and the bombing of a US military facility in Riyadh, Saudi Arabia, which Al Qaeda was believed to be responsible for. Even Japan was not an exception. A Tokyo subway sarin gas attack by members of Aum Shinrikyo, an offshoot of Japanese Buddhism, killed 12 people and injured thousands. Aum Shinrikyo also perpetrated a number of murderous attacks against its members and other people whom it considered its enemies. The case of Aum Shinrikyo surprised the entire Japanese society because of its very violent and aggressive character, but it also jolted the world by underlining the fact that a terrorist organisation based on fanatical religious belief may produce and use weapons of mass destruction without hesitation.

The phenomenon of religious terrorism has a very long history, including the case of the 'zealots' and the 'assassins' in the ancient and medieval ages. But in the modern age, terrorism had been perpetrated mainly by leftists and secular nationalists until the 1970s. In the early 1980s, however, the radical Lebanese Shi'ite movement Hezbollar started its suicide bombing operations, and since then religious terrorism has expanded and intensified all over the world. The 2004 version of *Pattern of Global Terrorism* lists up 80 terrorist organisations and a half of them can be classified as religious. Since the September 11, it has been found out that such terror has been globalised, as *National Strategy for Combating Terrorism* published by the Bush administration in February 2003 states, "the terrorist threat is a flexible, transnational networked structure, enabled by modern technology and characterized by loose interconnectivity both within and between groups."[1]

A significant majority of religious terrorist incidents are carried out by radical Islamic organisations and movements, particularly by Jehadists for the last couple of years, but extremists based on faith in Christianity, Judaism, Hinduism, and even Buddhism, have also adopted terrorist tactics as well.

The purpose of this paper is twofold. The first is to extract common characteristics from various types of religious terrorism for comparative purposes. And the second is to categorise religious terrorism into two types, territorialised and de-territorialised. Both types of religious terrorist groups have many common futures but defer from each other in measures they resort to and ultimate objectives they try to achieve. This author believes that the categorisation can be helpful to understand the nature of contemporary religious terrorism.

Characteristics of Religious Terrorism

To define religious terrorism is a very difficult task. There are more than one hundred definitions of terrorism, and probably a much wider variety of definitions concerning religion. As for the criteria to classify terrorism as religious or not, Bruce Hoffman suggests the following two elements; whether they are using liturgy or religious texts to justify or explain the violence or attract recruits, and whether there is some sort of clerical figures involved in some leadership roles. Based on these criteria, Hoffman does not consider the terrorism in Northern Ireland as religious.[2] Mark Burgess also argues that religious terrorism is primarily motivated by religion, thus, the Northern Ireland terrorism cannot be regarded as religious because terrorist groups there, both Catholic and Protestant, are overwhelmingly motivated by a political, not a religious, imperative.[3]

On the other hand, Mark Juergensmeyer asserts that violence in Northern Ireland is related to the renewed role which religion has come to play on both Catholic and Protestant sides, while he admits the fact that most observers question to what degree religion actually is at the heart of the dispute there.[4]

Probably it is almost impossible to define religious terrorism, as Aref M. Al-Khattar honestly states that his study on terrorism and the three monotheisms, namely Judaism, Christianity and Islam, fails to arrive at a single, unified definition of terrorism in the name of religion.[5] But he suggests the following as the closest to a unifying meaning regarding religious terrorism:

> Religious terrorism is a violent act against others (individuals, groups, or states) to coerce them to behave or act according to the perpetrator's (individual, group, or state) interpretation of a religion.[6]

In many cases terrorism is discussed and defined in the context of politics. But the moot question here is that is it really possible to separate politics from religion in an adequate manner. As we will discuss it later, most activists of religious terrorism believe that religion should be the centre of any domain in a society, including politics, based on their own interpretation of liturgy or religious texts. In this regard, Khattar's definition, although it may be tentative, could be very useful because it does not differentiate "religious" and "political" motives.

Unrestricted and very destructive violence is often referred to as one of the characteristics of terrorism based on religious motives. In fact, according to the Terrorism Knowledge Base data-set[7] provided by the National Memorial Institute for the Prevention of Terrorism (MIPT) based in Oklahoma City, USA, 32 terrorist organisations which can be classified as religious perpetrated 395 fatal attacks in 2004, each of which resulted in one or more deaths. This number represented 61 per cent of the total fatal cases carried out by identifiable terrorist organisations, but the ratio of deaths caused by religious terrorist attacks was disproportionately high, namely 84 per cent of the total victims, or 1,868 of 2,216 individuals.

Bruce Hoffman cites the following points as reasons why religiously motivated terrorism results in more victims[8]: firstly, religious terrorists conceive violence as a sacramental act or divine duty executed in direct response to some theological demand or imperative, therefore, they are unconstrained by the political, moral or practical constraints that may affect secular terrorists, secondly, religious terrorists seek to appeal to no other constituency than themselves, thus the restraints on violence that are imposed on secular terrorists by the desire to appeal to a tacitly supportive or uncommitted constituency are not relevant to religious terrorists, and thirdly, religious terrorists see themselves not as components of a system worth preserving but as 'outsiders', seeking fundamental changes in the existing order, and this sense of alienation enables them to contemplate far more destructive and deadly types of terrorist operations than secular terrorists.

Based upon these arguments, many observers warn that religious terrorists could bring about much more destructive and disastrous consequences once they acquire weapons of mass destruction, particularly nuclear weapons. For instance, Mohamed ElBaradei, Director General of the International Atomic Energy Agency (IAEA) cautioned against the danger of efforts by the Al Qaeda and other extremist groups to obtain WMDs, by saying, "I would be surprised if they did not try to acquire nuclear weapons and other weapons of mass destruction. That would be most horrible scenario because these extremist groups if they have the weapons, they will use it."[9]

These concerns are quite understandable, particularly in the post September 11 era. But it is worth examining whether any religious terrorist group operates without any restriction and constraint in the respect of degrees and manners of violence which they resort to.

Christian right wingers have been considered to be the most dangerous religious movements in the United States. They conceive their terrorist acts as divine duties because they "physically struggle with forces of evil against sin and other violations of God's law" according to a report of the Federal Bureau of Investigation.[10] In fact, for the last two or three decades they have carried out many terrorist attacks, including bombing, arson, shooting and others, targeting abortion facilities, federal buildings, homosexuality groups, and so on. But most of their attacks were well targeted and not indiscriminate, with a few exceptions, including the bombing of the Oklahoma City federal building in 1995, which destroyed almost an entire building and killed 169 people.

Terrorist attacks by Jewish fundamentalists are also to a certain extent perpetrated with some restraints. Since the late 1970s, Jewish fundamentalists have been quite active in the Israeli occupied territories and have executed many terrorist attacks against Palestinian Arab citizens and Islamic religious sites. But, according to Richard

Chasdi, their attacks are with less intensity than their Islamic counterparts because Jewish fundamentalists operate in so-called "friendly" areas.[11]

As Richard Chasdi argues, terrorist attacks by Islamic extremists may be more intensive than ones by their Jewish counterparts, but it is also a fact that many Islamic militant groups have shown their own restraints. Hamas, the Islamic militant group in Palestine, apparently takes into account their level of support in the Palestinian society, and has tended to refrain from terrorist attacks against Israeli targets, including suicide bombings, when their fellow Palestinians strengthen their criticism against its terrorist operations. Hamas also took part in the informal truce with Israel in the spring of 2005, when many Palestinians supported a peace settlement rather than 'armed struggles' after the death of Chairman Yasser Arafat in November 2004.

Ayla Schbley, who conducted interviews with about 40 activists of Hezbollah in Lebanon, concludes that their targets would most likely be well-defined, limited in scope and dimension, and they would not willingly use chemical, biological, or nuclear weapons, despite the fact that Shi'ite terrorist operations are mostly executed for fulfilling personal salvation by answering perceived divine message/will, or following directives from charismatic religious leadership, and terrorism provides some Shi'ite zealots with profound spiritual satisfaction and fulfilment.[12]

Therefore, it can be argued that there are two different types of religious terrorism concerning the degree of violence that religious terrorists perpetrate. One is not constrained by political and/or cultural situation and tends to execute unrestrained and indiscriminate violence. Another is constrained by political and/or cultural context in which it operates and the degree of its violence is calculable. What makes them different?

Territorialised and De-territorialised

As Gilles Andréani argues that in the post-September 11 world, "we face a ubiquitous 'de-territorialised' threat, without a clear set of territorial objectives or geographical bases,"[13] contemporary religious terrorism can be categorised into two subtypes, territorialised and de-territorialised.

Territorialised Religious Terrorism

Territorialised religious terrorism operates in the context of a particular territory or land, and it can be further divided into two subcategories: 'territory-liberation seeker' and 'theocracy seeker'. 'Territory-liberation seeker' seeks liberation of a certain territory or establishment of its own rule over a particular piece of land through violence and terrorism. Palestinian Islamic groups, such as Hamas and Palestine Islamic *Jehad*, are typical 'territory-liberation seekers'. They perceive Palestine as an indispensable part of the entire Muslim land, and have perpetrated many violent attacks, including suicide bombings, under a banner of liberating the whole Palestine from the Israeli rule. The article eleven of the Hamas Charter declares, "the Islamic Resistance Movement (i.e. Hamas) believes that the land of Palestine has been an Islamic Waqf throughout the generations and until the Day of Resurrection, no one can renounce it or part of it, or abandon it or part of it."[14]

Their Jewish counterparts, Kach and Kachne-Chai, are also 'territory-liberation seekers'. Extreme religious Zionists belonging to these Jewish groups claim that the creation of the state of Israel and its occupation over entire Palestine represent the beginning of the messianic era, and any Jew cannot concede even one inch of *Eretz Israel* (the Land of Israel). This was the justification that the young Jewish terrorist cited when he murdered Yitzhak Rabin.

Hezbollah in Lebanon, various groups in Chechnya, both Catholic and Protestant groups in Northern Ireland, Abu Sayyaf in the Southern Philippines, and many others can be categorised as 'territory-liberation seeker' type of territorialised religious terrorist groups because their main purpose is to liberate and/or dominate a particular territory. The Sikh religious extremists who perpetrated many terrorist attacks until the middle of 1990s fall under the same category, because according to Karandeep Singh they wanted to alter the population composition of the Punjab state in favour of Sikhs against Hindus by terrorising Hindu population and facilitating Sikh migration into Punjab and Hindu emigration from Punjab to other parts of India.[15]

'Theocracy seeker', on the other hand, struggles for maximalising roles of religion in a nation-state where they reside and an eventual establishment of a theocratic society governed by the law of God. Since the Enlightenment, the process of secularisation has expanded and deepened in many part of the world, and the principle of separation of church and state has been applied in the most nation-states, although with various degrees and in different manners. Bruce Lincoln, however, argues that there is an unresolved tension between the secularising character of the modern state and the potentially religious character of the nation, and the maximalists fiercely fight back against the secularisation process in order to restore religion to the central domain of culture.[16] Lincoln's argument resonates with John H. Garvey's analysis on the attitude of religious fundamentalists against the public/private distinction. According to Garvey, they see the public/private distinction as artificial, and believe that religion is inseparable from law and politics.[17]

Gamaat Islamiya and *Jehad* in Egypt, Front Islamique du Salut in Algeria, and Christian Identity movements in the United States can be classified as 'theocracy seekers'. All of them reject the idea of separation of church and state and try to topple the existing secular order in the nation-state which they belong to by violence and terrorism with the ultimate objective of establishing a theocratic society completely based on their own interpretation. Thus, they confront the nation-state but at the same time they are confined within the boundaries of a territorial state; accordingly 'theocracy seekers' can also be classified as territorialised terrorists.

De-territorialised Religious Terrorism

Rabei Osman Sayed Ahmed, an Egyptian, who was believed to play an important role in the simultaneous bombing attacks on trains in Madrid in March 2004, and was arrested later in Milan, allegedly said that nationality did not matter in holy war.[18] Indeed, the perpetrators of the Madrid bombings consisted of a variety of nationals, such as Moroccan, Algerian, Egyptian, Syrian, Spanish and Tunisian, while a core group was dominated by Moroccans. They had no intention to liberate any part of the Spanish territory or replace the present Spanish government in Madrid with an Islamic regime.

As the Madrid case exemplifies, since the September 11 we have seen many attacks carried out by de-territorialised terrorists, who are not bound to any particular state or land. They cross international borders very easily, and some of them come from the diaspora. It has been reported that a number of European nationals go to Iraq and take part in *Jehad* against American and other coalition troops although they had no specific reason for identifying themselves with Iraq.

De-territorialised religious terrorism has not emerged from nowhere. This new phenomenon is a result of the transformation of some territorialised groups into de-territorialised ones. One can trace the background of this process in the leaflet written by Ayman al-Zawahiri, an Egyptian terrorist and Osama bin Laden's deputy. Islamic extremists who have been inspired by Sayyid Qutb's *jahiliyyah* ideology have tried to topple a regime of the nation-state which they belong to, calling it a "near enemy." For instance, Muhammad Abd al Slam Faraj, a leader of the Egyptian *Jehad* group, stressed the importance to fight against a near enemy, and eventually the group killed Egyptian President Anwar al-Sadat as its "near enemy." But Zawahiri calls for opening a battlefront which includes the entire world. According to him, they should engage in "a battle that is not confined to a single region" because "the Jewish-Crusade alliance, led by the United States, will not allow any Muslim force to reach power in any of the Islamic countries" and "the Jewish-Crusade alliance will not give us time to defeat the domestic enemy then declare war against it thereafter."[19]

De-territorialised religious terrorism is mainly operated by Jehadist groups and individuals. Most of them may connect themselves by very loose networks. But some of them may not have any link or network, and perpetrate terrorist attacks in very responsive and reactive manners. They carry out attacks not because they are ordered or commanded but merely encouraged and inspired by Jehadist preachers and other terrorist attacks. The second bombings in London in July 2005, were a good example of this type because the attackers just followed the first suicide bombings which took place two weeks earlier.

In this sense, as Philippe Errera argues, the term 'decentralisation' may be misleading because we are dealing rather "with a phenomenon that has been more accurately characterised as an ideology, or even a social movement," and even the term itself "implies that we are still dealing with a well-organised core leaders, in full charge of the command-and-control system of the organisation."[20]

The dissemination of fatal violence inspired by de-territorialised Jehadist ideology certainly relates to the failure of many nation-states, which have not been able to provide any effective solutions to numerous problems in many fields, including political participation, economic development, human rights protection, social differences, and so on so forth. Although in the modern age the majority of Muslims consider the notion of establishing a new caliphate a utopian goal, for the last decade or so, the idea of establishing a *umma* under one caliph has attracted a significant number of Muslims who have strong grievances against a secular nation-state and feel themselves seriously humiliated by the current international circumstances mainly led by the United States and its allied forces. Of course the idea of *umma* is very much theoretical and nobody can offer any practical idea of how to establish it. But new information technologies, represented by internet, provide many young Muslims with a virtual community with which they can identify themselves as if it would be a modern *umma*.

De-territorialised Terrorism and Violence

As the September 11 has shown, de-territorialised religious terrorist attacks result in many victims and large-scale destructions. De-territorialised religious terrorists tend to resort to unrestricted and indiscriminate violence because they do not operate in a particular state or territory, and therefore, they do not need to be sensitive to general trends and opinions of a society in which they exist. For instance, in the video message, Mohammad Sidique Khan, one of the London suicide bombers of July 7, 2005, said that the public in London was responsible for the atrocities perpetuated against his "people" across the world because they supported democratically elected governments which carried out atrocities, and further he claimed, "Until we feel security, you will be our targets."[21] From his words it is impossible to trace any cue which indicates that Khan had a sense of belonging to the British society, although he had a British nationality and had been a teaching assistant at a primary school.

Another reason for their resort to unlimited violence is that purposes of de-territorialised religious terrorism may be different from territorialised religious terrorism. Territorialised groups mainly carry out terrorist acts for a supplementary measure in a conventional armed conflict and/or enhancing their political influence. Consequently territorialised religious terrorists tend to calculate costs and benefits which would result in terrorist operations.

On the other hand purposes of de-territorialised Jehadist groups are not clear. From various statements issued by Al Qaeda, Osama Bin

Laden and others, we can guess that they fight against "the Jewish-Crusade alliance, led by the United States" on the global theatre, but we do not know exactly what they want to achieve and what they want to construct after their "holy war." As R. James Woolsey said in 1994, "today's terrorists don't want a seat at the table, they want to destroy the table and everyone sitting at it,"[22] their purposes cannot be measured by conventional manners because they act only for their own interpreted God, thus, they escalate their violence without any calculation on costs and benefits.

Conclusion

National Strategy for Victory in Iraq, released by the National Security Council of the Bush administration at the end of November, 2005, categorises the Iraqi insurgent forces into three subgroups as follows:

- Rejectionists: the largest forces, most of them are Sunni Arabs who are against a new Iraq in which they are no longer the privileged elites.
- Saddamists and former regime loyalists: they harbour dreams of reestablishing a Ba'athist dictatorship.
- Terrorists affiliated with or inspired by Al Qaeda: they make up the smallest enemy group but are the most lethal and pose the most immediate threat.

National Strategy for Victory in Iraq underlines an importance of "isolating hard-core rejectionists," by arguing that the three groups "share a common opposition to the elected Iraqi government and to the presence of Coalition forces, but otherwise have separate and to some extent incompatible goals." It is the first time that the Bush administration classifies the insurgent forces into different types and recognises the necessity for taking different and nuanced approaches to each one in order to isolate a "hard-core" group. Before that, the administration did not make any differentiation among insurgent forces and treated them, instead, just as "terrorists." Ironically enough, such strategy, without any differentiating and nuanced approach, united different types of armed forces against their common enemy.

Thus, the categorisation of religious terrorism into the two types, territorialised and de-territorialised, provides us with a useful tool which could be used not only for academic purposes but also for policy practices. Groups of the two types, of cause, have many common characteristics and may work jointly under certain circumstances, but their ultimate goals are very different from each other and basically incompatible. In addition to that, as discussed in the previous section, means of terrorism and violence used by such terrorist groups differ too. Therefore, policy planners and strategists should take into consideration the categorisation and differentiation when they develop strategies for fighting religious terrorism.

Notes and References

1. *National Strategy for Combating Terrorism*, February 2003, p.8. *http://www.whitehouse.gov/news/releases/2003/02/counter_terrorism/counter_terrorism_strategy.pdf*, Accessed on August 10, 2004.
2. "Religion and Terrorism", Interview with Dr. Bruce Hoffman, *Religioscope*, February 22, 2002. *http://www.religioscope.com/info/articles/003_Hoffman_terrorism.htm*, Accessed on March 7, 2005.
3. Burgess, Mark (2004). *Explaining Religious Terrorism Part-2: Politics, Religion, and the Suspension of the Ethical*, Center for Defense Information, August 23. *http:/www.cdi.org/program/issue/ocument.cfm?DocumentID=2384&IssueID=138&StartRow=1&ListRows=10&appendURL=&Orderby=DateLastUpdated&ProgramID=39&issueID=1380a*. Accessed on Feb. 8, 2005.
4. Juergensmeyer, Mark (2000). *Terror in the Mind of God: The Global Rise of Religious Violence*, University of California Press, pp.36-43.
5. Al-Khattar, Aref M. (2003). *Religion and Terrorism: An Interfaith Perspective*, Praeger, p.95.
6. Ibid.
7. *http://www.tkb.org/Home.jsp* (accessed on July 10, 2005).
8. Hoffman, Bruce (1998). *Inside Terrorism*, Columbia University Press, pp.94-95.
9. "TV: IAEA Chief Says Al Qaeda Sought Nuclear Weapon", *Reuters*, April 9, 2005.
10. Federal Bureau of Investigation, Megiddo, p.15 (*http://www.odl.state.ok.us/usinfo/megiddo/megiddo.pdf*, accessed on December 20, 2005).
11. Chasdi, Richard J. (1997). "Middle East Terrorism 1968-1993: An Empirical Analysis of Terrorist Group Type Behavior", *Journal of Conflict Studies*, Vol.17(2), Fall.
12. Schbley, Ayla (2003). "Defining Religious Terrorism: A Causal and Anthological Profile," *Studies in Conflict and Terrorism*, Vol. 26(2), March-April, p.120.
13. Andréani, Gilles (2004). "The 'War on Terror': Good Cause, Wrong Concept", *Survival*, Vol.46(4), Winter 2004-05, p.38.
14. The full English translated text of the Hamas Charter can be found in Alexander, Yonah (2002). *Palestinian Religious Terrorism: Hamas and Islamic Jihad*, Transnational Publishers, pp.47-72.
15. Singh, Karandeep (1990). "The Politics of Religious Resurgence and Religious Terrorism: The Case of the Sikhs of India," Emile Sahliyeh (ed.), *Religious Resurgence and Politics in the Contemporary World*, State University of New York Press, p.260.
16. Lincoln, Bruce (2003). *Holy Terrors: Thinking about Religion after September 11*, The University of Chicago Press, pp.58-61, 63.
17. Garvey, John H. (1993). "Introduction: Fundamentalism and Politics," in Martin E. Marty and R. Scott Appleby (eds.), *Fundamentalisms and the Starte: Remaking Pololiti, Ecomomiec, and Moilitance*, the University of Chicago Press, p.13.
18. "The Talkative Terrorist on Tape: Madrid Plot 'Was My Project", *New York Times*, July 12, 2004.
19. al-Zawahiri, Ayman (2001). "Knights Under the Prophet's Banner", *Al-Sharq al-Awsat*, Dec. 2 (in FBIS-NES-2002-0108, Dec. 2, 2001). *http://www.liberalsagainstterrorism.com/wiki/index.php/Knights_Under_the_Prophet's_Banner*", Accessed on September 15, 2005.
20. Errera, Philippe (2005). "Three Circles of Threat", *Survival*, Vol.47, No.1, Spring, p.72.
21. *BBC*, "London Bomber Video Aired on TV," September 1, 2005.
22. Quoted in *National Commission on Terrorism*, "The International Terrorism Threat is Changing", 2000, p.2.

29

Water Resources Issue in the Middle East Conflict

BAKHTIYAR MIRKASYMOV

The problem of water resources in the Middle East is a catalyst for conflicts but it also has a potential to lay the ground for cooperation in the region. According to some estimates, the world population will exceed 8 billion by 2025-2035 with over 3.4 billion having no access to clean drinking water. The UN declared the decade, i.e., 2005-2015, as the "Water for Life" decade in order to draw attention of governments and NGOs to solve the problem of inadequate water supplies.

About 5 per cent of the world's population is living in the Middle East while the region's water resources account for less than 1 per cent of the Earth's total. The number of water deficit countries in the Middle East and North Africa increased from 3 to 11 in 1990 (which included Algeria, Israel with occupied territories, Qatar, Saudi Arabia, Somali, Tunisia, UAE and Yemen). Another six countries may join the list by 2025 (Egypt, Ethiopia, Iran, Libya, Morocco and Syria.[1] The main sources of water in the Middle East originate from the Tigris, Euphrates, Jordan and Nile river systems. One of the difficult problems in the region is the trans-boundary nature of the water sources. Most of the hydro resources from the northern part of the region are carried through rivers and underground sources. In many cases, water is carried over long distances to places where water is scarce.

Due to existing political disputes and unresolved conflicts, no comprehensive, equitable and basin wide water sharing agreements have been reached. Controversial nature of the international watercourse law makes it rather difficult to provide the basis for such agreements. National policies on water use are often conducted in clear disregard to the needs of the neighbouring countries or future generations. Such approaches cause tensions and may provoke clashes.

The UN Convention on the Law of the Non-navigational Uses of International Watercourses was adopted by the UN General Assembly in 1997. It stipulates that all riparian states should use their rivers in an equitable and reasonable manner which would allow sustainable development, optimal and long-term benefit, taking into account the interests of other riparian states and their natural, demographic and socioeconomic factors. Riparian states should prevent causing considerable harm to other watercourse states. In case of significant harm caused to another state, the responsible riparian should take necessary measures to eliminate or mitigate such harm, in consultation with the affected country. The watercourse states may establish a joint framework for the management of an international watercourse.

There is no mechanism to enforce this convention. Israel abstained from voting; Turkey was against it. A country's position with regard to the water law depends on its geography, socioeconomic situation, international situation and other factors. States controlling headwaters tend to favour absolute sovereignty over the river without obligation to other riparian states. However, most countries do not rely on this approach as it is not comprehensive and fails to settle disputes over water resources of trans-boundary watercourses.

Lower riparians insist on historically earlier use of river waters and are more likely to turn to international regulation or arbitration. Such states support the theory of absolute integrity of a river. It means that an upper riparian state has an obligation not to cause harm or in any way reduce the quantity or quality of the water reaching lower riparians. This theory also has drawbacks as it favours lower riparians at the expense of upstream states.

A limited sovereignty doctrine was evolved as a compromise. According to this theory, every riparian recognises rights of other riparian states to use water from the common source and undertakes to regulate its water use in such a way that it will not inhibit similar uses of other states. Recognition of mutual rights reflects growing demand in countries for guaranteed and uninterrupted access to water resources and their desire to avoid conflict situations. This theory has been widely used among riparian states.

Yet another theory, based on an ecosystem concept, considers management of a river basin as a single geographic and economic entity, irrespective of international borders. Riparians agree to joint exploitation of the river basin and participate in its development and protection in an equitable manner. In spite of its advantages, this theory remains rather utopian in modern international conditions and also due to lack of water in certain regions.

Historically, in the Middle East, water rights tended to be a matter of communal responsibility that took into consideration collective interests of the people. State provided the right to use water but not to own it. In the Middle Eastern environment, experts put emphasis on the 'needs' of riparian states rather than their 'inalienable right' on water.

As a result of the Arab-Israeli War, 1948-49, Israel occupied part of the land that was intended for the creation of an Arab state in accordance with the UN resolution. With the new boundaries in the region drawn sometimes with the use of force, every country started unilateral development of their water resources. This brought into collision interests of different states in the Jordan River Basin. At the beginning of the 1950s, Arab states discussed projects aimed at using water of the Hasbani and Banias tributaries of the Upper Jordan River. Israel was intent on changing the course of the Upper Jordan River and planned construction of a national water to transfer part of the water beyond the Jordan River Basin into the Negev desert.

In 1951, Jordan publicised plans to irrigate the East Ghor area with the Yarmouk River water. Israel stopped the water flow to the south of the Tiberias Lake and started drying the marshlands in the demilitarised zone near the border with Syria, provoking clashes. The construction of the Israeli national water carrier began in 1953 amidst Syrian protests over water being taken from the northern part of the Tiberias Lake. Israeli leadership's approach justified any actions supporting immigration even though they contradicted international law.

In that tense environment, the United States sent special representative Eric Jonston with the mission to prepare a comprehensive solution to the water resources sharing problem in the region. The plan sought to satisfy the minimum requirements of riparian Arab states as well as Israel. Due to political reasons, Jonston plan was not ratified.[2]

By 1964, the construction of the Israeli water carrier enabled Israel to transfer water from the Jordan River basin to the Mediterranean coast while Jordan was prepared to take the Yarmouk water for the East Ghor irrigation canal. In 1965, the Arab countries proceeded with the construction works that would turn the Litani River to Lebanon, the Banias river to the Yarmouk where it would be impounded by a dam for use by Syria and Jordan. Implementation of this plan would have reduced the capacity of the Israeli water carrier by 35 per cent and increased the salinity of the Tiberias Lake. Israeli army attacked construction works of Banias-Yarmouk canal in 1965-66. Border clashes in 1966-67 on land and in the air ended in a full scale Israeli attack in June 1967. Having occupied territory, Israel gained access to the Hasbani and Litani Rivers.

Israel depends on three sources of water: Sea of Galilee Basin, the West Bank Mountain Aquifer and the Coastal Aquifer. Overexploitation of underground water sources resulted in deteriorating water quality. Over half of the water consumed by Israel is taken from the occupied territories, thereby increasing the tensions between the Arabs and the Israelis. Israel consumes more water than is ecologically safe. Water sources are overused by 15 per cent and, according to some estimates, at the end of this decade this level will increase to 30 per cent, presenting a national security risk to Israel.

There is a great inequality in per capita water consumption and access to water resources between Israeli settlers and Palestinians. Several Israeli-Palestinian accords govern water control. According to 1993 Oslo Agreement, Israelis retain hydrological control of the West Bank as a means to limit use of underground water sources. Palestinians are allowed to dig a limited number of wells. 82 per cent of water from the West Bank is used by Israelis leaving one-third of Palestinians in the West Bank without reliable water supply. Israeli settlers on the occupied land have 3 to 5 times more per capita water than their Palestinian neighbours.[3] Inequality in water allocation results in Palestinians having 2.5 times less water than the recognised international minimum.

Despite the provisionary agreement of 1995 on the West Bank and Gaza, bilateral relations have had a conflicting nature. The Palestine Authority (PA) claims that Israelis have a privileged access to water at the expense of the Palestinians retaining proportions established after the 1967 war. For Palestinians living under Israeli occupation, water deficit is due to political rather than natural reasons. The West Bank provides 600-700 million cubic metres of water. Israelis limit Palestinian consumption to one-fifth of that amount. Palestinians are denied access to the Jordan River. It affects agriculture and economic development. On the other hand, Israeli settlers have no limits on water consumption. Having subsidies from the World Zionist Organisation, Israelis dug deep wells equipped with powerful pumps that provide 40 per cent more water than 300 more shallow wells dug before 1967.

About 3.7 million Palestinians, living in Gaza and the West Bank, consume about 260-290 million cubic metres of water including their agricultural industrial and household needs. 6.4 million Israelis consume around 2,129 million cubic metres a year.[4] Under international pressure, Israel agreed to compensate partially lack of water to the Palestinians allowing them access to some deep wells. Local Arabs have become completely dependent on the Israeli authorities for water supply.

In Gaza, 1.3 million people living there experience the most acute shortage of water in the Middle East. Underground water can provide up to 60 million cubic metres but over 150 million is actually pumped. Water shortage forces local people to resort to measures prohibited by the Israelis. 4300 wells are legal, but another 2400 have been dug illegally pumping another 70 million cubic metres from an already low level of underground water.[5] Excessive reliance on wells leads to salty water penetration and increase of nitrates above the permissible levels. According to Palestinian estimates, by 2010 water demand will increase to 300 million cubic metres exceeding by 75 per cent the water volume that is obtainable with safe ecological standards.

To deal with the situation, the PA demand damage compensation from the Israelis caused during the years of occupation. They are in favour of finding a solution to the water problems,based on international law through equitable and reasonable sharing of water resources. Desalination plants could be built with the money from international donors easing heavy reliance on underground water. Water resources allocation should be done based on international standards. The probability of negative consequences to ecology increases if one of the riparians is not subject to outside monitoring. Unilateral building of the Israeli separation wall may worsen the situation because 30 per cent of wells and water are expected to become inaccessible to the Palestinians after the Israeli pull out. The Israeli Institute of Water predicts worsening of the water shortage in

Israel after the separation because the Palestinians may use additional 50 million square metres of land without Israeli permission and water cost to them will be reduced.[6]

Israel plans to use more treated water for irrigation. Israeli, Palestinian and French researchers suggested mapping polluted areas of underground water. Computer modelling will help to determine directions of underground currents along Israeli and Gaza border. Then several scenarios of underground water exploitation will be proposed using mathematical models.[7]

Israelis would like to buy water from Arabs but the unresolved Middle East conflict reduces the scope of Arab-Israeli cooperation. Many rivers used by Israel, Syria, Lebanon, Jordan and Palestine originate from the occupied Golan Heights. The overall surface of the Golan Heights is 1,860 square km (1,250 square km are under Israeli occupation). Technical examination revealed underground water resources exceeding the surface water in the area.

The Israelis put forward certain conditions in their negotiations with Damascus over the Golan Heights. They want to ensure continuous uninterrupted supply of water.

Despite the fact that Lebanon is better off with water, by 2025, the country risks to experience its shortage. Lebanese authorities have to take measures to curb industrial and agricultural pollution, introduce more efficient modern irrigation, replace outdated drainage and running water systems. In 1999, the Lebanese government started implementing a 10-year plan including construction of water distribution systems in South Lebanon, and 30 new dams.

After leaving the 'security belt' in South Lebanon, Israel is very suspicious of any activity on the Lebanese territory that might reduce the flow of water to the Tiberias Lake. The yearly flow of the Hasbani River is estimated at 150 million cubic metres,[8] contributing 20-25 per cent of the water flowing to the Tiberias Lake.[9] The water level of that lake is crucial to the water supply of the densely populated Israeli territories on the Mediterranean coast and the Negev Desert because water from the lake is transferred to those areas via the Israeli water carrier. Lebanese efforts to supply water to several villages in the south of the country drew bellicose statements from Israelis in 2002. The Lebanese wanted to arrange a daily water supply of 10,000–12,000 cubic metres to 15-20 nearby villages.[10] Israeli authorities claimed that their northern neighbour tried to pump yearly 50 million cubic metres of water through a 16 inch diameter pipe from the Wazzani River. Independent observers from UNIFIL and USAID checked the diameter of the pipe which turned out to be 4 inches.

The Lebanese side claimed that they were entitled to 50-70 million cubic metres of water yearly from the Jordan River tributaries on their territory while they were using only 5 million at that moment.[11] As for Israelis, they consume about 150-160 million cubic metres of water from the South Lebanese tributaries to the Jordan River.[12] Lebanon was in favour of UN mediation. There were calls in the Israeli media to help in financing alternative development projects in South Lebanon on the basis of the Litani River and, if possible, water sources in Northern Lebanon.[13] It is unlikely though that Israel would participate in the Lebanese infrastructure projects.

The Kingdom of Jordan uses about 1,250 million cubic metres of water yearly while locally available water amounts to only 850 million cubic metres. Having one of the worst water availability in the region, Jordan has to manoeuvre between Israel and Syria in order to ensure continuous water supply. Forecast estimates predict a water deficit equalling one-third of the yearly consumption by 2010.

Syria is in a privileged situation compared to Jordan that is lower along the Yarmouk River. There were two agreements between Amman and Damascus on the use of the Yarmouk River waters. Despite the negative attitude towards the Israeli-Jordanian peace treaty of 1994, Damascus took steps to normalise relations with Amman. In 1996, the late Jordanian and Syrian rulers agreed to further implementation of the 1987 agreement and complete the construction of the Wahda Dam. In 2004, the new leaders of both countries participated in the ceremony of the dam construction. With the Wahda Dam fully operational, it will improve the Jordanian water supply by 10 per cent and increase Syrian electricity generation.[14]

After the conclusion of the Israeli-Jordanian peace treaty in 1994, the Jordanian leadership chose to settle problems with the Israelis at the negotiating table. According to that treaty, both countries have obligations to protect the Jordan River from pollution. Jordan has rights on most of the Yarmouk water while Israel becomes the main user of the Jordan River water. The treaty also regulates the operation and preservation of water sources during different seasons and envisions construction of a water desalination plant. According to

1997 agreement between Israel and Jordan, Tel Aviv promised to supply 50 million cubic metres of water.[15]

Their bilateral relations were not always smooth partly because Israel failed to fulfil obligations with respect to the Jordanian water supply. In March 1999, Israel decreased by half, the water supply to Jordan during drought. Having no water desalination plants, Jordan had to resort to water rationing. After a new Israeli government came to power, the water supply to Jordan was restored.

The critics of the Israeli-Jordanian accords claim that Jordan could not improve its water supply as the result of the peace treaty having to put up with less water than it was suggested in the Jonston plan in 1954. Israel retains control over the wells in the former occupied Jordanian territories. Jordan agreed to Israeli water transfer from the Tiberias Lake giving up control of the water quality. Despite that agreement, both countries continue to take up to one-third of their water requirements from the Jordan River and release untreated drainage into the river amounting to 20 per cent of the river flow.[16]

Having 225 billion cubic metres of water, 110 billion of which can be practically used, Turkey is rich in hydro resources in absolute and relative terms compared to its neighbours.[17] Turkey utilised only 38 billion cubic metres in 2000. Most of Turkish water potential is of internal origin while that of Iraq and Syria are of trans-boundary nature.

By the middle of the 1970s, Turkey experienced energy crisis, compelling the country's leadership to rely more on the country's hydro energy. Turkish Southeastern Anatolia Project (Guneydogu Anadolu Projesi-GAP) was planned as a huge project amounting to US $32 billion encompassing 6 provinces. The scheme includes construction of 22 dams, 19 hydro power stations, 13 auxiliary projects and increase of irrigated lands by 1.7 million hectares. It is hoped to increase the land productivity. Apart from purely economic goals, GAP was supposed to achieve social targets: improve population absorption in large regional centres, optimise economic structure and reduce regional imbalances and unemployment. The project was thought to be instrumental in easing the problem of Kurdish separatism.

The Turkish authorities decided to use fast economic development and large investments in the country's southeast with a big proportion of Kurdish population for their integration in the Turkish society. In reality, however, publicity for GAP project in 1984, coincided with the creation of PKK (Kurdistan Workers' Party). The party started an active struggle for independence from Ankara.

In 1987, Turkey initiated an agreement with Syria, according to which Ankara undertook to allow not less than 500 cubic metres per second of the Euphrates water flow at the Turkish-Syrian border. On an average, water passes at the rate of 1000 cubic metres per second in the river on the border.[18] Relations between the two countries were tense. Ankara accused Damascus of supporting Kurdish separatism. Syria insisted on her rights to the Turkish province of Hatay with a large proportion of Arab population.

Turkey's implementation of the Southeastern Anatolia Project (SAP) and impounding of the Ataturk Dam caused interruption of the Euphrates water flow to Syria in 1990. Massive hydro development in southern Turkey threatens to reduce Syrian water on the Euphrates by over 40 per cent, and Iraqi water by over 80 per cent.[19]

Agriculture has been quite important for Syria. In 1963, the Syrian government decided to commence construction of a huge dam called Tabqa on the Euphrates River. The dam was completed in 1975. Its name was later changed to Al-Thawra (revolution). The water reservoir in front of the dam bears the name of Assad. At first, the dam was supposed to produce 840 megawatts of electricity and irrigate 640 thousand hectares of land. However, the initial projections had to be scaled down because of hydrogeological difficulties. According to different estimates, irrigated land covers from 60 thousand to 400 thousand hectares while the electricity production is at 150 megawatts.[20] Despite the lower than expected economic output, the dam is very significant to the Syrian economy. The Assad Lake contains 14.2 billion cubic metres of water that is used for household and industrial needs of Aleppo. The rest of Syrian river water resources amount to slightly more than 6 billion cubic metres. Underground water from wells makes up another 2.67 billion cubic metres.

Syria implements irrigation projects on other rivers, among them, the Orontes River. Its headwaters start in Lebanon, then the river passes on the Syrian territory and flows into the Turkish province of Hatay. Two completed dams, Rustam and Hilfaya-Meharde, provide electricity and water to Homs and Hama cities. Turkish experts claim that Syria takes up to 90 per cent of the river's water leaving from the total river flow of 1.2 billion cubic metres per year only 120 million. Even that amount of water is under threat. Another two planned water flow regulating dams may reduce the water flow to Turkey to 25 million cubic metres per year. On the other hand, Lebanon and Syria signed an agreement in 1994 that allows Lebanon to use 80 million cubic meters of water from the Orontes River. Thus, the problem of the Orontes River water resources is an example of double standards in politics.

The conflict over impounding the Ataturk Dam in 1990, brought about an agreement between Syria and Iraq on the sharing of the Euphrates River water. The protocol stipulates that 58 per cent of water received by Syria from Turkey will be released to Iraq.[21] The positive effect of the accord is limited and depends on the amount of the Euphrates water coming to Syria from Turkey. In 2005, Iraqis and Syrians agreed to exert joint efforts to make Turkey fulfill earlier obligations regarding water allocation on the Euphrates, exchange information on hydrology and climatic changes.

Syria considers the Tigris and Euphrates Rivers as international watercourses with the trilateral use and hence water sharing based on quotas. Being in favour of international mediation, Damascus suggests finding the solution within the International Court of Justice or the Arab League. Iraq was also worried about the negative consequences of the dam building on the Euphrates accusing Turkey of the international law violation. Baghdad threatened to impose sanctions on companies having dam construction contracts in the Southeastern Anatolia.

Turkey insists on her sovereign rights over water as other countries have their rights over oil but the Turkish authorities say that they are ready to accommodate Syrian and Iraqi interests. The regions water resources provide Turkey with a considerable leverage over the neighbouring countries' policies.

There are signs that Turkey is moving towards accommodating Syrian and Iraqi concerns over the Euphrates River water flow. Ankara and Damascus agreed to conduct bilateral exchange and training of personnel, implementation of joint projects and exchange of new technologies within GAP–GOLD (General Organization for Land Development, Ministry of Irrigation, Syria) framework. But this is only a small step along the road to a comprehensive solution of the water resources problem in the Middle East. Some Turkish observers consider a possibility of Turkey signing the 1997 UN Convention on the Law of the Non-navigational Uses of International Watercourses to bring Turkish legislation in line with the EU standards in an effort to join the EU.

Water issues cannot be separated from or solved earlier than the political problems. Their parallel or simultaneous solution seems more productive. Experts suggest different ways to deal with the problem. Increased catchment of winter floodwater can add to the water budget. When it is possible to store water underground through artificial groundwater recharge, even more water is saved. New sources of water can be developed using technology. Projects like iceberg towing and cloud seeding do not seem to be feasible. The two most likely technologies to increase water supply are desalination and wastewater reclamation.

Setting up a region wide or a basin wide joint organisation would contribute to reasonable planning and allocation thus helping to end conflicts. Withholding of complete data on water availability and use should be perceived as a political liability, rather than an advantage. Making hydrologic remote sensing, and other geographical data freely available to all parties should be a major priority and could be done using electronic data handling systems and exchanges, such as is now occurring on Internet and related systems throughout the world. While new sources of water supplies may eventually be developed, cooperation over the existing water resources is essential; unless current water supplies are equitably and efficiently allocated and used, agreements to increase the overall pie will be jeopardised. Ways to reduce water waste and increase efficiency of water use should be studied and implemented.

"Virtual water" concept means that instead of spending water for certain uses or products they can be imported. Thus, 1000 tonnes of water is required for growing 1 tonne of grain that could be imported thus saving water for other uses. This concept can also be employed to reduce water demand but it seems that it has a limited application due to existing socioeconomic and political environment in the region.

Notes and References

1. Berman, Ilan and Wihbey Paul Michael. "The New Water Politics of the Middle East", *http://www.israeleconomy.org/strategic/water.htm.*
2. Wolf, Aaron T. (1996). "Middle East Water Conflicts and Directions for Conflict Resolution" International Food Policy Research Institute, Washington DC., March.
3. *CNN, http://www.cnn.com/2002/TECH/science/04/22/mideast.water/index.html.*
4. Nettnin, Sonia (2005). *Palestine Chronicle* Weekly Journal, Friday October 7. *http://www.palestinechronicle.com/story.php?sid=20051003233019336*
5. *http://www.metimes.com/articles/normal.php?StoryID=20050923-062237-4914*
6. *Al Jazeera. http://www.aljazeera.net/NR/exeres/92AE3C95-0213-43CA-A426-2789BD531E5.htm*
7. *Science Daily News* August 07, 2005. *http://www.sciencedaily.com/releases/2005/09/050907100405.htm.*
8. *BBC News,* September 16, 2001. *www.news.bbc.co.uk/2/hi/world/middle-east/2261060.stm.*
9. *Monday Morning,* No. 448, September 2002.
10. *http://www.iimes.ru/rus/stat/2002/15-10-02.htm*
11. *Jerusalem Report,* September 5, 2002. *http://www.iimes.ru/rus/stat/2002/15-10-02.htm.*
12. *Safir,* September 12, 2002.
13. *http://www.iimes.ru,* 15.10.05
14. *BBC News Online,* September 02, 2004. *http://news.bbc.co.uk/2/hi/middle_east/3473483.stm*
15. *Jordan Times. http://www.jordantimes.com/Tue/homenews/homenews5.htm.*
16. *BBC News Online. http://news.bbc.co.uk/1/hi/world/middle_east/4337341.stm*
17. Patrick, MacQuarrie (2004). "Wàtår Security in the Middle East. Growing Conflict over Development in the Euphrates Tigris Basin", Revised: 26 February 2004, originally submitted 15 September 2003, thesis, M.Phil, International Peace Studies Trinity College, Dublin, Ireland.
18. Kibaroglu, Aysegul, *Socioeconomic Development and Benefit. Sharing in the Euphrates-Tigris River Basin, Ankara, 06531, Turkey*
19. Patrick *op. cit.*
20. Greg, Shapland (1997). *Rivers of Discord,* pp.111-2.
21. Aysegul, Kibaroglu (2004). "Water Development and Cooperation", *International Expert Workshop,* Middle East Technical University, Ankara, Turkey.

Section VII

India and West Asia

30

India and West Asia

A Diplomats' Perspective

S.K. SINGH

I have spent many wonderful years working for India from 1956, in the Gulf region to almost the end of my diplomatic life in countries which were at peace, when they were at war, and even when they experienced civil war, in their years of prosperity and in the periods of lean resources, also during years of uncertainty and flux. It is a truism to say that flux persists in all our lives, whether as countries or as individuals or societies.

Truly civilisational links between India and West Asia go back through the millennia to Mohenjodaro, Harappa, and Mesopotamia. These links impact our cultures, languages, literatures, art forms, products, spices, craft, textiles, silks, wools, trade, commerce, and even weapons making. We in India feel that our interaction with West Asia has made us who and what we are today. The term "clash of civilisations" has therefore, little meaning for Indians but terms like coordinated growth and synergy have much meaning for me and for all of us Indians. The intense relationship of our two regions takes cognisance of the fact that we have both been evolving, growing, contracting, changing, and yet essentially remaining the same over the centuries.

As geographical, historical and cultural entities we have evolved in many complex ways. Both India and West Asia are not easy to define nor simple to describe. Over the decades and centuries, western authors and historians have referred to today's West Asia as the Near East, the greater Levant area, the Islamic world, the Middle East, etc. Post-independence India started defining the region as West Asia and North Africa. Even so, some of us have felt that this term perhaps failed to underline the importance to us of the Gulf region adequately.

As far as the entity called India is concerned, in a geographical, historical sense the periods of the Mauryan, Gupta and Kushan empries saw India too waxing and waning territorially and in political terms. In respect of its commercial, cultural, civilisational, religious, secular sense it has retained its sameness through the centuries. The same phenomenon persisted during the period of the literature which preceded the Sultanate of Delhi and later on the Mughal imperial era which was followed by the rise in India of the East India Company. In between was a period when the smaller *maharajas, rajas, nawabs* proliferated and the British colonial regime found their window of opportunity to establish themselves here. Until the departure of the British and the emergence of India into our partitioned independence, India defined itself variously during these eras and periods. In the years that preceded our Independence, the former colonial authority proceeded to disaggregate greater India which had once again tried to come together in the years preceding that, reuniting the pieces of scattered Indianness. They accomplished this surgery in the late 20s, 30s and right into the 40s. This has ensured that India they left in our hands was a much fractured entity with one separate India, two separate Pakistans, a separate Sri Lanka and a separate Burma, etc. The broad unity of Indianness however persisted and persists despite all these geographical games of separate sovereignties. Civilisationally, India has retained its multicultural, multilinguistic, multiethnic, multi-denominational, personality and its culturally and emotionally united, unique and singular character.

West Asia likewise includes the geographies and areas of 17 Arab states plus ones that are Persian, Turkish and Israeli, even though, in the latter too live more than a million Arabs. Within Turkey lived not only Turks but also the Kurds. Living in the Arab states of Lebanon and Syria are also some Druzes, and many Christians. Sudan too, has a non-Arabic speaking minority, also some Christians. Algeria and Morocco likewise have all along had their own respective Berber minorities. The Kurdish minority lives on territories and in areas that are included in Turkey, Iran, Iraq and Azerbaijan significantly speaking their own language Kurunji, and I am rather proud of saying that I was the only officer ever asked to learn Kurunji, which I did for almost a year and a half. The 17 Sunni states of West Asia and North Africa (WANA) region have been trying to add to their number a somewhat painfully emerging State of Palestine.

India boasts of having the second largest Muslim population in the world but India is neither Arabic nor Persian nor indeed Turkish but by either ethnicity or language even though over the centuries we have absorbed much from each one of them and structured one of our 18 national languages recognized under the Constitution–Urdu–which is perhaps the best of all understood languages throughout the breadth and length of India from North to South and East to West. Islam in India was for centuries, the religion of our rulers but not of the State, remained essentially a minority religion. Indian Islam acquired a distinct cultural, subcontinental genre of *qawwali*, devotional music, and absorbed specific characteristics of tolerant, mystique sufi Islam. In the poetic works of Malik Mohammed Jaisi, Ras Khan and Amir Khusro, the founders of the Hindi and Urdu languages of *Hindustani* written in Hindi, acquired a softness, gentleness, tolerance and an attitude underlining that all devotional commitment is all love—love of God and love of humans.

In the WANA region, Islam became a major unifying force but unity for them never meant total peace or total mutual trust. In West Asian lands and amongst West Asian people mutual trust and unity were absent in various periods and eras. While the Sunnis are the majority in an overall sense, including the Turkish Sunnis, but then in essence today's Turks do appear somewhat schizophrenic as they yearn at one and the same time to be considered both European and West Asian Islamic Sunnis. The Shiite segment of Muslims is a majority not only in the Islamic Republic of Iran, but also in Iraq and Lebanon. Shias have remained a significant community both in Syria and in Yemen. One can note *en passant* that 17 per cent of the Muslims of India, also about the same percentage of the Muslims of Pakistan, are Shias.

One issue that has affected, in practice, the security of West Asia as also international global security environment has been the Israeli-Palestinian problem. The ongoing peace process in the Middle East has often been punctuated with violence between Israelis and Palestinians. We may also consider the security challenges in the Mediterranean region and major questions of maritime security of the Persian Gulf region, significant both for global security and West Asia's security. Both these have come to be tied up with the security of the energy resources and safety of oil wells, gas and oil pipelines. Security of gas pipelines, which affects non-West Asian countries too, have gradually acquired great salience. The economic, political and security dimensions of Gulf oil and the sea lanes of communication sandwiching through the Hormuz Straits and around it have become major issues because a vast proportion of oil going to the most significant consumers of their energy resources pass through these routes.

Today nuclear weapons of mass destruction are a principal element in the security considerations of the region. WMDs of other kinds, chemical and biological, have already been used for years in this region along with conventional weapons. They have principally been of the chemical variety, there have also been threats of the biological ones. Questions regarding nuclearisation of the region started quietly with Israel's pursuit of this technology as a means to equate themselves and to protect themselves and deter their Arab neighbours who were without that resource.

The Israeli nuclear programme has emerged according to well-informed estimate as the world's fifth largest behind Russia, the USA, China and France, putting it ahead of UK, India, Pakistan and North Korea. Israel could perhaps congratulate itself that five Arab-Israeli wars, the wars of 1949, 1956, 1967, 1973, and 1982 could be fought without recourse to any WMD and without any significant nuclear threats and disasters. The international community, particularly the high decibel nuclear weapon states had been content to allow Israel neither to confirm nor to deny their nuclear weapons programme, its extent or its destructive capability. The quality of keeping total silence on Israel's weapons programme could not have been sustained in the absence of the superpower—great power tolerance for that.

The efforts of Iraq, Libya, Iran in the same direction have to be assessed against this background. The great power-superpower tolerance has also shown that they can maintain total silence and a policy of non-reaction even in the face of one private entrepreneur e.g., Dr. A.Q. Khan opening up a highly visible, mightily successful Walmart of nuclear weapons and ancillaries. The Iraqi nuclear facilities came in for aerial destruction. The current Iranian effort in pursuit of nuclear hardware appears heading towards problems also. The road could be a complex, narrow and winding one, which could be both interesting and perhaps dangerous. One thing we need to be conscious of is that Israel was the first in West Asia to raise the nuclear antenna. The rest of the world did not do any loud drumbeating against this, just as recently they have not done any drumbeating against Dr. A.Q. Khan's success as a nuclear entrepreneur.

But, in the case of Libya, Iraq and now Iran, pursuing the same objective has resulted in the decibel level of drumbeating by the big boys against them became significant. In these three cases problems of non-proliferation in West Asia have definitely become a global concern. Similarly, the cooperation between one of the five permanent members of the Security Council with our own close neighbour Pakistan to enable the latter to produce certain types of rockets and missiles on its own soil, also providing the major portion of wherewithal for producing and testing nuclear weapons including blueprints and all the technical know how too, has failed to attract the kind of opprobrium which would have been the lot of any other country and this shows the unfair nature of unequal criticism, unequal threat of punishment in the NPT world. One has therefore, to be conscious that the evolution of West Asia in the last 105 years has involved rapid changes in thought processes, ideology and theology. The availability or paucity, economic or highly inflated costs of a particular natural resource essential for modern technology produced very largely in this region has become critical for the security of the world. The electricity power and energy required for the existence and well being of today's *homo sapiens* all over the globe, their daily transportation, availability of water in urban habitats, etc. all, let us recognise, depends on this resource in a large measure.

Let us remind ourselves also how rapidly the world changes and keeps changing. It is interesting to recall that in 1901, the year Queen Victoria died and her coffin had to be carried into the Abbey Chapel by eight emperors themselves, or in a few cases by princes and statesmen representing some of them. These were the emperors of India, which was Queen Victoria's own son Edward VII, her grandson who was the emperor of Germany, the husband of her grand daughter, who was the emperor of all the Russias, the emperors of Austria, Iran, Japan, Ethiopia and the Ottoman emperor and Kalif of Turkey. Out of these today only one emperor has survived and that is the emperor of Japan. Let us therefore, understand how fast things can change and how fast they can be forgotten.

During this period, i.e., in the post-1901 era, much of West Asia had formed part of the Ottoman empire where Arab nationalism was beginning to experience its first stirrings of freedom. In the First World War, during which one saw that the principal imperial power of that period, Great Britain found it advantageous for itself to engineer the commencement of an Arab revolt against the Ottoman

Emperor. They promised liberation of Arab areas from the control of the Turks.

At the same time they sought to establish a Jewish homeland in the region. The French joined the British bandwagon to promote a 'homeland' for the Jews. We need to recall that the Jews had been oppressed and discriminated against in various countries of Europe for the last 600 or 700 years. Indeed, ever since the medieval ages, the Europeans in the process of providing a homeland for the Jewish community as well, proceeded to create a Palestinian problem introducing them into a land of the Palestinians.

Some Arabs got liberated during the First World War from the Turkish rule and not so gradually passed under British and French control through a recently devised system of mandate. This mandate system passed away after its utility in the Arab lands and the West Asian area was finished. These developments were not unconnected with the discovery in this area of oil. Iran saw a constitutional revolution against their Qajar emperor in 1905. Within two years an Anglo-Russian competition for influence ensued. This divided Iran into two spheres of influence, the northern one of Russian influence and the southern one of the British influence. In the southern sphere of influence, the British introduced massively the Indian railways, the Indian shipping lines through the Gulf and the Indian telecom lines. When I was in Iran in 1956, as a very young man I used to pay something like 900 pensions from the Indian exchequer to Iranian civil servants who had worked for either the railways or the telecom sector. It suited Britain post-1925 to ensure that the Iranian people went under the rule of an actual monarch who they had to invent out of nowhere out of an army regiment. During the 20s and early 30s, oil was discovered and its production commenced in Iran and Iraq. Britain, France, the Netherlands, Belgium and other Europeans as also some US companies were able to organise exploration, exploitation, distribution and marketing of oil in the region and outside in a manner that supported the West European and North American companies and states getting profits out of oil, mastery of its technology as well as structuring of the future of energy becoming a monopolistic enterprise.

Out of all this evolved over the years new technologies, the voracious automobile industry and the birth of air warfare and air travel. It was towards the end of the Second World War that Saudi Arabia and then Kuwait became major producers and suppliers of oil.

In the 50s and 60s, I was exposed to the open democratic and benign consultation system between the rulers and the ruled in the Gulf. It gradually brought about a tacit arrangement between the Sheiks and their people. The arrangement was simple. The Sheiks and the rulers would share the oil wealth with their people through a system of free education, free health services, free travel abroad for various purposes and the structuring of most modern roads, most modern buildings and sources for drinking water, housing, schools, hospitals and just total absence of taxation. In return the people would not insist on any interference with the total authority and autocracy of the rulers.

Sometime in the 70s, the system came under the pressure of global economics. This speaker had the very good fortune of being the Indian Consular Officer in the Gulf in 1956-1959 period. When I was young, the world was also young and simple. I used to travel by a small boat with just two small engines along with one assistant and one office functionary. I used to travel from Khurram Shahar to Basra to Kuwait, to Bahrain, Dubai, Sharjah right up to Muscat reviewing, watching and observing the welfare of the small Indian business community and their relations with the local community and rulers. I sat in many of those *majlis* myself in the evenings, they were simple and very open, very transparent. The currency of the gulf was still the Indian rupee. One used to be able to watch the concussion and charm of the Arab rulers and Sheiks towards their people and the interaction between all of them and their elaborate politeness towards the traveling, wandering Indian. The regional security and the impact of the global economy on the Gulf region and *vice versa* were then not anyone's concern at all. Life was much simpler then and the complexities of today's globalised world did not trouble the likes of us being a diplomat.

As I look back I do feel that the changing security dynamics in West Asia is but a reflection of a new world that has allowed itself to get swallowed by elemental greed, elemental anger, elemental fear and elemental hatred.

31

Changing World Order and West Asia

A View from India

ASHOK K. BEHURIA

The paper makes an effort to underline the defining elements of the evolving international order and situate the West Asian security dynamics within that context.

Big powers have big ambitions and the efforts to realise them often lead to problems for themselves and others around them. They are guided by their belief that their worldview is the most rational and just, which needs to be adopted by people in other corners of the world. This theme was inherent in the campaigns of Alexander, in the Roman conquests, in the conquests of Napoleon, the European Imperialists in the 19th and 20th Century and now the endeavour to work towards 'democratic peace' around the world.

One has seen Napoleon III bragging in late 19th century that he had an interest in the Trans-Danube region for France had an interest, "wherever there was a just, civilising cause to promote." The European powers had taken upon themselves the responsibility of civilising the colonies, which they called "the white man's burden." And now one finds US assuming global leadership and seeking to promote "the cause of political and economic freedom abroad."[1] The history of the world reminds us that such aggressive enthusiasm and missionary zeal often induce a sense of inhibition when applied with force rather than persuasion.

Such overconfidence of the big powers has been matched by the supercilious stubbornness of the smaller ones thanks to the anti-colonial inertia many of them retain in their worldviews. While one may argue that the disaffection of the people in such underdeveloped terrain is justified, the faulty channelisation of disaffection has led throughout history to actions like the one on 9/11, which have, more than anything else, provided the powerful with the excuse to use all potential at its command to carve out the world of its liking. As John

Lewis Gladdis writes, the acts of terror in the wake of 9/11 have unwittingly provided the US with a "chance to lead the world into a new era, together with the opportunity to do it, this time more seriously."[2]

The political leadership in the underdeveloped corners, rather than quietly justifying such acts of vengeance, should take note of this clear message of history that such steps only drive them deeper into underdevelopment and subject them to the sport of the big ones. They have to try their best to focus on developing their economies and societies and harness such disaffection to good use and divert it towards development of their resources.

9/11, A Departure?

There is an overabundance of literature in the field of international politics today emphasising the point that 9/11 or 'Black Tuesday' (September 11, 2001) has changed the contours of international politics. The scholars and analysts belonging to this school of thought argue that the conventionally called 'post-Cold War era' that started with the fall of Berlin Wall has ended rather too abruptly with the fall of the Twin Towers in New York.[3] The states in the existing international state-system are challenged from without (rather than only from within) and the enemy, the so-called war against terrorism is up against, is "amoeba like"[4] and the threat it poses is much more 'diffuse' and almost untraceable even with the most competent of intelligence penetration by satellite and radar systems.

They would even argue that they are faced with a "hybrid strategic landscape" which is part-terrorist and part-conventional. And the scholars, especially from the US would say it will require a Grand American Strategy that will involve clever use of "its military strength, its productive competitiveness and its diplomatic skills" to sail through such a landscape.[5] The emergence of terrorism, a non-state factor that seeks to wreck havoc with the existing international political order, has been recognised as a fundamental feature of the coming world order, that is yet to assume any definite shape, as the strategic thinkers in the US are becoming increasingly aware of the crevices within the global alliance against terror and trying their best to forge a lasting alliance until the menace is curbed.

A New Political Order?

Various elements of the new international political order are being recorded by social and political scientists the world over:

- Violence as a mode of political resistance is losing its legitimacy and appeal, even if it is being used increasingly as a mode of resistance.
- There is greater emphasis on democratisation of politics and liberalisation of economies in the developing and under-developed societies.
- A powerful multilateral consensus is slowly emerging among the states on the need to fight fundamentalism.
- The zeal of conservative Islamists is being replaced by the sober effort of the liberal thinkers within Islamic societies to rediscover the enduring appeal of Islam as a religion of peace and not of avenging suicides.
- The 9/11 has ushered in the "return of the state"[6] whose authority was on the wane in the wake of the overriding financial concerns of a globalising world.

Continuity or Break?

At another level, the horrendous prospect of coercive non-state actors emerging as chief determinants of civic (which includes the political) existence may have stirred the states—the monopolisers of (the so-called) 'legitimate' coercive power—into collective and cooperative action. But the march of history does not show any epistemological break even then. The continuities in world politics are there for all to discern—continuities understood as the culmination of a process that nourished what Benjamin Barber would call "tribalism" or "retribalisation"[7] in an overarching zeal to defeat and wipe out communism.

It is as if the self-fulfilling prophecy of Huntington has suddenly realised itself and the clash[8] between Islamic civilisation, surging with a relatively numerous young population driven by a 'consciousness without cohesion' and beset with economic crises and the Christian one, brimming with self-confidence after the end of the Cold War and in the driver's seat of the world economy, is going to define the context of the international politics in the days to come.[9]

As Mahmood Mamdani would assert, the fundamentalist violence that the US is so much perturbed about was a strategic weapon it had forged in its "crusade" against communism. He would say that "contemporary fundamentalism is a modern project, not a traditional leftover."[10] It reminds one of the analyses of Eqbal Ahmed who had pointed out that the terrorists of today were welcomed by US as "moral equivalents of America's founding fathers." Mamdani raises a very important point that if "terrorism was an official American Cold War brew," should we consider the states in South America, Africa and South Central Asia plagued by civil strife as "hosts of terror" or "hostage to terror"? The enthusiasm to join the mightiest nation in its avenging exercises must not cloud reason and the responsibility for the present mess lies squarely on the shoulders of the same nation that is waging a war against terror.[11]

There are many others who would look at the emerging reality in a more cynical vein. They would suggest that the present "social scientific theories are unsuited for the kind of political investigation" demanded by the emergence of what they would call "a military-industrial-media-entertainment network," a conceptual improvisation on the "military-industrial complexes" that Eisenhower talked about in 1961.[12] And there are profound observations such as that of Michael Ignatieff who would say that the "apocalyptic nihillism" exemplified by the suicide hijackers was indicative of efforts directed towards "the violent transformation of an irremediably sinful and unjust world." He would go on to say that such terror "does not express a politics, but a metaphysics, a desire to give ultimate meaning to time and history through ever-escalating acts of violence which culminate in a final battle between good and evil."[13] Philosophers like Jean Baudrillard would attribute such "mother of all events" to the 'natural' hatred for the "dominant and hegemonic world power," "among the disinherited and the exploited, among those who have ended up on the wrong side of the global order." He goes on to add that if "Islam were to dominate the world, terrorism would rise against Islam" and every world order dislikes any hegemonic domination and breeds its own counterforce.[14]

Even if there is a tendency to characterise 9/11 as the starting point or end point of an era, in the world that has survived the suicide attacks on the WTC and Pentagon, the continuities are more visible than the breaks. One has to bear in mind two most important lessons from the post-9/11 international politics—the preponderance of the

US in international politics and the dwarfing of United Nations as a multilateral forum. The nature of international politics at the moment suggests that the US will continue to remain the most powerful state and try its best to impose its will on others, through force or persuasion. It will go to any extent to protect its interests around the world. In fact, the multilateral foreign policy of the US threatens to replace the UN as an effective multilateral forum.

Breaks

As far as the breaks are concerned, many would suggest we are in for a major paradigmatic shift in international relations. Rather than revolving around the 'politics of the nations'[15], the contours of international politics will be determined by non-state actors, who are increasingly playing a more active and important role in international politics. In fact, they would say that the concept of international security has been redefined in the wake of 9/11, keeping these actors in mind. The efforts of the states—like US, Russia and China, who may not yet completely shed their cold war hangovers and continue their strategic agendas—to cooperate among them to uproot the menace of non-state or sub-state violence will characterise the nature and scope of international politics in the days ahead. Analysts have put it metaphorically: "Big shocks change perceptions. Actions and results follow. Today, thanks to Osama bin Laden, the world is in play.... Shake the system sufficiently, and new patterns emerge. Threaten it sufficiently, and historic enemies may sheathe their hatchets."[16]

And there are equally convincing theorisations that explain the phenomenon of violent resistance movements that lead to terrorism. A huge panoply of causes is isolated, analysed, understood and shared among the analysts to diagnose the phenomenon of terror. Factors like globalisation, communication revolution also feature in this construct along with more visible causes like pro-Israel policy of the US, its support to repressive and undemocratic regimes even when it championed the cause of human rights and democracy in its overt foreign policy posture, its military presence in the oil-rich west Asian region and the rising tide of unemployment among youth in the region that get attracted by the call of the Wahabites to join Jihad.

At another level, one finds the multiethnic multinational democracies devising the means now to handle internal dissent through peaceful means. A political consensus is now emerging slowly but surely over the need to attend to political dissent within state frontiers by means other than active and open coercion, even as social scientists apprehend that in spite of such devolved power sharing in various societies, democracy (defined as based on relatively broad and equal citizenship, binding consultation of citizens, and protection from arbitrary actions by governmental agents) will decline in quality across the world.[17]

The Case of Human Rights

It is also being apprehended by many that the thrust of foreign policy of the most powerful nation, the USA, which is very likely to determine the nature of the coming order in International relations, will shift its focus from Human Rights and thus international political landscape will have enough space for regional injustices. Analysts and thinkers like Michael Ignatieff would sound an alarm that the US has already reduced its emphasis on human rights issues in its endeavour to broad-base the alliance against terror.[18] And so have other powers like Germany, Australia and UK. Grieving that national security has trumped Human rights, he would say "divided between horror at the attacks and alarm at being enlisted as moral cheerleaders in a war on terror, many European human rights groups are sitting on the sidelines" and go on to suggest that the human rights activists "will have to engage soon in the battle of ideas" and have to "challenge directly the claim that national security trumps human rights. The argument to make is that human rights is the best guarantor of national security."

The war against terror has also brought about a sense of interest and moderation in scholars taking renewed interest in Islam more as a religion than as a corrosive worldview like the one championed by Mullah Omar and Osama bin Laden. While the intelligence agencies, the world over are gearing up for snapping the financial and material links that has sustained this terrorism, these scholars are engaged in the study of Islam as a religion of compassion and this has encouraged 'intercultural learning' which will certainly play a big role in bridging up the civilisational faultlines that provide theoreticians like Huntington, "the clumsy writer and inelegant thinker that he was"[19] with excuses to ignore the massive 'traffic' that moves across 'carefully maintained, even policed' cultural and civilisational boundaries with often 'terrifying ease' and build their grandiose self-fulfilling conceptual "gimmicks" that are "better for reinforcing defensive self-pride (of the West) than for critical understanding of the bewildering interdependence of our time," Edward Said would argue.

SITUATING WEST ASIA

It is in this context that politics in the region needs to be perceived in the context of the changing international political order. While the geopolitical and geoeconomic importance of the region will continue to invite international attention, it is certain that the following features will mark the politics of the region:

Energy Interests of the US and the World

US will protect its energy interests with a lot of caution in the region. In fact, many scholars from the US have already started arguing about the imprudent dumping of the Cold War strategy that laid an emphasis on ends rather than the means. They have urged the establishment to give due attention to 'shared interests' rather than 'shared values'. Energy interests, as ever, will continue to guide the Americans more vigorously in designing their policies towards the region. One saw the US issuing orders for its women workforce stationed in Saudi Arabia to abide by the Saudi regulations.[20] Through a mix of diplomacy and force, the US has tried to improve its image in the region and secure transmission of energy resources.

The evolving interrelationship among the states in the geo-politically important regions of Central Asia and West Asia can be facilitated by the US. The US interests in the region now expands beyond the terrain that is known as West Asia. Frank Vivian, a staff writer of the *San Francisco Chronicle* wrote on 26 September 2001, "The hidden stakes in the war against terrorism can be summed up in a single word: oil. The map of terrorist sanctuaries and targets in the Middle East and Central Asia is also, to an extraordinary degree, a map of the world's principal energy sources in the 21st century.... It is inevitable that the war against terrorism will be seen by many as a war on behalf of America's Chevron, Exxon, and Arco; France's TotalFinaElf; British Petroleum; Royal Dutch Shell and other multinational giants, which have hundreds of billions of dollars of investment in the region."

There is another argument that the US action against Iraq was not related to oil at all. Because, the world oil consumption is at present around 78 million barrels per day (mb/d), out of of which Iraq contributed 2 to 2.5 million in 2003. Regardless of who "controls" Iraq's post-invasion production, the volume of additional oil that Iraq could place on the world market would, in the short-term, be insignificant. Moreover, the reconstruction efforts would require more than $350 billion. Thus some would argue that, "despite the 'it's the oil, stupid' allegations of anti-war protesters, it is abundantly clear that there would be no room in Iraq's post-invasion economy for an unrestrained bonanza of oil revenues for the international oil companies."[21] However, this argument does not quite hold against the formulation that the US attack on Iraq was more to entrench itself deeper in the region and ensure its dominance there. The best way to do that would have been by flooring the most defiant of all regimes in the region.

Security of the States

There is a complex dynamic at work as far as the security of the states in the region is concerned. The unrepresentative character of the state structure, in most cases, has given rise to pressures against the system from within at one level causing internal security concerns. While at another level, chronic interstate suspicions have compelled the states to look for military ties with extra-regional powers, which again had implications for internal security as well. Saddam Hussein's attack on Kuwait, for example, had resulted in some states in the region welcoming external defence forces into their territory without any sense of inhibition, which in turn contributed to internal political disaffection. This is more so because there is a visible wide gulf separating the ruling class from the ruled. In the absence of popular participation in the decision making process, the efforts of the ruling class to secure the state from external attack is often being misconstrued as efforts to secure perpetuation of an unrepresentative and illegitimate system.

In this context, it is encouraging to find the incumbent powers in various states more circumspect in dealing with internal dissent. The case of export of terrorism from the region as elaborated in US newsmagazines in the case of Saudi Arabia seem to have compelled both the establishments in the region as well as external powers to advocate caution while dealing with internal resistance and internal dissent. The case of Saudi sponsored extremism as a policy alternative to divert domestic disaffection outward is being closely followed by the international media and states labouring under terrorism.[22] The small step forward taken in the direction of representative administration in Saudi Arabia is being regarded as a welcome measure which will ensure political participation at the internal level and this one believes will go a long way in containing the tide within their frontiers. However, as the recent electoral victory of the conservatives in Iran and Hamas in Palestine suggests, such phenomena may worry the forces seeking to initiate political reforms in these societies. However, it has to be remembered that any countermeasure will definitely cause more turbulence within some of the states in West Asia.

The relative importance of the region in World politics will depend on how the states in the region recast their domestic policies and whether they come together to address the issues that confront them. The states in the region will also have to strike a bond of unity to harness their collective energies to tackle the issue of fundamentalism at home and take care to explore the possibility of evolving a collective strategy to handle domestic dissent in a spirit of accommodation rather than confrontation. In their internal security matters, they should avoid external help.

On the other hand, keeping in mind the configuration of the non-state actors that are propelling international terrorism, the West will have to adopt a cautious policy towards the region. The door to freedom in the region will open through gentle nudges rather than firm pushes. The experiment in Iraq suggests that even if the US attack on Iraq dislodged the dictatorial regime of Saddam Hussein, it has been difficult to restore order and introduce genuine democracy in the post war situation. Leaving Iraq to Iraqis at the moment may lead to chaos and civil war. The fledgling democratic experiment in Iraq is thus in a critical shape.

Democratisation

The American policy towards West Asia aims at raising more participative political systems which can restrict dissent within territorial limits of the states concerned. The policy of the US thus has been, in the last few years, to convince the establishments about the use of expanding the base of popular participation within many of these societies.

The issue of initiating political reforms and addressing the problem of deficit of freedom in the region has thus been given prime importance in the post 9/11 international context. The political reforms initiated within the region suggest that the system of representative governance has been introduced in diverse ways in different countries like Morocco, Algeria, Egypt, Jordan, Kuwait, Qatar, and Yemen. The US Secretary of State has also come out with a statement last year in Cairo that the US was prepared to accept diversity within democracies. This is in continuation of Bush's statement earlier that, "working democracies always need time to develop...We've taken a 200-year journey toward inclusion and justice, and this makes us patient...as other nations are at different stages of this journey." All this suggests that the US is reconciled to a gradualist incremental approach to political reform in the region.

There have been others who would point out that the US approach in bringing liberal democracies within the region has been quite schizophrenic. The US is torn between its interests and the values it champions. They would like the US to adopt a more proactive approach and do away with "the institutions and ideologies of the Arab states" and help "Arab reformers escape the trap of liberalized autocracy."[23]

They would urge the US to shed its reluctance to author a change that would see the fall of the incumbent autocracies. They argue that the US policy and US aid programmes should be directed in part, at empowering groups and persons who support peace and democracy, rather than governments and leaders who have other agendas. Of the huge sums expended on aid, only a small percentage is spent on measures that can build up the civil society in these regions. The money spent on a few tanks and aircraft could, for example, build a network of peace schools, and fund a satellite TV station that brings objective news to the Middle East in the native languages. Otherwise, as some critics have pointed out, the monetary aid given to some of the countries in the region are channeled back to US through their policies of procuring defence equipments from US. "That aspect of the aid program is actually a giant subsidy to US defense industries, good for the US military-industrial complex, but bad for the aid recipients. Aid dollars should be expendable anywhere in the Middle East. That will build local economies, promote regional economic interdependence and help local economies grow," some would argue.[24]

The ruling classes in different states in the region have so far adopted a policy of openness only in so far as it does not trump their hold and control over the states they rule. They have to do more. The best way to handle political dissent is through accommodating it within the political system. In this context, one can say that the rulers in the region will have to think beyond their survival and their formulation that they are in the throes of *marhalla intiqaliya* or an eternal transition and take steps to introduce popular representation.

The Palestine Situation

One of the most encouraging effects of 9/11 has perhaps been the surge in intellectual activities in the United States and other Western countries to identify the potential bases of antipathy towards the US and the West. And there has been a consensus among the scholars and policy analysts that US policy towards the Palestinians has been one of the main reasons of the crystallisation of popular antipathy against the US. This has led to some proactive policy on the part of the US towards resolving the Palestinian issue. The West Asian peace process, with the decision of Israel to pull out of West Bank and Gaza were decisions which were unthinkable some years back.

The recent victory of Hamas in the January 25, Parliamentary Elections in Palestine has come up as a big surprise for everybody around the world. Some commentators have called it "a big political earth quake." Benjamin Netanyahu, leader of the opposition Likud Party called it the rise of "Hamastan," "a representative of Iran and in the image of the Taliban." President Bush came out with a statement that "United States does not support political parties that want to destroy our ally Israel." Condoleezza Rice remarked, "It's a terrorist organization; we're not going to deal with it. It has certain obligations if it wishes to govern in accordance with international standards, and we will see whether it's prepared to meet those obligations." Reactions from the neighbouring Islamic countries have been rather sober, urging the international community to give Hamas a chance.

As all these comments pour in from all directions, after Hamas' unexpected victory in the January elections, they reveal the dilemmas in the democratic commitments of the so-called free nations. Preach democracy in theory but shun it in practice. There is a precipitate doubt in the minds of scholars and analysts in US and other western countries that democracy in Middle East and Gulf will bring Islamists to power, making the region more dangerous than it is today.

If Hamas has won such a huge popular mandate, it has to be accepted and given a chance. Power has its own dynamics and it induces moderation in the people that wield it. Hamas leaders pouring their venom against Israel before the elections may become an entirely different kettle of fish when they settle down to conduct the affairs of the government.

On the contrary, if the external pressures mount on them even before they start, it is very likely that they will revert to their extremist methods. To corner the Hamas at the moment would not be a wise policy at all. Ms. Rice's statement is rather welcome in the sense that she just spells out the principles and leaves the rest to the Hamas. She perhaps understands that "journey to democracy is rocky and difficult" as she reflected recently: "I come from a part of the United States in which my parents were not even guaranteed the right to vote until 1965, when I was 10 years old."

If democracy is to succeed, such popular verdicts will have to be honoured. If Hamas has won a mandate from the people, it cannot also disregard the overwhelming popular support for peace among the Palestinians. To apprehend its victory as a throwback to the earlier days of uncompromising *intifada* is to close one's eyes to reality. The people of Palestine today are tired of the years of confrontation and want the process of peace to stay the course. If Hamas strays from the track, they should know, they will have to pay for it in the next elections.

Conclusion

From the perspective of political geography the region will remain as important as ever. But the region could enhance its influence in the international politics in the following ways: (i) by forging stronger interstate bonds across the ethno-national and cultural divides that weaken the unity among them, (ii) by consciously breaking out of the sense of dependence on the West by forging new bases of legitimacy to the existing regimes in the region, (iii) by allowing political dissent a legitimate role in internal politics, and (iv) by integrating themselves with the evolving world order. It is time for the states in the region to consolidate themselves internally and promote 'distributive justice' within their frontiers and create conditions for peace and harmony which will shield the prospect of external interference. This will require a Jehad-e-Akbar or Greater Jehad on the part of the people within the region. The US cannot bring democracy and freedom to the region. The people in the region will have to work sincerely towards it.

Endnotes

1. Thus one finds scholars coming up with formulations like 'New American Century', 'American Internationalism' etc.
2. Lewis Gladdis, John (2001). "And Now This: Lessons from the Old Era for the New One", in Strobe Talbott and Nayan Chanda (ed.), *The Age of Terror: America and the World after September 11,* Basic Books, Yale Centre for the Study of Globalisation, New York, p.21.
3. Ibid. p.3.
4. Paul Kennedy characterises forces of terror like this in his article, "Maintaining American Power", in Strobe Talbott and Nayan Chanda (ed.), *The Age of Terror: America and the World after September 11,* Basic Books, Yale Centre for the Study of Globalisation, New York, p.60.
5. Ibid. p.62-63.
6. John A. Hall, "Return of the State", *http://www.ssrc.org/sept11/essays/hall.htm*
7. Barber, Benjamin (1992). "Jihad as McWorld", *Atlantic Monthly, http://www.theatlantic.com/politics/foreign/barberf.htm.*
8. Huntington, Samuel P. (1993). "The Clash of Civilizations and the Remaking of World Order", *Journal of Foreign Affairs.*
9. Also see Huntington's interview with Michael Steinberger in *New York Times,* 20 October 2001, where he says that the youth in the Muslim world "are intelligent, ambitious young people who aspire to put their educations to use in a modern, developed economy, and they become frustrated by the lack of jobs, the lack of opportunity. They are cross- pressured as well by the forces of globalization and what they regard as Western imperialism and cultural domination. They are attracted to Western culture, obviously, but they are also repelled by it."
10. See Mamdani's article "Good Muslim, Bad Muslim: An African Perspective" *http://www.ssrc.org/sept11/index.htm.*
11. Eqbal Ahmed's address at Colorado University which was later reproduced as an article on "Genesis of Global Terrorism", in *Daily Dawn,* 23 September 2003 Later, based on this lecture, he wrote a chapter, "The Culture of Violence", in a book edited by Zia Mian and Iftikhar Ahmad, *Making Enemies, Creating Conflict: Pakistan's Crises of State and Society,* Mashal Pakistan, Lahore, 1997.
12. Derian, James Der (2004). "Before, After and Inbetween", Social Science Research Council", *www.ssrc.org/sept11/essays/der_derian.htm.*
13. Ignatieff, Michael (2001). 'It's war-but it doesn't have to be dirty', *Guardian,* 1 October.
14. Baudrillard, Jean (2002). *The Spirit of Terrorism and Other Essays,* Verso, London and New York, 2002, pp.13-15.
15. Hans J. Morgenthau's book of the same name, *Politics Among the Nations,* projected a realist perception of international politics and such realism was founded upon the post-second-world-war state system that seemed to stabilise as the most desirable, if not perfect, system in the realm of international politics. The very nomenclature 'international politics' (rather than inter-state politics) did leave the space open for new nationalities to emerge as political units and this has in turn punctured the notion of the inviolability of the existing state system. However, that will require a completely different discussion.
16. Achilles Zaluar and Richard J. Zeckhauser write that "predominantly Muslim nations shifted strategy, attacking or at least shunning Islamic extremism rather than appeasing it. Quiet efforts by the United States and Cuba turned the Al Qaeda holding pen at Guantanamo Bay into an olive branch. President George W. Bush became a formidable world leader.... it is now recognized that terrorist groups have the potential to undermine any nation, including those that have hosted or supported them..... This common enemy and common threat generate a strong confluence of interest among nations, even among traditional adversaries. Armed conflicts between Russia or China

and the United States now seem more remote.", in their article "Sensible Patterns Since September 11: A Shaken World", *International Herald Tribune*, February 14, 2002.

17. Charles Tilly, in his article "Predictions", 2003.
18. However Ignatieff has some consoling points when he says that "Human rights has gone global by going local, anchoring itself in struggles for justice that can survive without American inspiration or leadership. The movement does not have its headquarters in Washington" but still expresses his dismay at America's latest vacillation and admits, "but, if Washington turns away, the movement loses the one government whose power can be decisive in stopping human rights abuses.", Michael Ignatieff, "Is Human Rughts Era Ending?", *New York Times*, 5 February, 2002.
19. Said, Edward (2001). "Adrift in Similarity", *Weekly Al-Ahram* (Cairo), October 11-17.
20. The highest-ranking woman pilot in the US air force, Lieutenant Colonel Martha McSally, posted in November to the Prince Sultan Air Force Base sued the defence secretary of US for orders that make her wear the abaya and head scarf when she goes out of her base in Saudi Arabia, prohibits her from driving, and directs that she should always be escorted by a male. The pilot argues that the instructions given to her amount to asking her to abandon American values "that we all raised our right hand to die for". (Tahir Mirza writes about it in his column in daily *Dawn (Karachi)*, January 10, 2002.)
21. Hepburn, Donald F. (2003). "Is it War for Oil?", *Middle East Policy Journal*, Volume X(1), Spring.
22. In fact, Saudi Arab has been characterised by many commentators in the West as a crucible of Islamic terror. As this article was being sent to press, media reports suggested that Americans are pulling out of Saudi Arabia. This is perhaps to accommodate the voice of dissent within Saudi society rather than the demands of the Saudi monarch. In fact, Saudi monarch would have wished American security machinery to stay on in the country strengthening their hold on power. But now the Saudis will have tough time handling dissent within their country. It will be interesting to see how the Americans, the self-proclaimed champions of democracy and human rights, will contribute to the Saudi efforts of management of dissent within Saudi society and stall the process of exporting Islamic wrath to other countries in the region.
23. Brumberg, Daniel (2003). "Bush Policy or Bush Philosophy", *Washington Post*, November 16 and also see writings by Thomas Carothers and Marian Ottaway available in the web at *http://www.carnegieendowment.org*.
24. Ami Isseroff, "American Policy in the Middle East: Getting What You Pay For", available on *http://www.mideastweb.org/americanpeace.html*

32

Managing Foreign Workforce in the Gulf

Redefining the Rules of Engagement

N. JANARDHAN

The overwhelming presence of foreign workers in the Gulf Cooperation Council (GCC) countries is no longer just an economic issue. It is now being viewed as a national security threat, with political implications, by the labour-importing countries amid international calls for equal rights for expatriates. On the premise of pursuing a more rational policy, the GCC countries are planning a "zero-tolerance" approach, aimed at resorting to expatriate labour only when there are no local or regional alternatives.

With about 12.5 million expatriates in the six countries—up to 70 per cent of the workforce—the implications for the GCC countries are profound in the social, political, economic and cultural spheres. Though there is no real evidence of a security threat, some are warning that the "guest workers" are a "time bomb waiting to explode." Compounding their worries is a hitherto unheard practice–workers protesting for not receiving salaries on time.

Coinciding with these developments are a number of changes on the ground during the last five years—a combination of 'indirect' taxes, inflation and new employment visa regulations have made life harder for a large section of the expatriate population. The GCC countries even unsuccessfully negotiated a controversial deal to curb the duration of stay of unskilled expatriates to a maximum of six years.

Concurrently, as a means of intensifying economic diversification beyond the oil sector and containing the practice of capital flight in the form of remittances, the GCC governments are increasingly adopting liberal economic measures, including granting the expatriates access to local stock markets, and, more importantly, the right to own property, which guarantees residency even without employment.

However, despite redefining the rules of engagement, there is no denying that foreign workers form the cornerstone of economic progress in the GCC countries by being ready to consider any job, which many nationals are still unprepared for, despite growing unemployment. But the task of replacing foreign labour is easier said than done. While job substitution may be easier to implement, job creation and relevant vocational education are likely to be far more critical. The strategy to develop local human resources and boost competitiveness in the public and private sectors has been far from successful till now, though the possibility of a more dynamic and indigenous workforce emerging in the next decade is realistic.

All these policies have an impact on the labour exporting countries, including India, which has about four million people in the region. Viewed from an Asian, and, more specifically an Indian perspective, this paper discusses the demographic changes in the region; changing patterns of the labour force; characteristics of expatriates in selected countries; prospects and challenges for foreign workers; mismatch between educational qualifications and employment available to nationals; absence of a strong work ethic among them; changing ground realities for the blue-collar and white-collar expatriate workers and their implications; governments' views and policies on labour issues, including the nationalisation drive; impact of free trade agreements on labour policies; implications of the new 'Look East' policy adopted by the GCC countries; as well as the changing image of India in the region, which makes the future of GCC-India relations pregnant with possibilities.

Figure 32.1

Indian Diaspora in GCC Countries (December 2004)

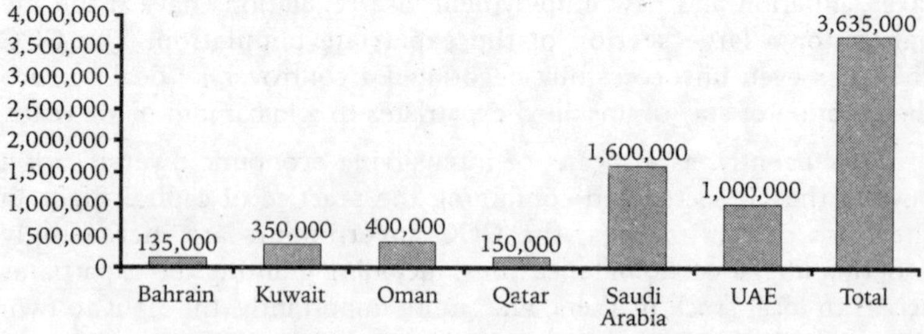

Note: In total, there are over four million Indians in the GCC countries, including those residing without any legal status.

Source: Ministry of External Affairs, India, and independent sources.

Demographic Pressure and Nationalisation of Workforce

The percentage of expatriates in the GCC countries grew from 22.9 in 1975 to 38.5 in 2002. Reliance on foreign labour has resulted in expatriates currently constituting 88 per cent of the workforce in the UAE, 83 per cent in Qatar, 81 per cent in Kuwait, 72 per cent in Saudi Arabia, 55 per cent in Bahrain, and 54 per cent in Oman (Source: *Middle East Economic Digest*, 2004). However, the composition of expatriates has changed periodically. Starting with Arabs, the GCC countries later preferred Asian workers because they were considered relatively less expensive to employ, easier to dismiss, more efficient, and not difficult to handle. One of the more important factors that favoured Asian workers was their willingness to live and work in the region without their families, a trait unavailable among the Arabs.

Table 32.1

Major Expatriate Communities in the GCC Countries (2002–in thousands)

	Bahrain	*Kuwait*	*Oman*	*Qatar*	*S. Arabia*	*UAE*	*Total*
Indians	100	295	300	100	1,400	1,000	3,195
Pakistanis	50	100	70	70	1,000	450	1,740
Egyptians		275	15	35	1,000	130	1,455
Yemenis					1,000	35	1,035
Bangladeshis		160	110		450	100	820
Sri Lankans		160		35	350	160	705
Filipinos		60		50	500	120	730
Jordanians/ Palestinians		50		50	270	110	480
Syrians		95			170		265
Iranians	45	80		20		40	145
Indonesians					250		250
Sudanese					250		250
Kuwaitis					120		
Turks					100		
Bidoons		70					70
Total	**280**	**1,475**	**630**	**420**	**7,000**	**2,488**	

Note: Numbers of only major expatriate communities in each of the GCC countries are included in the table. As a result, the figures in the columns do not add to reflect the total.

Source: Tattolo Giovanna, "Arab Labour Migration to the GCC States," *http://jmobservatory.eco.uniroma1.it/users/wp%20work%20progress/Todisco%20 movimenti%20 popolazione/ARAB%20LABOR%20MIGRATION%20TO%20THE%20GCC%20STATES.doc*

Table 32.2

GCC Workforce–2002

	National Population	Expatriate Population	Total Population	Population Growth Rate (per cent)	National Workforce	Expatriate Workforce	Total Workforce
S. Arabia	15,500,000	7,300,000	22,800,000	3.3	2,800,000	5,300,000	8,100,000
Bahrain	400,000	270,000	670,000	1.6	100,000	200,000	300,000
Kuwait	900,000	1,500,000	2,400,000	3.3	200,000	1,100,000	1,300,000
Oman	1,900,000	600,000	2,500,000	3.4	300,000	600,000	900,000
Qatar	200,000	500,000	700,000	2.9	30,000	250,000	280,000
UAE	900,000	2,700,000	3,600,000	1.6	200,000	1,700,000	1,900,000

Source: Middle East Economic Digest, April 9-15, 2004

Due to demographic changes, characterised by a population growth of three per cent or more in all the six GCC countries, there is now increasing pressure on the governments to accommodate nationals not only in the public sector, but in the private sector too. About 40 per cent of the Saudi population is under the age of 15 and another 40 per cent aged between 15 and 39; 27 per cent of Qataris are less than 15 years old; unemployment rate in Oman is estimated to be 15 per cent; and the unemployment rate among nationals in some of the other GCC countries ranges from 9 to 20 per cent.

Policy Changes and Impact

In order to meet the challenge, the GCC countries are taking various measures, including nationalisation of the workforce. Jobs in many sectors are no longer open to expatriates, especially in Oman and Bahrain, which have lesser oil revenues, growing unemployment and a willingness among nationals to accept occupations dominated so far by foreigners. In Oman's private sector, the number of expatriates has fallen by 24.4 per cent since January 2003. Statistics reveal that more than 134,000 expatriates left their jobs over the last three years.

By 2004, the country's single largest industrial enterprise, Oman LNG, achieved 73 per cent Omanisation of its staff through a plan to initially use foreign expertise and then train Omanis to take over. The total number of expatriate workers in the private sector with valid work permits stood lower at 415,135 at the end of July 2005 compared with 549,338 at the end of January 2003.

The government has banned expatriates from selling and transporting fruit and vegetables in several regions of the country. Expatriates cannot engage in frankincense trade or in the sale of mobile phones in certain cities. Running beauty parlours and photo studios is now for nationals only and even the sale of 'abayas'–the traditional Arabic full-length robe for women–is reserved for Omani women. The government also intends to restrict the selling of perfumes, gifts, textiles and accessories and the running of cyber cafes to its nationals. Since January 2003, the number of expatriates, engaged in agriculture and fishing sector, fell 32 per cent, from 66,000 to 45,000 in July 2005. More than 59,000 expatriate workers left the wholesale, retail trade and car repairs sector, making the total number of expatriates working in these areas as low as 89,461 as against 148,516 in January 2003.

According to an official study, up to 20,000 locals were unemployed in Bahrain. Further, a survey revealed that almost 70 per cent of Bahraini men and women have no qualms about taking on domestic work. The survey found most willing to work as babysitters, nurses for the elderly, drivers or gardeners. However, there was a strange twist–only 10 per cent of respondents were willing to employ Bahrainis in those positions due to inadequate skills, cultural problems and "security threat."[1]

Saudi Arabia is no exception either. Facing a population explosion and growing unemployment, the government decided in 2003 to reduce its foreign workforce by more than half in 10 years. Following a study, the Interior Ministry's Manpower Council announced a 20 per cent ceiling on the number of expatriate workers and their dependents in the Kingdom. The directive also meant that workers from a single country would account for no more than 10 per cent of the workforce by 2013. When fully implemented, the decision is aimed at reducing the number of foreigners to around four million, nearly half of what it was in 2003. Even if all new recruitments are stopped, this means that around three million foreigners will have to leave the country by 2013.

According to Saudi Interior Minister Prince Nayef, "The employment of expatriates should not be at the expense of Saudi job-seekers. More than 100,000 young Saudis enter the job market annually." He added that expatriates were also increasing the government's burden in terms of "pressure on infrastructure facilities and increasing crimes."[2]

As a follow up, the government announced plans to create about a million jobs for Saudis by 2010 through a combination of Saudiisation and the creation of new jobs. Accordingly, the government has reserved 22 job sectors–in addition to 34 sectors that already existed–for its nationals. New jobs reserved for Saudis include administrative managers and their assistants, procurement managers, secretaries, car salesmen and public relations jobs.[3]

Starting February 2004, Saudi travel firms were instructed to employ only nationals. Many of the employers who did not find replacements were shocked to learn that their expatriate recruits were jailed and humiliated by shaving their heads. The 'Saudiisation' of the travel business followed similar steps to ensure 100 per cent Saudi employment in the Kingdom's gold and jewellery business. In 2001, gold shops were told to prepare for complete Saudiisation, which has been largely successful.[4]

Cultural Diversity Policy

As part of its 'cultural diversity policy', the UAE announced in 2003 that it intends to scrutinise visa applications of Asian workers because of official concern about their growing numbers in the country, its reliance on foreign workers and its changing demography. Officials stress that there is no specific quota for any nationality, but, when a company applies for an employment visa for a new worker, the ministry checks its records. If it found that the majority of those waiting to be recruited are of the same nationality, the company is asked to recruit nationals from other countries.

Such was the concern that the then UAE president, Sheikh Zayed bin Sultan Al-Nahyan, said, "This imbalance continues to pose a grave problem which threatens the stability of our society and the prospects for future generations." In line with this approach, officials even suggested that they were not averse to making life less comfortable for expatriates to stem the flow.

Many of the unwritten benefits are presently being tapered out. Earlier, companies agreed to provide inflated salary certificates so that workers, otherwise ineligible for family visas, could still get their families over. But now they are not as forthcoming. Employment permits granted to Arabs increased after secondary school certificates—which many Asian workers didn't have—were made a minimum requirement for getting work permits. This rule applies to all workers except labourers in the construction sector and Arab workers, and officials believe this will make for a more multicultural demography.[5]

Measure of Success

However, till now, the drive to nationalise the workforce in the GCC has had limited success. Efforts to replace expatriates are complicated by the fact that privately-owned businesses tend to prefer expatriates as they are cheaper, more skilled and easier to dismiss. The Emiratisation process, for instance, sets quotas for the minimum level of nationals a banking institution must employ, but according to the National Human Resource Development and Employment Authority, almost three quarters of all banks have not met the quotas. However, the government has not penalised any private companies for non-compliance yet.

In September 2002, Saudi officials declared that within six months, all expatriate taxi drivers—some 90 per cent of the total—would be replaced by their nationals. However, this deadline was not met and it was later announced that the substitution process would take a further three years to implement. Further progress is yet to be recorded.[6] Reflecting that the plans are not working too well in the other sectors, Saudi Arabia announced plans to recruit about 50,000 Vietnamese workers in 2006, including housemaids.[7]

FTA Demands and Security Threat Perception

The overwhelming presence of foreign workers is now being perceived as a national security issue amid international calls to grant equal rights to expatriates. Further, the GCC governments are worried over demands made by the United States during negotiations over free trade agreements (FTA). In particular, they are opposed to demands for permanent settlement of expatriates, giving them salaries equal to nationals and greater rights in the areas of education and health, and therefore keen on implementing a 'zero-tolerance' policy towards violators.

In the case of the UAE, the US demanded that it allow the creation of trade unions, amend labour laws in compliance with international laws and revise sponsorship rules. It also demanded that US citizens be treated equally with UAE nationals in terms of ownership.[8]

As a defence mechanism, the GCC ministers of labour and social affairs recommended a maximum six-year uninterrupted legal stay, tougher recruitment conditions, deporting surplus expatriate workers and making renewal of residence permits more difficult. However, the first mentioned proposal was sidelined owing to pressure from the private sector. Among those openly discouraging the plan was the Riyadh Chamber of Commerce and Industry which stressed the importance of expatriates for the ongoing development programmes. Instead, the GCC labour ministers proposed a quota system to limit the number of foreigners permitted to work in the region, which is now under consideration.[9]

The GCC countries are also worried about the political impact on the domestic populace following the electoral processes undertaken by Iraqi and Palestinian expatriates, and possibly Indians in the future.[10]

The possibility of a new threat emerging was highlighted after France experienced riots by immigrants in late 2005. Highlighting the issue, President of the Arab American Institute James Zogby said, "France and the rest of Europe are learning now that 'guest workers', in their third generation and still denied justice, are not only a shame that eats at the moral fibre of a society, they are also a time bomb waiting to explode."

Zogby added, "In this region, workers are trapped in horrible conditions, denied justice and their basic humanity...It hurts not only them, but the image and the moral fibre of the countries which host them. We must do better. They clean your offices, build your cities and yet remain invisible. You must see them, incorporate their rights into your vision and defend them."[11]

In a similar vein, Kuwaiti intellectual, Dr. Shamlan Yousef, commented, "Can a lesson be learnt from the events in France? No one can deny the fact that policies of discrimination and oppression are being practiced in the Gulf countries against the foreign workforce—low salaries, long working hours and absence of rights..."[12]

Inflation

High oil prices and the continuing economic boom have had a negative impact as well. For example, in 2005, petrol prices went up by a whopping 32 per cent all over the UAE, rents by about 34 per cent in Dubai, and prices of goods and services by 10 per cent. Dubai moved 10 places to become a more expensive city to live in than the likes of Washington. According to a survey, almost 70 per cent of those questioned felt that the cost of living had increased by more than 20 per cent over the year, while 26 per cent felt it was higher by 40 per cent or more.[13]

Worse still, this inflation has not been matched by a commensurate increase in salaries. In the UAE, public sector pay increased by 25 per cent for nationals and 15 per cent for expatriates. In the latter case, however, implementation has not been uniform, with skilled workers and professionals benefiting more than the semi-skilled and unskilled workers. As a result, some expatriates have had to take extreme measures to balance their budgets, including sending their families back to their home countries.

Apart from inflation, measures that can be likened to indirect taxes are in vogue in countries like the UAE. Over the last five years, health services, including consultation, medicines, hospitalisation and surgeries—which were otherwise free for expatriates–now come at a price. In addition, rising electricity, water, sewage and municipality charges have ensured that making both ends meet is a daily struggle for many families, leading to loans, debts and accompanying problems in several cases.

No wonder then, expatriates in Saudi Arabia heaved a sigh a relief in 2003 when the Shura (consultative) Council rejected a proposal to impose 10 per cent income tax on foreign workers who earned more than $800 a month.[14] But tax-free environs in the region may soon become a thing of the past with a UAE proposal calling for the imposition of value-added taxes under consideration. The initiative suggests that initially taxes could be imposed on tobacco and other 'harmful' products.

Friendly Policies

In the midst of these developments, some of the countries have taken some positive measures to protect the rights of migrant workers. For example, Saudiisation initiatives have also been backed by giving foreigners the right to litigation. In 2003, the government approved the formation of the Saudi Human Rights Committee (SHRC) and the National Human Rights Association, an independent human rights monitor. One function of the SHRC is to look into violations against low-skilled foreign workers, who are vulnerable to abuse, not only by their employers, but also by recruiting agencies that either illegally deduct money from their salaries or recruit them for non-existent jobs by charging exorbitant visa fees.

In 2004, the Saudi Shoura Council even took the extreme step of announcing that foreigners would be allowed to apply for citizenship. However, this legislation allows only those meeting strict language and residency requirements to be eligible, including fluency in written and spoken Arabic, adherence to Islam, and a minimum residency period of 10 years.[15]

The UAE government intends to implement a 13-point package to improve the plight of private sector workers. The proposed changes include permitting workers to establish labour organisations and increasing maternity leave to 56 days from 45, during which time a woman cannot be dismissed or threatened with dismissal. The authorities are also preparing to implement a new prerequisite for immigration and employment–by requiring legal contracts between employers and agencies on one hand and domestic workers on the other.[16]

'Name and Shame' Campaign

The increasing frequency of workers marching along the roads and blocking peak-hour traffic in protest against unpaid wages and poor living conditions has led the UAE government to announce that it will 'name and shame' defaulting employers. An estimated 10,000 expatriate labourers protested in 2005. These measures and other steps aimed at repatriating child jockeys came in the wake of a US report in June 2005 that criticised some of the GCC countries and threatened sanctions for failing to fight human trafficking.

Other measures include the introduction of a minimum wage system and substitutes for the sponsor system and expansion of the role of trade unions. Bahrain and Kuwait already have trade unions for nationals, where expatriates can be non-voting members. Qatar has also introduced a law last year granting workers the right to form trade unions. The Saudi Shura Council has allowed 'workers committees', while Oman plans to allow trade unions.[17]

As a means of containing the practice of foreign workers transferring home about $27 billion in remittances annually, the GCC governments are liberalising the economy. Bahrain has allowed expatriates with more than 15 years working experience and a healthy bank balance to reside in the country even without work permits. Dubai has allowed expatriates to own real estate and trade on stock exchanges, which ensures that money which was hitherto sent home as savings will be invested investments in the GCC countries. This will begin to alter the edifice of the relationship between the GCC countries and expatriates because the fruits of engagement will no longer be shared by just the labour exporting countries, but will be considered mutually beneficial.

Tracking India's Resurgence

It is important to mention here the changing perception of India and Indians in the Gulf. After being the major source of unskilled and semi-skilled workforce to the Gulf for decades, India is now ready to increase its share of skilled and professional workforce in the region, which is bound to alter the composition of the Indian workforce in the

Figure 32.2

Total Transfers by All Expatriates between 1993 and 2002

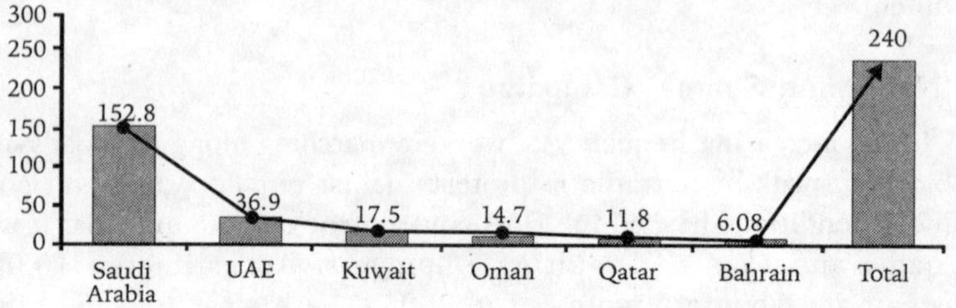

Note: (1) The funds accounted for mearly nine per cent of the GCC's combined gross domestic product.
(2) Given that foreigners in GCC countries slphoned out nearly $ 28 billion in 2002, India and Pakistan.

Source: Adapted from the website of the Arab Monetary Fund, Abu Dhabi.

region to some degree.[18] The wealth of resources India has on offer in the realm of the knowledge economy is an encouraging prospect for the GCC countries—more than 380 universities, 11,200 colleges and 1,500 research institutions; as well as the increase in the IT workforce–from 6,800 in 1986 to 650,000 in 2003 to over two million by 2010.

While characteristics such as India being the world's largest democracy, home to 150 million Muslims, fourth largest economy in the world in terms of purchasing power parity and world's largest provider of peacekeepers are well documented, India's strategic importance is being viewed in what the country is shaping to be—fourth largest consumer of energy by 2010, which is indicative of the economy's momentum; hi-tech plans to develop super-computers, complete nuclear fuel cycle facilities and placement of own satellites in orbit; replicating IT success in biotechnology, biogenetics and pharmaceuticals; and more importantly, a strategic partner of the United States, which is willing to share nuclear technology for civilian use and even share roles in joint space missions.

It is now evident to the GCC countries that the beneficiaries of India's knowledge economy include the United States and Europe, which will help India play a bigger role in the region's service industry. Further, the expanding private sector in the region has given Indians a level playing field over Western expatriates, and an opportunity to showcase their professional skills.

Prominent Gulf Arabs are taking note of changes in India and voicing their opinion. Comparing Indian youth who were finding easy employment in the services sector as part of the outsourcing drive in the last few years, Khaled Al-Maeena, Editor-in-Chief of Jeddah-based *Arab News*, wrote, "In the same period, I wonder how many of our young people found jobs. I also wonder how many of our young people have a good command of English and the necessary IT skills. I am afraid of the answer, and I am really afraid of finding out how many of our young people have good work ethics."[19]

Bahrain-based intellectual Abdullah Madani wrote in his column about the Gulf's "recognition of India's advancement in numerous vital areas and its political willingness to learn and benefit from India. This is a new Gulf rhetoric, given the prevailing preconceived idea in the region that India is no more than a backward nation that is only capable of supplying poor quality goods and unskilled labour. In fact, there is now a sound base for the two sides to establish a long-term strategic partnership."[20]

Of course, there is a pervasive feeling all around that economic sense is common sense. The GCC has finally realised the significance of reinforcing and expanding cooperation with India, while maintaining its strong ties with Pakistan, a key element that was missing in its diplomacy armour in the past. This marked a big leap in the GCC's new "Look East" policy. In April 2004, the GCC chambers of commerce and industry called for prioritising economic cooperation with Asian countries with large populations and developing economies, especially India and China.

Figure 32.3

Exports Increasingly Dominated by Asia

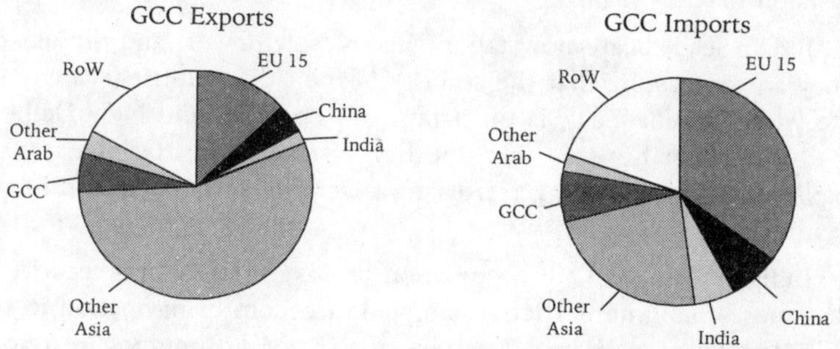

As a result, after years of hesitation, ideology—in the form of Islamic solidarity—took a backseat and the GCC dealt with India and Pakistan on a level playing ground by signing a framework agreement on economic cooperation that will ultimately help create free trade areas between the GCC and India and the GCC and Pakistan. Talks are under way to sign a GCC-India Free Trade Agreement. The GCC has also granted India 'dialogue partner' status—the first from the developing world and only the fourth in the world other than the US, European Union and Japan to get that privilege.

While having a president and prime minister who are not Hindus, as well as King Abdullah bin Abdulaziz's visit—the first by a Saudi head of state in 60 years—have contributed positively to India's image, more importantly though, the Gulf is taking note of India's renewed interest in the region. While energy security—to ensure stable prices and uninterrupted fuel for its economic growth—is certainly a factor, India is now willing to showcase its power and influence in the region. This expanded security perspective is driven by necessity, ambition and opportunity. The desire to lead coincides with its rise as a major power with continental aspirations. Former premier Atal Behari Vajpayee and Prime Minister Dr. Manmohan Singh have called for looking beyond the immediate neighbourhood.[21] After Pakistan, China, Russia and the US, the Gulf is the focus to ensure against any maritime or landward threat to it from the region, serve as a base to pursue India's interests, confront terrorism and extremism, as well as tap the investment potential.

By focusing on the Gulf and restoring traditional linkages with the immediate and extended neighbourhood, India is seeking to address its 'four deficits' in the historical, security, economic and global decision-making realms.[22]

Indian leaders are now talking about 'soft power' and diplomacy. They are suggesting that the security of the Gulf countries, as well as the wider Middle East, is of 'paramount concern' and New Delhi is ready to contribute to the stability of the region by sharing its experience in combating terrorism, maritime security and military training.[23]

India is keen on Gulf cooperation to ensure safety and security of sea-lanes of communications, safety and freedom of navigation in the shipping lanes and trade routes, positively influencing religious extremism and/or transnational terrorism, narcotics trafficking and proliferation of weapons in the region, and, most importantly, achieving peace in the subcontinent, given Pakistan's Islamic bond with the region.

In order to highlight its potential and achieve its objectives, the Indian military services are undertaking a major build-up of conventional arms, creating ways of delivering nuclear weapons and defending against them, planning construction of warships, enhancing military logistics in Central Asia and negotiating with the US for an Asian version of NATO. All these come in the wake of India's existing maritime security involvement involving Asian, African, European and Gulf countries, as well as Russia and the US.

Conclusion

Going back to demographic and labour issues, incorporating population policies into overall development strategies continue to pose a challenge for the GCC countries. This includes discrepancies between the high rates of population growth, emergence of new global challenges, cultural threats and damage to the national fabric. The discourse over reducing the gap between the number of nationals and foreigners in GCC societies and the nationalisation of labour force has not had the desired results.

The GCC countries need to rid themselves of their obsessive preoccupation with defensive demographic strategies and introduce new policy instruments to tackle their problems. They must consider that demographic variables, if properly utilised, could be the cornerstone of the region's comprehensive development march.

The truth, however, is that with the advent of globalisation and the accelerated pace of integration of the GCC economies into the global economy, the flow of migration will basically be determined by economic factors, which will necessitate granting rights that the GCC countries are not used to. For example, the continuation of the policy of double standards in salaries for the same kind of work done by the GCC nationals and expatriate workers would be difficult under the obligations of international agreements and increasing globalisation.

There is no doubt that the demand for foreign workers is not going to reduce drastically, but the regional governments will be proactive in trying to limit the dependence on expatriate labour in sectors where the indigenous population was hitherto either unqualified or unwilling to serve, but is now ready to do so. The intensity will, however, depend on factors such as the number of young nationals entering the labour market, the effect of the nationalisation of labour markets, the economy's capacity to generate new jobs, the employment qualifications of national labour in relation to the requirements of the job market, the willingness of nationals to take low-prestige jobs, competition from other groups of foreign workers, especially Asians, as well as political and security considerations. Labour-exporting countries like India can definitely take heart from Bahraini Minister of Labour Dr. Majeed Al-Alawi's assertion that the size of expatriates in the region is expected to grow by 5 per cent every year to reach 18 million by 2017.

Endnotes

1. Janardhan, Meena (2005). "Indigenisation policies hit expatriates", *Inter Press Service* (IPS, Bangkok), 14 December.
2. Employment for locals is an economic safety valve that helps let out political steam. It is a fairly full-proof political diversion, apart from being a sound economic plan.
3. Janardhan, N. (2003). "Saudi nationalisation drive targets expatriates", *Inter Press Service*, 5 February.
4. *Gulf News* (UAE), 12 March 2004.
5. Janardhan, Meena (2003). "Welcome Mat shrinking for Asian workers in UAE," *Inter Press Service*, 3 October.
6. For more, read Rutledge, Emilie (2005). "Are the GCC's Labor Nationalization Policies Working?" *www.gulfinthemedia.com* (UAE), 25 July.
7. *Arab News* (Saudi Arabia), 20 January 2006.
8. Al-Tayer, Ahmad Humaid (2004). "GCC should talk to US on free trade", *Gulf News*, 8 December.
9. *Gulf News*, 22 December 2005.
10. Bowing to longstanding pressure from Indians in the Gulf, the Indian government announced in January 2005 that it was considering voting rights for expatriates.
11. *Aljazeera.net*, 7 November 2005
12. "France and the Gulf", *Al-Seyassah* (Kuwait), 15 November 2005
13. Rutledge, Emilie (2005). "Pleasure of high oil prices brings pain of inflation to Gulf countries", *The Daily Star* (Lebanon), 14 October.
14. Janardhan, N.(2002). "Saudi income-tax proposal scares Gulf expatriates", *Inter Press Service*, 10 May.
15. Pakkiasamy, Divya (2004). "Saudi Arabia's Plan for Changing its Workforce," Migration Policy Institute, 1 November 2004
16. Janardhan, Meena (2004). "Workers struggle with laws, wages in UAE," *Dawn (Pakistan)*, 13 October.
17. Janardhan, Meena (2005). "Labour authorities to name and shame defaulting employers", *Inter Press Service*, 12 October.
18. The Indian embassy in the UAE estimates that 25 per cent of the Indian working population can be categorised as unskilled, 50 per cent as semi-skilled/skilled and the remaining as professionals and businessmen.
19. "14,000 jobs and a bitter lesson in skills," *Gulf News*, 23 February 2004
20. "The GCC finally takes a note of the Indian giant," *Gulf News*, 5 September 2004
21. "PM keen on building strong ties with Gulf states", *www.sify.com*, 7 August 2005
22. Statement by Indian Minister of Defence Pranab Mukherjee at the Carnegie Endowment for International Peace, Washington, DC, 27 June 2005
23. Statement by India's Special Envoy to the Middle East Chinmaya Gharekhan at the Second Gulf Security Conference in Bahrain, 2-3 December 2005

33

India and West Asia

Past and Future

A.K. PASHA

Before and after Independence India extended support to the Palestinians and other Arab National Liberation Movements (NLM's) in West Asia. In the process India came to establish close ties with secular nationalist regimes especially with Egypt under Nasser. Soon India also came to befriend, encourage and extend support to those countries which opposed the Baghdad Pact. Friendship with progressive Arab leaders to some extent neutralised the Pakistani propaganda against India, especially over the Kashmir issue, as also during the 1965 Indo-Pak War. Indian leaders (Nehru even visited Saudi Arabia, Iran, Turkey and other states) were not only concerned about implications of developments in the region on India's security but also on its secular policy, which came to be appreciated gradually even by the so-called conservative States in the region.[1]

New Directions in India's West Asia relations

The Bangladesh crisis and the Indo-Pak War in 1971, made a major impact on India's ties with West Asia, as Pakistan sought to get closer to the Arab States. With great skill India managed to convince the Saudi and other Arab (Muslim) leaders about its desire to live in peace with Pakistan and respect its territorial integrity. The 1972 Shimla Agreement and the release of Pakistani POW's helped the atmosphere as Arab concern for Pakistan's integrity had become a major issue in India's bilateral ties. Egypt's posture during the 1971 crisis in the sub-continent brought home to India the harsh reality of little to choose between the so-called progressive and conservative Arabs in their stand towards such a crucial issue affecting Indian security. In the light of new regional developments, India had to reorient its policy from one of heavy reliance on Cairo to other power centres, which had emerged in the region particularly Iraq, Iran and the Arab Gulf monarchies. Iran, Turkey, Jordan and others continued to give diplomatic and political support to Pakistan. Israel's refusal to withdraw from Arab lands made several of Israeli supporters abandon it just before and after the 1973 Arab-Israeli Conflict (AIC) and the subsequent oil embargo by OAPEC against USA and other supporters of Israel. India did not have to make such opportunistic changes in her foreign policy towards the Arabs. The Arab states due to India's unequivocal support for their causes assured New Delhi of uninterrupted supply of oil and treated India as a friend. Soon it opened significant possibilities for exports and also of Indian workers in the Gulf area.[2]

India and the GCC States

The Gulf crisis over Kuwait when Iraq marched into the oil rich Emirate on 2 August 1990 posed India a major challenge. Due to India's close political and economic ties with Iraq, an impression was created in the initial stages of the crisis that New Delhi tended to support Baghdad. This led to misunderstanding between Kuwait (and also with other GCC states) and India. To be correct, India strongly disapproved the Iraqi invasion of Kuwait and called it illegal and should be vacated but stopped short of condemning the invasion. Since India did not openly take sides, Kuwait and other GCC states assumed India's sympathies lay with Iraq. India by neither condoning nor condemning the Iraqi takeover of Kuwait justified its position by saying that, "condemnation is not part of the Indian nature." India, of course by simply 'regretting' the invasion did not win the gratitude of either party. It was in fact misunderstood by both the parties and by others as well. The Kuwait crisis affected India in many ways. New Delhi mainly due to domestic turmoil displayed no apparent enthusiasm to resolve the crisis. The minority government of V.P. Singh appeared to be primarily concerned with the evacuation of more than one lakh Indian citizens from Kuwait and 25,000 in Iraq. This lack of interest was partly due the economic compulsions flowing from the situation in the Gulf which pushed India to seek IMF loan where US support was crucial. The Indian government partly shared the fear of the West in particular the US, that Saddam Hussein would control bulk of the world's oil reserves in the region, if he were not pushed out of Kuwait. No wonder India allowed (initially) the-refuelling of US military planes and went along with the sole superpower at the UN. Room for India to manoeuver in the crisis was rather limited, given the resolve of the West and the GCC to contain

Iraq. But as a leading member of NAM, India did take some initiatives, but these came too late, partly because of deep differences within NAM, especially Yugoslavia's one-sided stance. India's economic linkages with the GCC had increased during 1970s and 80s, especially due to growth in India's oil imports. Also the oil rich GCC states as seen earlier, embarked on ambitious economic development, which led to demand in GCC States for labour. India with its surplus labor became a major source of supply. The four million plus Indians currently in the GCC States send about $12 billion (US) dollars as remittances annually. Indian banks also mobilise huge deposits and investments from Gulf Indians. The GCC states provide good market for India's exports. During 2004-05, India's exports to GCC were $9536 million and imports $6900 million. The bilateral two-way trade exceeded $17 billion.[3] With a trade turnover of $11 billion in 2004-2005, the UAE was India's biggest trade partner in the GCC. India's exports to UAE in the same year were $7.1 billion and its imports were $4.5 billion. UAE was the first GCC State to provide aid to India through Abu Dhabi Fund for Economic Development (ADFAED). A large number of India's public and private companies have taken active part in the industrialisation of the UAE. They are either contracts or joint ventures. UAE has about 20 joint ventures mostly in private sector in India especially in Gujarat and the offshore and onshore oil drilling platforms plant in Maharashtra is significant. The stand adopted by the UAE on Kashmir was remarkably restrained. Some in the Indian bureaucracy and mass media continue to portray UAE as a country (Dubai) which is haven for Indian terrorists besides giving it negative publicity in India. This image needs to be corrected as this might lead to unpleasant problems in future.[4]

In the aftermath of the 1991 Gulf War, India through its skillful diplomacy succeeded in clearing the misunderstanding with Kuwait. Nevertheless, ties deteriorated since Kuwaiti leaders publicly expressed anguish and surprise at India's ambivalent attitude during its occupation by Iraq. The Kuwaiti foreign minister, during his visit to India in February 1992 had said, "We will forgive, but not forget." Not surprisingly, Kuwait, at a number of OIC meetings after the war, had launched a scathing attack against India and had supported Pakistan's stand on holding a plebiscite in Kashmir. Despite Kuwait's unhappiness, a number of Indian firms were given contracts in the rebuilding of Kuwait. India has shown interest in Kuwait's downstream petrochemical sector whereas Kuwait has evinced interest in Indian petrochemical complexes. India has also shown interest in a gas based fertiliser plant in view of huge quantities of gas in Kuwait and a refinery on the West coast of India. The Kuwaiti Amir Sheikh Sabah, visited India in June 2006 and bilateral ties have improved significantly.[5]

While political relations have been far from bonhomie as a result of Cold War alignments and Saudi-Pakistani affinity, Indo-Saudi economic cooperation took a more realistic form and flourished. Mrs. Gandhi's visit in 1982, led to the setting up of an Indo-Saudi joint commission to promote economic and technical collaboration. The assumption that a combination of Indian technology, skilled manpower and Saudi money could work to strengthen South-South Cooperation yielded tremendous economic gains during the last 25 years. The commission has been meeting regularly since 1983 in Riyadh, and in New Delhi. India's exports to Riyadh during 2004-05 stood at $1379 million and imports at $1264 million excluding crude oil and petroleum products. These figures are among the highest between India and an Islamic country. Saudi Arabia is not only a major economic partner of India; it is also a major power in the Gulf region. Saudi Arabia has emerged as the 13th largest market for Indian exports and a source for 5.5 per cent of India's total imports. India is the fourth largest market for Saudi goods and ranks 10th in the Saudi imports market list. Riyadh is the largest source of crude oil to India supplying half a million barrels of crude oil per day and a total of about 24 million tonnes annually, which is nearly a quarter of India's total crude requirements. However, much of this adverse balance is made up by the flow of remittances by over 1.8 million Indian workers in the Kingdom.

Moreover, Indian companies, both public and private, have taken and completed numerous lucrative projects in the Kingdom and earned huge profits. The Saudis have given loans to India through the Saudi Development Fund for a number of projects. Since the emergence of militancy in Kashmir in 1989 and the demolition of Babri Masjid in 1992, India worked hard to improve ties with the Kingdom. The opportunity to cement that effort came following 9/11, when Saudi Arabia responded to Indian overtures, realising that both countries face the common scourge of terrorism. Since then, Riyadh has condemned all forms of extremism and terrorism, including the attack on Indian Parliament in 2001, while making its fervent appeal for peace in the Indian subcontinent. The healing process was demonstrated by the quick aid dispatched to the victims of the Gujarat earthquake the same year. The Saudi King was the chief guest at the Republic Day celebrations in New Delhi on 26 January 2006. He became the first Saudi king to get that honour. For the last half century, no Saudi ruler has visited India as the Kingdom enjoyed close ties with Pakistan. Saudi Arabia is playing a key role in the talks for a Free Trade Agreement between GCC and India. Based on its new diplomacy of promoting cooperation between importers and exporters of crude oil, India and Saudi Arabia are encouraging cross investments in the industry, mining and energy sectors. Clearly relations are likely to see upswing in the coming years. In the broader context of rapidly improving Indo-US ties, closer Indo-Saudi ties are likely to be upgraded which could include security cooperation to meet common challenges. King Abdullah's visit should kick-start a process that will look beyond religion, oil and employment, as well as remain independent of its close ties with Pakistan.[6]

In a smaller or larger measure, Bahrain, Qatar and Oman have either expressed disinterest or toned down their earlier stance on Kashmir. Bahrain which takes a balanced view of "UN resolutions and framework of Shimla" position, has removed UN resolutions and kept the Shimla framework from its published text in the context of Indo-Pak relations for a number of reasons. Indo-Bahrain trade turnover in 2004-05 stood at $262 million, with Indian exports being $150 million and imports of $112 million. Bahrain's PM Sheikh Khalifa bin Salman Al Khalifa's visit in January 2004, the first to India boosted the cooperation between the two countries. Both signed an extradition treaty, mutual legal assistance and judicial cooperation in civil, criminal and commercial matters, bilateral investment promotion agreement and avoiding double taxation. The Bahrain King is likely to visit India at end of 2006 or early 2007. There is a wide gap in Indo-Qatar trade with India's exports in 2004-05 being $203 million and imports $640 million and is reflected in adverse and growing deficit in balance of trade for India. Qatar's Rasgas has also tied up with Petronet LNG for supply of 7.5 m tonnes per annum of LNG for Petronet's plants at Dabhol, Maharashtra and at Cochin, Kerala. There are about 200,000 Indian workers in Qatar who send valuable foreign exchange, which is quite helpful in reducing the balance of trade deficit faced by India in its trade with Qatar. For the last 5 years both countries had been building the necessary infrastructure to receive the LNG. Petronet imported 2.5 mt in 2004 and in 2005 it got 5 mt of gas from Qatar. India hopes to get 25 per cent of its gas requirements from Qatar. The Qatar Emir and his Foreign Minister have visited India recently and Qatar investments have recently increased. India's relations with Oman are very old and historic. During the 1971 Bangladesh crisis, Oman was the only Arab (Muslim) country that did not completely side with Pakistan. In recent times high level visits have taken place between the two states. PM Narashima Rao visited Oman in June 1993. Oman has decided to participate in two oil refineries in India. The Indian Minister of Chemicals and Fertilizers visited Muscat on 4-5 December 2001 and signed key agreements relating to the Oman-India Fertilizer project in Qalhat south of Muscat during the visit. Oman-India Fertilizer Company (Omifco) is a $1 billion project and is producing 1.65 million tonnes of granulated urea and 0.25 m tonnes of surplus ammonia per year which India imports. President S.D. Sharma visited Oman in October 1996, and the Oman Sultan came to India in April 1997 and in August 1998, Prime Minister A.B. Vajpayee visited Oman. There is modest military-security cooperation between the two states. Indo-Oman trade improved from $96 million in 1990 to $157.47 in 2001-02; $212.44 in 2002-03 and in 2004-05 India's exports were $258 million and imports $19.82 million only. There are a number of joint ventures which Indian companies have taken up in Oman. Both signed an extradition treaty and agreed to intensify defence cooperation as also in the IT, tourism, education and HRD, healthcare and small and medium industries sectors.[7]

India after having consolidated its ties with the ASEAN countries is gradually focusing on the GCC states. The key factor driving its policy is India's quest for energy security. Despite global search for oil/gas resources, the focus remains on the Gulf region to ensure adequate supplies to sustain India's high economic growth. The Gulf region has also become an important foreign exchange earning zone from the standpoint of growing trade, workers remittances, investment opportunities and tourism. One can notice numerous steps being taken by both to strengthen bilateral ties in all of these areas. The economic and commercial content in India-GCC relationship is growing and has more potential to expand for mutual benefit. The new development is that India would be working in tandem with the US in the Gulf region to further promote its core interests. India by continuing to support the just causes of the Palestinians and other Arabs and also to work for closer ties with

Israel and the US has shown the unique balance one can maintain by having good ties with both the parties.

Many argue that India's ardor for the Arab cause has waned and become muted. Most Arab states including the GCC states despite having very close ties with the US and EU have been unable to restrain the US and others in their support to Israel. Israel's continued defiance of UN Resolutions, violation of international law, contempt of world public opinion and opposition to the establishment of an independent Palestinian state is mainly due to this solid support. Not only due to the collapse of USSR and the Kashmir issue but also due to a variety of other factors, India has been working hard to improve relations with the US. If the US has become India's largest investment and trading partner and cooperation in military, nuclear and security matters is increasing, it is due to the desire of both the states to achieve it, which is helped by the growing role of private sector in this relationship. Despite overwhelming opposition from the public in the GCC states, the US is looked upon by the GCC leaders as a security and defence partner. India in cooperation with such like-minded states has been striving hard to remove the factors creating instability in the Gulf region. Both the GCC states and India face the menace of terrorism and religious extremism and there is urgent need to combat such forces. A modest beginning has been made for cooperation in the field of security with the GCC states which has a rich potential field ahead. Many also emphasise the need to fulfil the growing aspirations of the Gulf people to play a greater role in the sphere of political participation and decision making process. Political systems which allow more space for civil society, and work for enhancing their legitimacy would prove to be more enduring as also develop stake in peace and stability. Of course, external support will continue to be a major factor in Gulf politics but it is internal support, which will eventually decide the final outcome especially major issues concerning peace, security and stability.[8]

Growing Cooperation with Israel

Nehru was convinced towards the end of his life that his decision to oppose partition of Palestine and not to exchange diplomatic personnel with Israel was right. Since then successive Indian governments highlighted the absence of full diplomatic ties with Israel as evidence of India's support for the Palestine cause. Apart from other reasons, it must be emphasised that the Kashmir problem propelled India not to establish diplomatic ties with Israel as perhaps that would have provided Pakistan an additional lever to undermine India's interests in the area. In the wake of the Kuwait crisis, end of Cold War, disintegration of USSR, Madrid Peace Conference in October 1991, and India decided to establish diplomatic ties with Israel on 29 January 1992. Things changed dramatically after 1991, when the Arabs and the Israelis after more than 40 years of hostility, wars and boycotts decided to sit down and negotiate their problems at the Madrid Conference.

With this new development India was encouraged to break the ice with Israel. The Jewish state was no more a *pariah* as before. There were also major changes as far as PLO was concerned. After Yasser Arafat (PLO leader) in 1988, accepted UNSCR 242 and renounced violence and recognised Israel, the result was in 1991, PLO had more than 92 embassies and 28 Lower Level foreign representations. India was the first non Arab country to recognise PLO in 1975, as the sole and legitimate representative of the Palestinian people and gave the Office of the PLO in New Delhi full diplomatic status of an Embassy in 1980. In 1988, India was one of the first countries to recognise the State of Palestine. Under the changed atmosphere India voted in favor of resolution 46/86 in December 1991, at the UNGA which rescinded its 10 November 1975, Resolution 3379 equating Zionism with Racism. This was the first indication of change in India's West Asia policy especially towards Israel. Soon thereafter, meetings between Israeli and Indian officials at the UN, US and elsewhere took place.

Apart from India's support to UNSCR 687 of 3 April 1991 imposing terms of ceasefire on Iraq, this was a major shift. Some have even called this a reversal in India's policy towards the region. India's decision was accompanied by many statements highlighting the benefits India stands to gain with its new policy towards Israel. India's Defense Minister (on 23 February 1992) Mr. Sharad Pawar said it paved the way for drawing on Israel's successful experience to curb terrorism. He said that such an exchange would be of mutual advantage to the two states, adding that India would also like to get acquainted with Israeli experience in developing technology for anti-terrorist operations. It was also mentioned how India could play a significant role in the Arab Israeli Peace Process.

Additional areas of cooperation were also cited, i.e., defense cooperation and cooperation in agriculture. India's trade with Israel has increased manifold. Israel has emerged as a major supplier of weapons to India. Major bilateral agreements have been signed in almost all fields and high level visits have taken place. Initially, Indian Chief Ministers and other state ministers visited Israel. From the Israeli side, former Prime Minister and foreign minister Shimon Peres has been to India several times followed by Ezer Weizmann the Israeli President. However, the most controversial has been the visit of Israeli Prime Minister Ariel Sharon during the NDA Government. L.K.Advani (the then Home minister) and Jaswant Singh (the then External Affairs) also visited Israel and generated lot of controversy over cooperation in nuclear field, terrorism and internal security.

There was no doubt about the NDA giving primacy to Israel and US in its foreign policy. But the emergence of UPA government under Manmohan Singh has apparently not made much of difference to Indo-Israeli ties. They remain as warm as before with India importing large quantities of arms from Israel. Although the MEA is cautious but the Defense Ministry has been receiving delegation after delegation from Israel. Despite the call by left parties for a review of ties with Israel and a ban on import of arms in view of continued occupation of Palestinian lands the Congress led UPA appears to continue the ties started by P.V. Narashima Rao in 1992. The Indo-Israeli relationship is flourishing with two-way trade surpassing the $ 3 billion mark with no end to military exchanges. After the Madrid Conference and the Oslo Peace Accords of September 1993, between the PLO and Israel, UN members and people all over the world expected Israel to withdraw from Arab lands leading to peace between Israel and its neighbours and the formation of a Palestinian State. But the assassination of Israeli Prime Minister Yitzhak Rabin on November 4, 1995 and the emergence of Benjamin Netanyahu as Likud Prime Minister of Israel and his opposition to the Oslo peace process including a Palestinian state and acceptance of US neocon agenda calling for a Clean Break with the Oslo process while addressing US Congress on 10 July 1996, gave a severe jolt to the Peace Process. Many accused Arafat of rejecting in July 2000, the Israeli peace offers in Camp David talks with President Clinton and Israeli Prime Minister Ehud Barak. There are no reliable reports as to what was offered to Arafat. But what is clear is that Arafat was willing to concede 77 per cent of historic Palestine and wanted Israel to withdraw from the 23 per cent of Palestine i.e., Gaza, West Bank and East Jerusalem. Israel offered Gaza and parts of West Bank but refused to dismantle Jewish settlements and rejected the right to return of the Palestinian refugees and reiterated its stand that Jerusalem will remain the eternal and undivided capital of Israel.

The US proposed Tenet and Mitchell Plans and later the Road Map for Peace with UN, EU and Russia but so far no progress has been made. Meanwhile Israel has reoccupied the Palestinian territories, especially after the withdrawal from Gaza and the election of Hamas government in January 2006, and has carried out indiscriminate arrests, killings of the Palestinians, imposed curfew on most towns and cities, carried out house demolitions, and humiliating check points to prevent free movement of Palestinians. The result has been that deprivation, destruction and death rained on the Palestinians. Growing Indo-Israeli ties continue to impact on India's ties with some Arab states.[9]

Indo-Iranians Relations

Iran is one of the most important countries in the West Asian region with which India has maintained fairly extensive relations in diverse fields. India did not ignore Tehran even though the Shah of Iran was close to US and Pakistan. Indo-Iranian ties developed as reflected in several high level visits, expanding bilateral trade and investments. But India saw the emergence of the Islamic Republic of Iran in 1979, under the leadership of Ayatollah Khomeini as a positive development. Khomeini was described as the father figure of the Iranian revolution, by A.B. Vajpayee (then, India's minister for external affairs) who also said, "We are waiting for the day when we can welcome Iran in the Non-aligned Movement." More significantly, "India viewed the revolution in Iran as a reflection of Iran's quest for identity and national self-assertion and a desire to charter an independent course without outside big power influence." Iran's preoccupation with Iraq in the Eight-Year War (1980-1988) and the new Islamic regime's penchant for taking up Islamic causes led to strain in India's ties with Iran. Iran's stand on the Kashmir issue, the Babri Masjid and Indian Muslims introduced new factors in Indo-Iranian relations.

Indian response to Iran's attitude was muted because of the importance New Delhi attached to economic ties, Iran's role in the Gulf, Central Asia and Afghanistan, and also due to the conviction that the realists ruling Iran wanted to do business with India. The

1990-91 Kuwait crises, Soviet withdrawal from Afghanistan, the disintegration of the Soviet Union together with the end of the Cold War and the emergence of Pax Americana in the region and identical security threat perceptions brought India and Iran closer. US attempts to isolate Iran under its dual containment policy, economic sanctions and American influence over the GCC states and Iran's bitter experience with Iraq compelled Tehran to strengthen relations with India and China. These common political and strategic concerns led to extraordinary developments in Indo-Iranian relations, including numerous high level visits and a large volume of bilateral trade. Prime Minister P.V. Narasimha Rao visited Iran in September 1993, where he told Ayatollah Khameini's that the destruction of Babri Masjid was an isolated incident and did not impinge on India's secular character. He assured him that the Indian Muslims were very much a part of the national mainstream and the decision making process in India. The Indian Prime Minister also made it clear that India was firmly committed to protect the interests of minorities and that New Delhi could deal with such issues ably and competently without outside help.

More significantly, during the UN Human Rights Commission (UNHRC) meeting in Geneva in February-March 1994, Iran played a crucial role in persuading Pakistan to withdraw the resolution it had tabled on Kashmir. Iranian President Rafsanjani visited New Delhi in April 1995, and emphasised the need for strategic cooperation to ward off outside interference and domination in the region. The election of Mohammed Khatami as President in 1997, and the favourable statements to expand cooperation on the basis of mutual respect augured well for future ties. All of this notwithstanding the Iranian support for the participation of Hurriyat leaders in the OIC Tehran Summit in December 1997. During the Khatami period relations improved in all fields and concrete steps were taken to push the gas pipeline issue also. Mahmoud Ahmadinejad won the Iranian Presidential election in 2005 and the new President has followed the line of Khameini even more closely than his predecessors. The formidable western (especially the US) naval presence in and around the Gulf led to pressures on Iran and India, making it clear to them that both could cooperate on this issue affecting their security. India had invited President Khatami as the chief guest at the Republic Day in 2003. But relations have deteriorated since September 2005, when India voted twice against Iran at the IAEA thereby casting a shadow on the gas pipeline and LNG imports.[10]

India and Turkey

Due to obvious ideological and foreign policy divergences, Indo-Turkish relations grew *albeit* slowly although Turkey was one of the first country to establish diplomatic relations with India in 1947, with an Embassy in New Delhi since 1948. As early as on 14 December 1951, India and Turkey signed Treaty of Friendship. Maulana Azad became the first major Indian leader who visited Turkey when he was India's Education minister and signed several agreements in the fields of education, culture and scientific cooperation. He also signed the cultural relations agreement with Turkey on 29 June 1951. Even though India's Prime Minister Jawaharlal Nehru had serious reservations over Turkey's pro-west foreign policy and its support to Pakistan on the Kashmir and other issues but he was, attracted towards Turkish policy and commitment to secularism. It is this crucial reason more than anything else which prompted him to visit Turkey in June, 1960. Turkey's close alliance with the West and its membership in NATO and CENTO was cause for deep concern in India. No wonder the bilateral military cooperation pacts with the US and support to Turkey and Pakistan and domination of Turkish political system by the military and other factors led to a drift in Indo-Turkish relations. But it must be stressed that there were no fundamental bilateral problems between India and Turkey.

The real breakthrough in ties came with the Turkish Prime Minister Turgut Ozal making his first ever-official visit to India on 9 April 1986. Mr. Ozal brought with him a strong delegation of nearly 200. The unusually big size of the delegation was intended to impress India that Turkey is keen on establishing closer relations with India in mutual interest—as one journalist put it. Ozal assured India that Turkey was keen on developing mutually beneficial political and economic links with India without prejudice to its close relations with Pakistan. In return Prime Minister Mr. Rajiv Gandhi visited Turkey from 17 July 1988, on a three-day official visit. He became the second Prime Minister from India to visit since Jawaharlal Nehru visited Turkey in 1960. Undoubtedly, there had been a significant improvement in Indo-Turkish relations thereby giving a new direction to bilateral ties. Bilateral trade turnover, which was a meager \$5 million in 1985, exceeded \$120 million in 1987. Many experts felt that even this increased trade does not reflect the true potential for commercial exchanges between the two countries. Since Mr. Ozal's visit to India in April 1986, more than 20 Turkish delegations, including those at the ministerial level, have visited India.

During Rajiv Gandhi's visit three agreements were signed: (1) Maritime Agreement; (2) Mutual Judicial assistance in civil and commercial matters; and (3) Mutual judicial assistance in criminal matters. In addition, an agreement was also signed between IRCON and the Turkish National Railways for a US $25 million railway electrification project. It was also agreed that: (1) a banking delegation led by the Governor of the Central Bank of Turkey will visit India in October-November; (2) further negotiations for finalising the Avoidance of Double Taxation Agreement will be held in Delhi in December 1988; and (3) Turkish side will send an agricultural delegation to India in early spring 1989. Apart from this warmth, apparently aimed at setting the stage for a new turn in ties, what may have discomfited Pakistan were the announcements regarding the Turkish ban on exports of nuclear inverters and the lifting of the ban on screening Sir Richard Attenborough's film, 'Gandhi' in Turkey. During the visit Gandhi welcomed the Turkish decision on the ban and formally thanked Mr. Ozal for it.

India's trade with Turkey has been modest. Turkey's imports from India in 1989, 1990, 1991 were $43, $78 and $87 million dollars respectively. But India's exports in 2005, jumped to $1.274 billion and imports were $219.7million. A trade protocol was finalised in February 1992 and the establishment of a Business Council envisaged in the Protocol of the Turkish-Indian Joint committee for Economic and Technical Cooperation signed in 1992, paved the way for widening contacts between the business partners of the two countries. Ties made gradual progress despite some chill in the political ties due to Turkey's stand on Babri Masjid and on the Kashmir issue as member of the OIC contact group in addition to the stand taken during the 20th OIC Foreign Ministers meeting in August 1991, at Istanbul. Despite these developments a three year cultural exchange programme (CEP) was renewed in January 1992, under which exchanges of teachers/professors, scholars, artist's takes place. Film weeks have been regularly held under the CEP. The ICCR has been sending music and dance groups and Turkish groups have also performed in different cities of India which have proved to be popular and increased people to people contacts between the two countries. End of Cold War in 1991 had its clear impact on Turkey's security perceptions as the disintegration of USSR was a major factor, Turkey's *raison d'etre* for its membership in NATO virtually collapsed. Having had close ties with Pakistan since 1950s, Turkey was disturbed by the Babri Masjid's demolition by BJP in December 1992.

Reference was made to the demolition of the Mosque by Turkey in the UN General Assembly but mostly it called for reconstruction of the mosque at the earliest. Turkey being a prominent member of the OIC took part in the meeting of the OIC Bureau which in an extraordinary meeting in Dakar, Senegal on January 11, 1993, especially called to discuss the expulsion of the Palestinians by Israel; also discussed at Pakistan's request the Babri Masjid demolition issue also. A very negative resolution was adopted by the OIC even in the face of hectic lobbying by India with key OIC members including Turkey. It was a disappointment for India that Turkey had joined hands on this issue at OIC with Syria, Morocco, Pakistan, Senegal, Indonesia, Kuwait and Saudi Arabia. Turkey being a member of the OIC contact group consisting of Pakistan, Saudi Arabia in addition to the OIC Secretary General sought to paint the situation in Kashmir as posing a threat to peace and security in the region. Pakistan had planned to raise this issue in the draft resolution tabled for the UN General Assembly in 1994 but it suffered a setback as it could not muster enough supporters to its resolution at the first committee dealing with security and disarmament issues. At the 20th meeting of the OIC foreign ministers held in Istanbul (Turkey) 4-8 August 1991, they expressed concern at the alarming increase in the indiscriminate use of force and gross violations of human rights committed against innocent Kashmiris. Reactions in Turkey to events in India–the Babri Masjid demolition and the Bombay riots were low key compared to reaction from Iran. Except for critical articles in the lively Turkish press and a single but deadly attack against an Indian diplomat. Turkey's Ambassador to India Mr. Yusuf Buluc pointed out that Turkey had been using its influence to bring moderation to discussions and resolutions in the OIC.

In order to consolidate the political and economic relationship with India, Prime Minister of Turkey Mr. Bulent Ecevit came to India on a three-day visit from March 31 to April 2, 2000. During Ecevit's visit both signed three accords including one on cooperation in agriculture; a memorandum of understanding (MoU) for establishment of a mechanism for political consultation between the foreign ministries of the two countries; and another MoU extending cultural exchange programme between the two countries. Growing

Indo-Turkish relations have become a matter of concern to Pakistan as it wanted Mr. Ecevit to visit Pakistan at least briefly before or after his visit to India, but reportedly the Turkish Prime Minister refused to yield. It must be pointed here that the Pakistani military ruler General Parvez Musharraf chose Turkey as the first country for his visit abroad after he seized power in a military *coup d'etat* from the civilian government of Prime Minister Nawaz Sharif in October 1999.

The joint statement issued at the end of Ecevit's three-day visit said, the two sides proclaim that the shared values of "secularism and democracy" are the foundations of a new political relationship between the two nations. Both also declared their "conviction" that the suppression of international terrorism, "regardless of its origin and motivation" is an "essential element" for maintenance of peace and security. Prime Minister A.B. Vajpayee visited Turkey from 16-19 September 2003 in the continuing effort to consolidate bilateral relations. The visit was marked by mutual warmth and a desire to add further all round substance to the relationship. The signed agreements were: On cooperation in the field of Science and Technology and on Extradition; A Protocol of intent of cooperation in the field of IT and Computer software; Protocol for setting up of a joint working group on combating terrorism. Both have been cooperating to combat terrorism. It has been mutually agreed that suppression of international terrorism, regardless of its origin and motivation, is an essential element for maintenance of international peace and security. Turkey supports India's initiatives to get the UN adopt a comprehensive convention on international terrorism, which would be an important step towards strengthening international legislative framework for combating this menace.

Like Turkey, India also had to reorient its foreign policy in the wake of the momentous developments after the end of cold war, demise of USSR, war over Kuwait, and the emergence of Pax Americana in the West Asia. India lost its longtime military supplier and principal diplomatic supporter, i.e., USSR especially when militancy in Kashmir escalated. India initiated a gradual but necessary change in its policy towards the West Asian region while at the same time continuing its basic policy. India could not ignore the fact that the US had emerged both during and after the 1990-91 Kuwait crisis as the undisputed leader of the West. For Israel, the end of the Cold War removed one strategic motive for US support for Israel, but US aid and support to Israel continued. After the "Madrid and Oslo peace process", Israel initiated strategic partnership with Turkey in the region. Both Israel and Turkey became closer on Syria, Iran, Kurds, terrorism and other issues. Many Israeli analysts began saying that strategic partnership is emerging between Israel, India and Turkey and that this potential entente is an opportunity to the US to think 'outside the box' about building new security structures in the West Asian region.[11]

Indo-Iraqi Ties

India came to have close ties with Iraq until the latter invaded Kuwait in 1990. India's succeeded in cultivating Iraq without alienating Iran. India's deepening relations with Iraq also facilitated the subsequent process of political understanding and fruitful economic cooperation between India and the Gulf States. During the Kuwait crisis India voted in favour of the "mother of resolutions" passed by the UN Security Council (687) on 3 April 1991. India under the NDA initially was considering the idea of sending Indian troops to Iraq for stabilisation work but due to opposition from Congress and left parties it had to drop the idea as the Parliament enacted a resolution opposing the war. India's special envoy to West Asia C.R. Ghare Khan visited Iraq in May 2004 and offered assistance in the drafting of Iraqi constitution. He also offered help in the reconstruction of Iraqi infrastructure. India is a member of the Donors Group for Iraq and has made modest contribution for the Iraq fund. India has also helped Iraq in setting up field hospitals and other infrastructure projects.

India under the NDA government (as seen above) initially due to pressure from the US and also with an eye on the lucrative economic projects in Iraq was considering the idea of sending Indian troops to Iraq for stabilisation work (i.e., to suppress the Iraqi resistance). In fact the RSS on June 11, 2003 said, "There is no question of RSS opposing move to send Indian troops." But on 14 July 2003, India officially rejected the idea of sending Indian troops to Iraq. There were also massive protests throughout the country not only against the US invasion and occupation of Iraq but also against sending Indian troops to Iraq. Once the Indian Parliament deplored the US invasion and occupation and also the realisation that the Iraq crisis would be long drawn out with mounting Iraqi resistance, there was no question of sending Indian troops. India participated in an international conference on Iraq sponsored by the US and EU at Brussels on 21-22

June 2005. External affairs minister K. Natwar Singh took part in the meeting. He told the conference that India had been deeply distressed by the suffering of the Iraqi people especially for the past many years. He pointed out that what Iraq needed was security and stability and for that India was prepared to provide assistance in capacity building for security management.

It might be recalled that the US has been seeking Indian participation in training Iraqi police force in countries close to Iraq. The Minister praised the Iraqi Prime Minister Ibrahim al Jaafari for his detailed presentation on developments in Iraq at the Conference. He said, "They have analysed the challenges that their government faces and the steps that it is taking for political and economic reconstruction of the country. On behalf of India, I wish all success for these efforts." He also said, "We in India have been closely following recent political developments in Iraq. The restoration of sovereignty to the Iraqi people is essential for the establishment of peace and stability in the country. In this context, the holding of elections in Iraq on January 30, 2005, was a noteworthy and encouraging development. The UN played a useful supporting role in that process. India has welcomed the formation of the Transitional Government in Iraq. India would like to see a stable, peaceful, prosperous, united and democratic Iraq." He also told the Iraqi Prime Minister that Iraq could draw upon India's experience in drafting a Constitution for its multicultural society. Earlier in May 2005, when the new Iraqi government was formed under the Prime Ministership of Ibrahim al Jaafari, India's Prime Minister sent a message of congratulations to Dr. Jaafari, wishing the people of Iraq success in their efforts towards building a peaceful, stable and prosperous Iraq. The official statement said: "In keeping with India's long tradition of close political, economic and cultural ties with Iraq, the PM conveyed that India values and would like to strengthen further the bonds of friendship between our countries and peoples, and would be happy to assist in the reconstruction of Iraq and cooperate in areas of mutual interest."

India's the then external affairs minister Mr. Natwar Singh told an international conference in Brussels in June 2005, sponsored by US-EU on Iraq, "we condemn the actions of disruptive forces in Iraq and the activities of terrorist's elements." It is well-known that most if not all the Iraqi resistance to the US is led by the Sunnis. This is also clear by the fact that most Sunnis boycotted the January 2005 elections to the National Assembly. As seen earlier the US has developed a close relationship with the Kurds since 1991, and now after its occupation of Iraq it's wooing the Shi'is. Sunnis were deliberately targeted and excluded from the new power centres. Can India remain silent to the ruthless policy of killing and destruction pursed by the US against the Sunnis in Iraq? Most Sunnis believe they are waging a national liberation struggle for freedom which the US is trying to crush. How can India forget the fact that Iraq is essentially an Arab country as Sunnis, Shies and most Kurds speak Arabic? What the US is doing in Iraq is leading to a split between different groups and clearly the Sunnis are being crushed and suppressed. The US is going about destroying one Sunni city after another and killing hundreds and thousands in the name of containing the challenge of terrorism. India's silence to the US use of all military means on an almost daily basis including planes, helicopters, tanks, napalm, cluster and phosphorous bombs and indiscriminate arrests of hundreds and thousands of Iraqis is to be seen in the background of rapidly warming ties with the US. Not only Indo-US economic ties are growing but defense, nuclear cooperation is also intensifying.

What the NDA started the UPA is taking forward by strengthening, widening and deepening Indo-US ties. Since India did not send troops to Iraq as the US desired, now the US wants Indian support to its policies in Iraq. In effect, India is indirectly approving what the US is doing in Iraq. Once the US fully dominates Iraq and through it much of the Gulf region (with the cooperation of Israel) it will aim to extend its influence for broader West Asian region. The day will not be far when the US will directly apply pressure on India itself. Iran remains a stumbling block hence the pressure for regime change in Tehran. Already the US has plans to supply Iraqi oil from Kirkuk to Haifa in Israel via Syria (hence, pressure on Assad) and Jordan which would add to the complications in the region for India. US opposition to Iran-Pakistan-India gas pipeline and US suggestion to India to talk to the US for its (India's) energy security is primarily due to Pax Americana in the Gulf region. In future, the US along with Israel and Pakistan are likely to put more pressure on India. The simple truth is that the US has established itself so close to India in the Gulf region (India's extended neighbourhood) that there is bound to be spillover of its military success in the region on India in the coming years.[12]

Violence in Iraq: Implications for India's Security

Most secular Arab regimes are on the defensive due to internal challenges and India has been supporting conservative Arab regimes lately than the former. For New Delhi to fight terrorism in Kashmir and other parts of India and in the larger context of South Asia, it needs cooperation with West Asian states. One can argue that India cannot make much of a difference in the Arab-Israeli Peace equation but upgrading of ties with Israel since 1992 has undoubtedly led to greater Indo-Israeli military and political cooperation and especially under the enthusiastic support of the NDA. Muslim groups in the region have adversely reacted to this development especially since India has traditionally maintained close ties with Arab states. Both privately until recently and now publicly, growing number of political leaders in the Arab/Islamic World have expressed concern and growing apprehensions about the direction of New Delhi's domestic and foreign policies, religious extremism at home and closer identification with Israel and undue emphasis on fight against terrorism.[13]

The BJP's communally biased policies against the Muslims led to internal turmoil and economic disruption which affected the entire country. This had an impact on India's interaction with the Islamic world as well. Moreover, Pakistan continues to paint India in dark colours at international foras particularly UN and OIC on violation of human rights and killings in Kashmir, treatment of Indian Muslims and of India being hegemonic in South Asian affairs.[14] The NDA government in New Delhi and the BJP in Gujarat and their short sighted policy towards minorities especially the killings of Muslims and Christians (elsewhere as well) has provided a handle for outsiders to meddle in India's internal affairs. Only sound secular domestic policy and vigorous support to secular forces in the region and opposition to mounting outside military interventions may prevent the spread of terrorism to India. It has also become clear that placing undue emphasis on Israeli collaboration for fighting terrorism in India is unwise, as it would send a wrong message regionally and domestically.

The time has come for India to vigorously support the creation of an independent sovereign state of Palestine at the earliest and impress on Israel to implement the innumerable UN Resolutions and peace accords signed in this connection. Continued denial of this basic right in the face of universal support and growing concern for the sad plight of the Palestinian people, is an issue, which India has to address on a top priority basis. If this issue is resolved to the mutual satisfaction of both the parties one of the root causes for the growth and spread of violence and terrorism in West Asia would have been addressed. It has also become obvious that both Israel and the Palestinians are incapable of reaching a solution to their conflict, which very often threatens peace and security in the region. Moreover, other regional problems and contentious issues are getting interlinked with this problem thereby posing greater danger of terrorism acquiring a much larger regional dimension, and perhaps transnational dimensions as well. Since India has come to establish good relations with the US, it has become imperative to find an external solution to the vexed AIC.

The consensus which exists among all the major powers towards the establishment of a Palestine state could be the starting point. To start with, India must not only impress on the Israeli government to halt its brutal and oppressive policy towards the Palestinians but also mobilize global support in this direction. In this effort, the focus obviously has to be on the US which is the backbone of support to Israel giving over $3 billion a year, besides providing diplomatic and political support to Israel. Any delay in the resolution of this conflict has the potential to adversely affect peace, security and stability in the region with consequent implications for India. Although India has had historical ties with Iraq and now vital interests in the Gulf region, US military adventure against Iraq would have adverse consequences both on India's domestic and foreign policies as also alienate India from the masses in the region. Our support to Egypt during the 1956 Suez Crisis is still fondly recalled by Arabs all over the region. It has by now become obvious that the US wants to redraw the map of the region in order to keep the beleaguered state of Israel as a predominant unchallenged state and add Iraqi oil to US control and influence. It also wants to divert US public attention from the troubled US economy.

India must be with the people of the region and support their just aspirations for greater participation in the decision making process and on the broader issue of democratic reforms and also safeguard its vital interests. US domination of the region is slowly elbowing India out from the region as witnessed by sanctions on Iraq (and now chaos, civil war and threat of disintegration), Iran, and Libya, Sudan so on which has led to reduced trade, winding up of joint ventures, return and exodus of Indian workers and declining investments from these countries for sometime. India must work with the secular and democratic forces in the region as also the dominant world powers which have vast stakes in the region to maintain security and stability so very essential to protect our vital interests and also to prevent the spread of violence and terrorism from this volatile region to India in future. Hence India needs to enhance its interaction with the countries to seek greater role in cooperation with the states vital to our interests especially to enhance our energy security.

As briefly mentioned earlier, India should actively encourage the process of political change and modernisation because once the Arab regimes become democratic and responsive to popular demands, not only their legitimacy would be enhanced but more significantly their dependence on outside states especially the US would come down and this may lead to reduction in outside military and political intervention which has periodically undermined peace, security and stability in the region thereby adversely affecting India's interests as well. Genuine political reforms would give the politically oppressed people a stake in the running of the political system as also much needed political stability and smooth transition. The abundant oil resources could then be used to advance economic development so as to engage the Arab people in the process of political and economic progress. Undoubtedly, for India the threat of religious extremism and terrorism from the region would be contained if this democratic process takes shape.

Along with these steps towards democratic transformation, steps to resolve the thorny AIC would also be debated, discussed, at all forums thereby allowing Arab leaders to take initiatives for meaningful steps along with serious US-Israeli steps (which are very essential). Both these steps would reinforce the momentum for change and India could then look forward to greater peace, security and stability in future. In certain Arab countries like Lebanon, Bahrain, Jordan, Morocco and elsewhere the Islamic parties have gained substantial seats in legislature or come to share power through democratic process but such groups have learnt how to be responsible and accountable. It is far easier to control terrorism and religious extremism in liberal democratic political systems which have high level of legitimacy and strong domestic control than the present political set up where the rulers want to remain in total control of all state affairs. The long overdue change has now become inevitable but if its not coming soon internal and regional terrorism may be directed against such autocratic regimes unwilling or unable to initiate reforms. Instability, religious extremism and terrorism could spillover into South Asia as security and stability in both the regions are mutually interlinked through a number of factors.

The most credible option to tackle this twin menace which is posing a new challenge to India's composite culture, liberal democratic structures and also national security is to adopt a political course that prevents alienation of any segment of the population "especially religious [Shia] minorities-so that they do not take recourse to unlawful activities or violence. In India certainly the rising tide of Hindu radicalism has led to the emergence of certain Muslim radical groups as happened in Sri Lanka in which Israel also has a hand in it."[15]

Future Ahead

As seen before, US-Israeli policies in West Asia are undermining peace, security and stability leading to periodic threats of military intervention or wars or to prop up threatened fragile regimes against their own population who are demanding political participation. US invasion and occupation of Iraq and the continued instability (fear of split) in many parts of Iraq, in addition to US accusations of Iranian interference in Iraqi affairs is used by the US to put pressure on Iran to cease its ideas and programmes. This has also put pressure on Indo-Iranian ties. US attempts to reshape Afghan affairs to increase Pakistani influence also strained Indo-Iranian ties. US determination to stay in Iraq and its attempts to put pressure on Iran for regime change is partly linked to help project its power into Central Asia, due to the oil and gas reserves. From the Indian perspective, increasing bilateral ties with West Asian states is not only important due to trade, oil and gas, but also due to the unique proximity of these states to events in Afghanistan, Central Asia, and the Gulf region.

The common political and strategic concerns of India and these states will reinforce and strengthen mutual cooperation. If India, in cooperation with the US, reassures the GCC states of its desire to work for peace, security and stability in the region, outside interference could be contained, which many perceive to be the root cause of much of the problems which India and the GCC states face. 1990s saw the reassertion of regional identities in the face of external interference. The critical space is widening and it is in this context that India and GCC can create opportunities for mutual benefit and jointly prepare to meet emerging domestic and external challenges in the Gulf region. Instability in the region poses a greater threat than ever before, as India's reliance on Gulf oil is growing. Although the issue of terrorism is also important and imported arms from Israel are vital for India's security, there is no need for India to place undue emphasis on bilateral relations between India and Israel as the NDA did.

The United Progressive Alliance (UPA) government's common minimum programme stated that it will maintain the independence of India's foreign policy stance on all regional and global issues even as it pursues closer strategic and economic engagement with the US. But the reality is that in the context of West Asia, US policy is actually affecting India. Pax Americana in West Asia has serious implications for India. US control of Gulf oil may in the future affect our quest for energy security, especially as India imports 70 per cent of its oil from the Gulf. US support to undemocratic regimes in the region is encouraging the Islamic groups to use violence in the face of reluctance of the rulers to initiate political reforms. The security and future of Indian workers in the region in the wake of continued wars and instability has become an issue of concern. Peace, security and stability in the region are crucial for India's own security and prosperity and these calls for new initiatives, dynamism and recognition of the traditionally close historical relationship with the Islamic world. India's economic ties are unlikely to suffer with these states in the immediate future, but one needs to be cautious in developing closer ties with Israel until the process of reconciliation gets under way between the Arabs, especially Palestinians and Israelis.

The UPA coalition should attempt to balance the traditionally close ties with the West Asian with Israel. Security and stability in the region is linked with early resolution of the Iraq crisis, (withdrawal of US forces) establishment of a viable Palestinian state and encouraging political reforms and democratisation which alone can contain the growing slide towards religious resurgence, violence and terrorism leading to instability in the region. But what we are witnessing are record oil prices of over $78 per barrel. US policies to put pressure on Syria and Iran and the July-August 2006 Israeli invasion of Lebanon, have made the situation much worse. Not only oil prices have gone up but the region is sliding towards more extremism, hatred and growing animosities which are bound to affect security, stability in the region. India's oil demand is expected to grow from 2.4 million barrels a day (mbd) in 2003 to 6 mbd in 2025. India's import of oil and products costs the economy over Rs. One lakh crores and imports from GCC countries alone are over $ 6 billion. As seen above any disruption of oil supplies or increase in oil price due to uncertain situation in the Gulf will adversely affect Indian economy both in the short and long term. [16]

US-Israeli policies in the wake of occupation of Iraq and attacks on Lebanon and more particularly US support to authoritarian regimes has, allowed Islamic groups to grow but this is also allowing these groups to use the Palestine issue in order to resort to greater violence. US talk of further fragmentation of the area into smaller states based on sectarianism and ethnicity and the emergence of so-called Shia crescent is likely to further destabilise the already volatile region. Growing US domination or military presence in the region and control of oil resources and blind US support to Israel is leading to religious extremism and most resort to terror not only against the US but also against US allies in the region. India should actively encourage the process of political change and modernisation because once the Gulf regimes become democratic and responsive to popular demands, not only their legitimacy would be enhanced but their absolute (abject) dependence on US would be reduced.

India can help the GCC states in establishing the elections commissions, train officials for elections, help in drafting electoral laws, preparing constituencies, and prepare mechanism to resolve electoral disputes, so on. This would lead to reduction in US military presence and thereby maintain security and stability so very essential to protect India's vital interests as also contain the growing violence in this volatile region from spreading to India in future. The big question is: Are the GCC monarchies willing to share power with their own people and reduce their dependence on US. The other big question is: will the US allow them to become truly democratic and independent from US influence.[17] The UPA soon after assuming charge on 24 May 2004, among other things emphasised the importance of India's relationship with the Islamic world that went back many centuries. MEA said, "We greatly value our relations with our Islamic friends. There are over 50 countries (in OIC) whose religion is Islam. A large number of our people work and prosper in West Asia...Our foreign policy is not linked to any doctrine or dogma and we take decisions keeping in mind our vital national interests."

The absence of reference to Israel clearly suggested that there would be a reassessment of foreign policy on this front. The Road Map for Peace in West Asia has failed to take off. Much of the onus for this failure lies with the Israeli side. Israel is yet to withdraw from the Palestinian towns, villages and cities. The release of few hundred prisoners, which were first of all illegally arrested, was seen as a major concession but this gesture did not impress either the Palestinians or world public opinion. Palestinians are facing enormous hardships and feel humiliated in their own homeland. The whole situation is caught in the midst of US inaction, desire of the Palestinians to go through the peace process and negative approach of the Israeli authorities to genuinely implement the Road Map.

The Palestinians need India's support to get justice for their long sufferings. The Palestine issue has become one of the most important flash points which have led to repeated conflicts and violence should be contained. India should speak out in favour of the Palestinians and condemn Israeli atrocities which are taking monstrous proportion. This is essential not only to repair our ties with the Palestinians but also to restore ties with the Arab/Islamic World. Another issue which is impacting our foreign policy in the region is the US invasion and occupation of Iraq. The fact that the US has continued to keep it full forces in Iraq is mind boggling. The daily killings of innocent Iraqis by US and other forces and destruction in holy places are fuelling the Iraqi resistance. The Iraqi resistance to US occupation has not only intensified but the US has confirmed that from being sporadic attacks by the old Baathist elements, a full scale guerilla war and now a well organised sophisticated resistance is going on against US forces. India did a wise thing by not sending troops to Iraq at US request.

The anti-US sentiments sweeping the Islamic world are likely to intensify in the coming years due to US war against some Muslim states and groups. US-Israeli policies are spreading hatred and extremism and this is the prime reason for spreading violence and terrorism. India in cooperation with Iraq and other like-minded states in West Asia should work for speedy end to US occupation of Iraq and also speak out on the atrocities committed by US occupation forces. Although the UPA Government's CMP stated that it will maintain the independence of India's foreign policy stance on all regional and global issues even as it pursues closer strategic and economic engagement with the US but the reality is in the context of

West Asia, US policy is actually affecting India in a gradual way. Pax Americana in West Asia has serious implications for India: (1) The US control of Gulf oil sources may in future affect India's quest for energy security especially as we import 70 per cent of our oil from Gulf; (2) the US support to undemocratic regimes in the region is encouraging the Islamic groups who are increasingly using violence in the face of reluctance of the rulers to initiate political reforms; (3) the security and future of Indian workers in the GCC states in the wake of continued wars/instability has become an issue of concern; and US desire to involve Pakistan in its strategy in West Asia has implications on India's security and influence in the region. India's close ties with Iran and Syria has come under pressure as the US intensifies its pressure on these two states while doing nothing to restrain Israel, which seeks to increase its influence in the region with US help.

It is imperative to maintain close ties with Iran in the context of Afghanistan, Central Asia, gas from Iran and the Pakistani factor. India's relationship with Iran is also vital for Iraq and stability in the Gulf region. President Mubarak of Egypt was awarded the 1995 Nehru Award for International Understanding in 1997, but so far he has not visited India to receive the award? The answer lies in India's pro-Israeli tilt which has become a burden and its high time India pursues a more vigorous policy to strengthen its relations with the Arabs. Peace, security and stability in West Asia are crucial for our own security and prosperity and this calls for new initiatives, dynamism and recognition of the traditionally close historical relationship with the Islamic world.[18] At the moment Pax Americana in West Asia is there to stay. In fact, the US military presence (naval) is bound to increase. US pressure on Iran and Syria is likely to intensify in future. The 2006 January victory of Hamas in Palestine and the negative reactions from US/West and Israel is bound to complicate the search for peace in West Asia.

The Israeli invasion of Lebanon in July 2006, in order to eliminate the Hezbollah and US complicity in it by airlifting cluster and other bombs used in Iraq and by opposing an early ceasefire has only strengthened the religious group. Also, Israel by arresting and imprisoning half of the Hamas cabinet members, Speaker of the Palestinian Parliament, and over 20 MPs and other top Palestinian officials has only strengthened the Hamas religious group. The area of armed resistance to US-Israeli policies has only widened with violence spreading to new areas in the region. US and Israel would like to tackle the peripheral groups first before launching all-out attacks on Iran and Syria–two states still holding out in West Asia–a region which has largely been subjugated to US-Israeli domination. The developments in Iraq especially growing violence and sectarianism are pushing the country (until recently secular) towards religious chaos and perhaps fragmentation. US pressure on Iran may lead to greater violence in Iraq. Together with the lack of infrastructure, growth of religious groups and their militias and US inability to handle the developments, the implications flowing from Iraq are likely to impact the region as a whole if not beyond. Shia-Sunni tensions are raising in the GCC states especially Saudi Arabia, Bahrain and Kuwait, in view of developments in Iraq. If US or Israel attacks Iran, Tehran may retaliate on oil tankers in the Gulf or Strait of Hormouz. Oil terminals, platforms may also become vulnerable thereby pushing oil prices. If violence spreads in the Gulf region, Indian workers who are nearly four million in GCC states may have to be evacuated from some places.

India should prepare contingency plans to move the workers to nearby safe places. India must also keep an eye on the impact of religious groups preaching extremism on Indian workers. US ability to dictate terms to India on our energy security like its doing in the case of Iran and Syria are bound to increase in future. As the US pressure on Iran mounts, Tehran will urge the GCC states to remove or reduce US presence like bases and troops. India need not share Iran's view on this issue entirely but actively work to have autonomy for Iran and the GCC states. India's energy needs are growing and everything possible has to be done to have the closest ties with oil producers in West Asia. This has become vital for India's growth and prosperity. Of course India must have good ties with the US but our policy in West Asia should be autonomous of US and not subservient to the dictates of the dominant power in the region.[19]

After 9/11 it has become obvious that some countries in West Asia are looking to the East especially Saudi Arabia and they find India an attractive partner in all areas. India should extend support to such countries which are willing to stand up to US hegemony as the US doctrine of preemption is likely to destabilise the region still further. US domination is clearly restricting the role of India and the room for autonomous engagements is narrowing. Under the notion of unlimited war on terror, the US is justifying its military intervention to contain terrorism. All of this is mounting pressure on the states in the region as also on India. Not only the West Asian rulers have become nervous due to US incited developments but despair and discontent among the unemployed youth in the region is also mounting. India should quietly urge the leaders in the region to move ahead with democratic reforms leading to political participation in the decision making process, thereby enhancing regime legitimacy. This will help reduce the Arab regime's dependence on outside powers, and tackle the dependency syndrome.

It is clear that most regimes are going through severe loss of sovereignty and threat of disintegration, civil war and identity crisis. While continuing the fight against terror, problems like Israeli occupation of Arab lands and ending US occupation of Iraq and other local concerns need to be addressed. It is surely possible for India to strive for good ties with US and Israel (without of course, identifying too closely with their agenda in West Asia) and also maintain close ties with the Arabs, Turks and Iranians. Indian diplomacy needs to rise to the occasion as it has both challenges and opportunities in energy rich but unstable West Asia.

Notes and References

1. Pasha, A.K. (1990). "Nehru and the Crises in West Asia", *Detente*, Vol.IX(1), May- June 1990, pp.4-9.

2. Pasha, A.K. (1999). *India and West Asia: Continuity and Change*, Gyan Sagar Pubs, Delhi, pp.36-51.

3. Ibid., pp.61-66; Pasha, A.K. (ed.) (1992). *The Gulf in Turmoil*, Lancers Books, New Delhi, pp.70-126.

4. Pasha, A.K. (2003). "India and the GCC States: Challenges and Opportunities", *World Focus*, Vol. 24(2), February, pp.9-15; Pasha, A.K. (1989). "Prospects for Peace in the Gulf", *Detente*, May-June, p.23; Pasha, A.K. (1998). "India and GCC" in Riyaz Punjabi and A.K. Pasha, *India and the Islamic World*, Radiant Publications, New Delhi, pp.34-48.

5. Pasha, A.K. (ed.) (1999). *India and Kuwait: Strategic, Economic and Political Developments*, Gyan Sagar, Delhi; Pasha, A.K. (1994). "Communal Revivalism in India: Its Impact on Ties with West Asia and North Africa", in Muchkund Dubey (ed.), *Communal Revivalism in India: A Study of External Implications*, Har Anand, New Delhi, pp.54-88.

6. Pasha, A.K. (ed.) (1999). *Perspectives on India and the Gulf States*, Detente Publications, New Delhi, pp.208-227; Pasha, A.K. (2006). "Indo-Saudi Ties Ready for a New Dawn", in *GCC-India Research Bulletin* (Dubai), Issue No.1, January, pp.6-7; Pasha, A.K. (1997). "Perspectives on Indo-Saudi Relations", *Asian Affairs* (London), Vol. 1(12), September 1997, pp.24-26; Pasha, A.K. (1992). "Indo-Saudi Arabian Summit is Long Over Due", *The Pioneer*, 3 August; Pasha, A.K. (2000). "India and West Asia: Continuity and Change", in Nalini Kant Jha (ed.), *India's Foreign Policy in a Changing World: Essays in Honor of Professor Bimal Prasad*, South Asian Publishers, New Delhi, pp.240-269.

7. Pasha, A.K. (ed.) (1999). *India, Bahrain and Qatar: Political, Economic and Strategic Dimensions*, Gyan Sagar, Delhi; Pasha, A.K. (ed.) (1999). *India and Oman: History, State, Economy and Foreign Policy*, Gyan Sagar, Delhi; and Pasha, A.K. (2006). "India and Oman: Emerging Security Partnership", *GCC-India Research Bulletin*, Issue No. 2, June, pp.10-11; See also Azhar, Mohammed (1999). *Contemporary Gulf Economies and Indo-Gulf Relations*, New Horizon, New Delhi, pp.179-198.

8. Pasha, A.K. (1999). *Aspects of Political Participation in the Gulf States*, Kalinga Publications, New Delhi, pp.1-16.

9. Pasha, A.K. (1993). "India and Israel: Growing Cooperation", *World Focus*, Vol. 14(8), August, pp.19-22; see also Pasha, A.K. (1993). "India's West Asia Policy: Continuity and Change", *Strategic Analysis*, Vol. XVI(6), September, pp.783-804; Pasha, A.K. (2004). "Arafat, Palestinians and the Future", *Mainstream*, Vol. XLII(50), December 4, pp.7-12.

10. See Pant, Girijesh, P.C. Jain and A.K. Pasha (eds.) (1996). *Contemporary Iran and Emerging Relations*, Neelkanth Publications, New Delhi, pp. 145-157; Pasha, A.K. (2005). "Iran and its Regional Influence", *World Focus*, Vol. 26(8), August, pp.3-8; Pasha, A.K. (2000). *India, Iran and the GCC States: Political Strategy and Foreign Policy*, Manas Publications, New Delhi, pp.262-277; Pasha, A.K. (1988). "India, Iran and the GCC States: Common Political and Strategic Concerns", *Encounter*, New Delhi, March-April, pp.66-74.

11. Pasha, A.K. (2006). *India and Turkey: Past and Emerging Relations*, Academic Publishers, Delhi; Pasha, A.K. (1995). *India and OIC: Strategy and Diplomacy*, Centre for Peace Studies, New Delhi; Pasha, A.K. (2003). "South India and the Gulf: Trade and Diplomacy during the late 18 th Century", in N.N. Vohra (ed.) (2003). *History, Culture and Society in India and West Asia*, Shipra Publications, New Delhi, pp.237-249;

12. Pasha, A.K. (ed.) (1999). *Contemporary Gulf: State, Society, Economy and Foreign Policy*, Detente Publications, New Delhi, pp.261-330; Pasha, A.K. (1995). "The Recent Gulf Crisis: India's Options", *Strategic Analysis*, Vol. XVII(10), January, pp.127-186; Pasha, A.K. (2003). *Iraq: Sanctions and Wars*, Sterling Publications, New Delhi.

13. Pasha, A.K. (2006). *India, Iraq and the Gulf*, Academic Publishers, Delhi; Pasha, A.K. (2004). "The Iraq Crisis: An Overview", *World Focus*, Vol.25(1), January, pp.3-6; Pasha, A.K. (2005). "India and the Gulf States", in Rajen Harshe and KM Seethi (eds.),

Engaging with the World: Critical Reflections on India's Foreign Policy, Orient Longman, New Delhi, pp.389-424.

14. Pasha, a.k. (2005). "Iran and Iraq: Are they becoming Friends"? *South Asia Politics*, Vol. 4(5), September, pp.29-34; See also Khan, Javed Ahmed (2005). *India's Energy Security and the Arabian Gulf*, Arise Publishers, New Delhi, pp.99-132; Pasha, A.K. (2005). "Iraq and its Future: An Overview", *World Focus*, Vol.26(2), February, pp.3-6.
15. Pasha, A.K. (2002). "Arab Response to Peace Process", *World Focus*, Vol. 23(3), March, pp.14-19.
16. Pasha, A.K. (20050. "Post-Arafat Palestine", *World Focus*, Vol.26(5), May, pp.7-10; AK Pasha (2006). "Pax-Americana in the Gulf Region: Implications for India for India's Energy Security", in SN Malakar (ed.), *India's Energy Security and the Gulf*, Academic Publishers, Delhi, pp.116-122; Pasha, A.K. (2004). "Palestine after Arafat: Struggling to Survive", *Hard News*, Vol.2(1), December, pp.60-61; Pasha, A.K. (2000). "Conflict Resolution and the Palestinian Peace Process", in Arun Kumar Banerji (ed.), *Conflict Resolution in the Post-Cold War Era*, Minerva Associates, Calcutta, pp.72-91.
17. Pasha, A.K. (2004). "Foreign Policy of UPA Government: India and West Asia", *World Focus*, Vol. 25(10-12), October-December, pp.41-46.
18. Pasha, A.K. (2005). "Asian Security : Flash Points in the Gulf Region", *World Focus*, Vol. 26(10-12), October-December, pp.25-33; Pasha, A.K. (2004). "Why is the Islamic World Important for India?", *Hard News*, Vol.1(7), June, pp.56-57.
19. Pasha, A.K. (2006). "India's Policy Towards the Arab World in the 21st Century", in V.D. Chopra (ed.), *India's Foreign Policy in the 21st Century*, Kalpaz Publications, Delhi, pp.215-250.

Index

C

D

E

F

G

H